Wave, Wind and Current Power Generation

Scrivener Publishing
100 Cummings Center, Suite 541J
Beverly, MA 01915-6106

Publishers at Scrivener
Martin Scrivener (martin@scrivenerpublishing.com)
Phillip Carmical (pcarmical@scrivenerpublishing.com)

Wave, Wind and Current Power Generation

Victor M. Lyatkher
and
Ziaur Rahman

Scrivener Publishing

WILEY

This edition first published 2022 by John Wiley & Sons, Inc., 111 River Street, Hoboken, NJ 07030, USA and Scrivener Publishing LLC, 100 Cummings Center, Suite 541J, Beverly, MA 01915, USA

For more information about Scrivener publications please visit www.scrivenerpublishing.com.

Wiley Global Headquarters
111 River Street, Hoboken, NJ 07030, USA

For details of our global editorial offices, customer services, and more information about Wiley products visit us at www.wiley.com.

Library of Congress Cataloging-in-Publication Data

ISBN 9781119829300

Cover image: Green Turbine Background Image: Korn Vitthayanukarun | Dreamstime.com
Wave: Leigh Prather | Dreamstime.com
Cover design by Kris Hackerott

Set in size of 11pt and Minion Pro by Manila Typesetting Company, Makati, Philippines

Printed in the USA

10 9 8 7 6 5 4 3 2 1

Contents

Preface

The book deals with modern problems of using wind energy – low-level and high-level (jet) currents in the atmosphere and the energy of water flows of rivers (without dams), ocean and sea currents, including those caused by tides. These areas (air and water flow energy) taken together do not solve the problem of using renewable, environmentally friendly energy sources, but they form an essential part of it. The energy of these flows in elemental, natural conditions passes from one form to another and dissipates into heat. The engineering measures discussed in this book aim to convert some of the energy of flows into electricity, thereby reducing the natural rate of energy dissipation of these flows. It turns out that by reducing the rate of dissipation of atmospheric surface low-level currents by only 1%, it is possible to provide all of humanity with energy at rates twice the per capita consumption of the wealthiest countries.

The physical laws underlying the considered flow energy conversion schemes are equally implemented to water and air flows. Therefore, it is natural to combine the issues of calculations and modeling of wind and hydroelectric units into a single section. There is a lot in common in the schematic diagrams of power plants. In contrast, specific design solutions due to the considerable differentiation in densities and flow rates are different. In addition to general approaches, the book describes real projects in the field of wind and tidal energy. Lastly, this book discusses the future of wind energy: high altitude turbines (HATs). Unlike the jet stream turbine it can harness wind energy from a wider wind band. HATs have been on trial for the last few years with limited commercial success. We describe HAT technology in detail with two case studies. It has proven added benefits over conventional wind turbines and all other renewable sources of energy.

1

Renewable Energy of the World

"All over the hill... there were wind turbines of gigantic dimensions... the wheels of wind turbines rotated noisily in the gusty wind..."

Herbert D. Wells, *When the Sleeper Wakes,* 1899

"Wind power promises a clean and free source of electricity... The U.S. Department of Energy (DOE) aims to see 5% of our electricity produced by wind turbines in 2010...A little research, however, reveals that wind power does not, in fact, live up to the claims made by its advocates..., that its impact on the environment and people's lives is far from benign, and that with such a poor record and prospect, the money spent on it could be much more effectively directed."

A Problem with Wind Power
by Eric Rosenbloom
http://www.aweo.org/ProblemWithWind.html

Renewable energy sources known to modern mankind are associated with solar radiation reaching the Earth, and with the mutual asymmetry of the arrangement of the orbits of the planets, primarily of the Moon relative to the Earth and the Earth relative to the Sun. Solar radiation can be directly converted into thermal or electrical energy, but, in addition, it forms all biological energy sources, as well as atmospheric and oceanic thermal flows and the cycle of river and ground waters. The asymmetry of the planetary orbits provides a harmonic change in gravitational forces, causing deformations in the earth's crust, long waves in the atmosphere, seas and oceans, manifested in the regular rise and fall of the water level, especially noticeable in bays and straits. These changes in the water level are naturally

Victor M. Lyatkher and Ziaur Rahman. Wave, Wind and Current Power Generation, (1–58) © 2022 Scrivener Publishing LLC

accompanied by flows filling the respective bay in one phase of the tide, and emptying it in the other phase.

Wind energy and water energy directly related to solar radiation and the presence of the earth's atmosphere are the most ancient energy sources mastered by human society. If Noah's Ark really existed, then it probably moved under sails and oars. The mill driven by the force of the wind or the flow of water has been grinding grain in Persia as early as 200 years before the Common Era. Even earlier, wind turbines and water wheels were used in China. In Europe, wind turbines appeared in the 8th century. The practical development of wind energy as a source of electrical energy began only at the end of the 19th century. The first 12-kW wind turbine (Fig. 1.1) was built by Charles Brush in 1888 in the Cleveland area (Ohio, USA). It worked without any problems for almost 20 years.

This unit echoed the ideas of using active wind pressure, incorporated into the design of old windmills. They worked almost independently on the shape of the blades (wings) moving slower than the wind, and the efficiency of using wind energy was very low.

Fig. 1.1 Charles Brush's Wind Turbine. Cleveland City, USA, 1888.

The rapid development of aerodynamics at the beginning of the 20th century, associated primarily with aircraft construction, affected the theory and practice of using the energy of free water and air flows. The simple, but very important results of German scientist A. Betz and Russian professor N. E. Zhukovsky determined the efficiency limits of wind turbines and indicated the ways of their optimization—the blades should have well-streamlined profiles and move faster than the flow. A 100-kW first large wind turbine having a special aerodynamic design was built in Crimea near Balaclava City in 1931 (Fig. 1.2).

By 1936, a new project, a 5000-kW wind turbine, was designed in Russia, which, however, was not implemented, and the wind energy line in Russia itself was assigned to the section of agricultural production, where

Fig. 1.2 High-speed wind power unit. 100 kW. Balaclava City, Crimea, USSR, 1931.

it stalled. The situation was not much better in the United States after the failure with the Putnam-Smith project; the said project led to the construction of a 1250 kW wind turbine in 1941, which collapsed on 23 March, 1945, because of having produced wind lower than what was estimated. Consequently, the interest in wind energy was lost for almost 40 years.

The energy of tides compares favorably with the energy of wind and that of oceanic or sea flows of other origin, by its high predictability. Although the height of the tides and the shape of the smoothed seagram at each fixed point do not remain constant, these changes are well predicted by modern methods. Attempts to harness the energy of tides in a variety of forms have been made since ancient times. Engineering initiatives in the last 100 years have been aimed at creating and using the static pressure occurring on the barrier blocking the mouth of the sea bay with high tides. Relevant projects based on the experience of river hydropower were put forward in Canada, the USA, England, France, and Russia. Finally, in 1936, the construction of a tidal power plant (TPP) of Kwodi (USA) began with a traditional dam, a hydroelectric power station, spillways and all the rest of the set of structures and equipment typical for a river hydroelectric power station. This attempt, dubbed "economic madness," was soon abandoned. In the next 50 years, four more TPPs with a traditional set of hydraulic structures were nevertheless built. These were TPPs Rance (France, 1967), Kislogubskaya (USSR, 1968), Annapolis (Canada, 1984), Jiangxian (China, 1985), which, despite various new technological methods of construction, turned out to be not as effective as estimated and very expensive. As a result, practical interest in this field remained stifled for a long time.

A noticeable revival in the field of tidal energy has been observed in recent years, when the Russian technology for the use of floating power units almost completely equipped in the factory and transported to the installation site has been generally accepted; experience has been accumulated in the study of orthogonal turbines, the operation of which does not depend on the direction of flow and, most importantly, there were ideas to move away from the traditions of pressure river hydropower and directly use the energy of tidal flows.

It is important that, with the new technology of using tides, a system of several TPPs can provide a guaranteed basic power even without external storage devices, although storage elements are needed to fully utilize the energy of the tides.

According to this principle, with free-flow collinear turbines resembling traditional wind turbines, the first TPPs of a new generation have already been built in Devon Coast (England, 2003, Fig. 1.3) and in Hammerfest strom (Norway, 2004) with 300 kW turbines each.

Fig. 1.3 Hydraulic unit MCT in repair and maintenance position. Devon Coast, 2003. 300 kW.

The costs of these pioneering facilities turned out to be so high (in Norway, for example, about 11 million US dollars per vehicle) that plans to expand work in this area and create a large TPP (Fig. 1.4) have not yet been set into motion.

We hope that the situation will change with new, orthogonal turbines in combination with a new technology of using tidal energy and floating block technology for the construction of TPPs.

The large-scale use of any renewable sources was constrained by a number of general technical and economic reasons:

— high degree of dependence on natural factors;
— lack of confidence in the perfection of available technologies;
— large initial capital investments;
— low level of world prices for fossil fuels;
— relatively low share of these sources in the balance of energy resources consumption.

The development of alternative energy sources was understood by many people as a kind of additional source for a niche market, such as for the agriculture industry. In addition to that, progress in this sector was also

Fig. 1.4 Tidal power plant project based on the unit at Hammerfest strom.

hampered by the insufficient purchasing power of the population and the lack of funds of local authorities for investing in facilities, which was most typical, for example, for Russia during almost the entire 20th century.

However, the undeniable advantages such as lack of fuel costs, and environmental safety which the inexhaustible resources have, are a favorable basis for the recognition of alternative energy and assure investors a quick return on promising projects.

The search for an environmentally friendly alternative to the traditional energy industry prevailing today in the context of special public attention to the problem of environmental protection is of great importance also because energy facilities operating on natural fuel negatively affect the environment by polluting the air and water basins, causing acid rains, etc.; this leads to their making a significant adverse contribution to worsening the greenhouse effect.

The use of renewable energy from water currents on a large scale has taken place since the mid-19th century in the form of the construction of dams and diversion hydroelectric power plants, and since the mid-20th century in the form of the construction of tidal power plants with a traditional pressure front cutting off the bay from the sea. In the USA, Italy, South Korea, Russia, and Canada, a number of projects have been implemented to use tidal flows, without creating a pressure front. The authors also have proposals for Bangladesh. We usually considered layouts with collinear turbines, similar to traditional high-speed wind turbines with a horizontal axis, as well as layouts with orthogonal units of the Darrieus type. A comparison of economic efficiency, which was carried out for turbines of the same diameter and at a tidal flow velocity of up to 2 m/s, turned out to be in favor of the traditional solution with a pressure front. Blocks of tidal power plants with orthogonal turbines were proposed by V. Lyatkher in 1981; independently and in a slightly different form, similar layouts were proposed by Barry Davis and promoted by Blue Energy Canada Inc. Their detailed studies were carried out in Russia and small, full-scale samples were built. Orthogonal turbines in floating blocks can qualitatively change the situation with the development of tidal energy, but an even greater effect is expected from the use of multiblade, multistage orthogonal turbines installed in oceanic or river flows without creating a pressure front. These turbines can have linear generators with opposite motion of inductors and a squirrel cage rotor. For water turbines, this is especially important, since the flow rates and, accordingly, the speeds of the turbine blades are small here, and the doubling of the speed of crossing the magnetic fields in the generator, achieved by the counter rotation of adjacent turbine tiers, significantly increases the efficiency of the machines.

The material consumption of modern conventional hydropower construction is, on average, 82 kg/kW of metal (mainly steel) and 1.5 m^3/kW of concrete. The material consumption of promising power plants using the energy of air or water flows is not expected to exceed the indicated values. With the development of ways to improve the quality of the energy generated, attention to these sources and the scale of their use in new technological forms should steadily increase.

Over the past 25 years, wind energy has been the fastest-growing energy sector. For example, in the USA, it was the case that the wind power industry has developed at the fastest pace since 1990—the output of wind power plants has increased, on average, by 22.6% per year. From July 1998 to June 1999, wind power plants with a total capacity of 1,075 MW were commissioned in the United States, including 895 MW at new ones, and 180 MW at those under reconstruction. This required a capital investment of about one billion dollars—at the time about $1000/kW. In 2006, 2,454 MW was commissioned (an increase of 27%) with an investment volume of about $4 billion ($1,630/kW). The total installed capacity of US wind farms at the end of 2006 was 11,603 MW. The average annual output of these wind farms was expected to be about 31 billion kWh which corresponds to an average capacity of 30% of the installed capacity. The actual average capacity of actually operating wind farms in California, which started the mass construction of wind farms earlier than others, is only 24%. The observed rapid development of systemic, large-scale energy, had a real economic basis—the cost of energy on the bus bars of wind farms in areas with moderate wind is now in the United States from 3 to 6 cents per kWh. This is almost the same as the cost of energy at gas stations (3.9 to 4.4 cents/kWh) which is slightly lower than coal-fired power plants (4.8 to 5.5 cents/kWh) and significantly lower than all other types of modern power plants. According to the forecasts of the American Wind Energy Association (AWEA), the cost of energy at wind farms could drop to **2.5** cents/kWh. Globally, the growth rate of the wind power park in the first years of the new century turned out to be even higher than in the United States; in 2005, wind turbines with a total capacity of 11,769 MW were installed, which was 43.4% greater than in 2004. The total capacity of wind farms by the end of 2005 reached 59,322 MW which was 25% greater than the total capacity as at the end of 2004. The year 2006 was a record year for the world wind power industry. The growth of installed capacity reached 32%, and the total wind power plant capacity at the end of 2006, according to GWEC, was 74,233 MW. Sales of equipment for wind power plants in 2005 reached $14 billion, so the average cost of wind turbines (excluding transportation, installation, and commissioning costs) was about $1,200 per kW of installed capacity. There are GWEC forecasts, according to which after 2020 12 to 16% of the world's energy needs will be covered by wind energy using those design solutions that are already widespread at the present time. These wind farms are expected to have a total installed capacity of over 1,254 GW with an average annual capacity of 27.8% of the installed capacity.

Wind power continues to develop rapidly worldwide. About 6,500 MW new wind turbines were installed around the world in 2001, with their annual sales of about $7 billion. This was a large international increase in wind energy facilities, well above the results in 2000 (3,800 MW) and 1999 (3,900 MW). At the end of 2001, the global wind power capacity was about 24,000 MW. Germany alone has set an international and national record by introducing new wind power capacities exceeding 2,600 MW during 2001. In 2005, this figure was the second highest in the world (after the USA), but the commissioned capacity amounted to only 1,808 MW (it was 2,431 MW in the USA). By the end of 2003, the installed capacities of wind power plants in Germany exceeded 12,000 MW; by the end of 2005, 18,428 MW and, by 2020, energy generation from wind turbines should cover up to 20 to 30% of the country's total energy demand. The experience of Germany, Denmark, and Spain clearly shows that wind energy can reliably provide from 10% to 25% (or more) of the energy supply of a particular region or country.

In the United States, 2001 was a breakthrough year for wind power with previous national records left far behind; nearly 1,700 MW of new capacity was commissioned at a cost of $1.7 billion. New installations introduced in the United States at that time accounted for nearly a third of the world's wind power capacity commissioned in 2001.

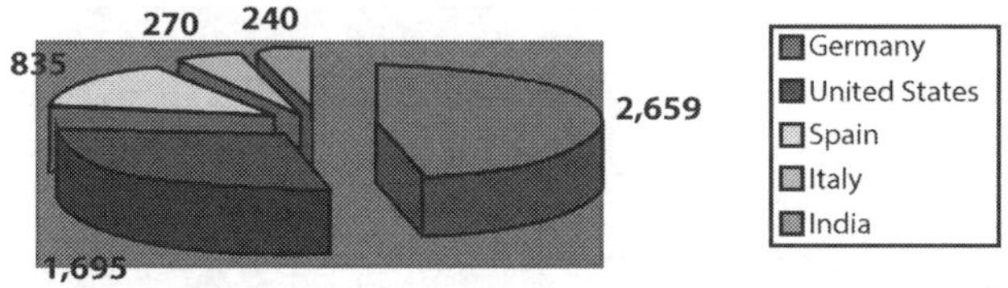

Increase in power capacity in 2001 (in MW)

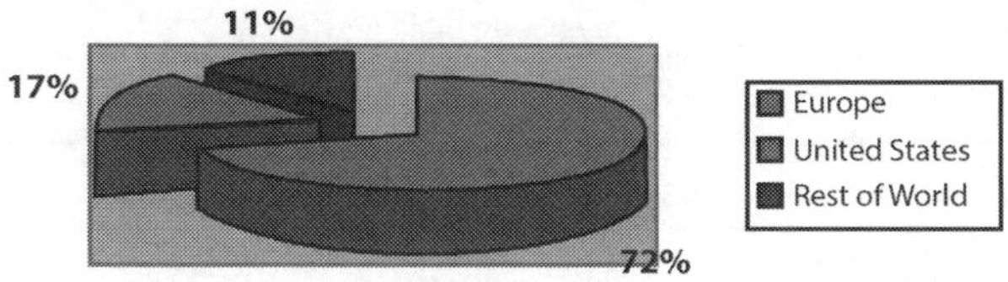

70% of the world's wind energy now comes from Europe. European countries accounted for 2/3 of the 2001 growth.

In a couple of years, the commissioning of capacities in the USA significantly exceeded the indicators of other countries; however, by 2006, the total capacity of US wind turbines was half of that in Germany. This important difference was due to the temporary withdrawal of the US administration from some of the benefits provided to wind turbine manufacturers and wind farm owners. After the restoration of the "tax umbrella," the rapid growth of wind farms in the United States continued. The same happened in Denmark. The government's refusal to subsidize wind energy has virtually halted its development. The global wind energy market is still dominated by the "big five" countries, each of which has a capacity of over 1,000 MW: Germany, USA, Spain, Denmark, and India. In 2006, there were ten such countries.

The number of countries with wind power capacities of several hundred megawatts was increasing, and by 2006 it had already approached two dozen. Table 1.2 shows the increments and cumulative capacity in the five largest wind user countries mentioned in 2000 and 2001, and at the end of 2005.

The data provided below show that the fantastic triumph of wind energy in the world is still, to a large extent, based (alas!) on the political decisions of the leadership of the countries mentioned. The state of matters in small

Table 1.1 Countries where the installed capacity of the wind farm exceeds 1,000 MW.

Name of the country Installed capacity of WPP, MW (% of the amount worldwide)
Germany 18,428 31
Spain 10,027 16.9
USA 9,149 15.4
India 4,430 7.5
Denmark 3,122 5.3
Italy 1,717 16.9
United Kingdom 1,353 2.3
China 1,260 2.1
Japan 1,231 16.9
Netherlands 1,219 2.3
Total for 10 countries 51,936 87.5
Worldwide 59,304 100

Table 1.2

Main wind energy markets (by capacity commissioned, MW)	**2000**	**End of 2020**	**2001**	**End of 2020**	**2005**
	Growth	**Grand total**	**Growth**	**Grand total**	
Germany	1.669	6113	2.659	8.750	18428
USA	53	2566	1.695	4.261	9149
Spain	713	2502	835	3337	10028
Denmark	552	2300	117	2417	3122
India	90	1167	240	1407	4430

countries like Denmark simulates a possible situation in the world — **the termination (or absence) of political and financial state support immediately stops the development of wind energy.**

The average increase in capacity (MW/year) over 5 years (from 2001 to 2005) and the increase in 2005 for the indicated countries are as follows:

Country	Average growth over 5 years	Growth in 2005
Germany	1,935	1,808
USA	977.6	2,431
Spain	1,338	1,764
Denmark	141	4
India	604.6	1,430

Let's take a closer look at the situation with wind energy in developed countries.

Germany. As reported by German Bundesverband Windenergie (BWE) (http://www.wind-energie.de/informationen/informationen.htm), in 2001 Germany set a record of 2,659 MW. This country retains its world leadership in wind energy, having 11,500 wind turbines by 2003, with a total capacity of 8,750 MW, and by 2006, 18,428 MW. Wind power employs 35,000 people. In 2003, Germany received just under 3.5% of its total electricity from wind power, but the adverse effects of pulsating wind power were not noticeable. The largest number of wind power plants was located in the northern state of Lower Saxony. In 2003, there were 2,426 MW plants operating there. Schleswig-Holstein (1,555 MW) remains the leading region in terms of the proportion of electricity generated by wind energy (28%), followed by Mecklenburg-Western Pomerania (21%), Saxony-Anhalt (11%), North Saxony (10%), and Brandenburg (9%). Among the new wind farms introduced in Germany, the 105 MW Sintfeld wind farm near Paderborn ranks first as the largest onshore facility. It consists of 65 1.6 MW turbines and will generate electricity for 70,000 typical European homes. Germany is also taking steps to harness its offshore wind energy potential. Turbine manufacturer Nordex has finalized an agreement with Winkra-Energie GmbH for the supply of seventy 5MW turbines for the 350MW offshore facility Nordliche Pommersche Bucht located east of the Rugen Island. The first phase of construction was completed in 2006. Two more development stages of the same size are planned, as a result of which the capacity of the entire project will reach almost 1,000 MW with a capacity of 3.6 billion kilowatt-hours of electricity annually. The total investment will be approximately EUR 1.53 billion including all costs (EUR 1,530/kW). The rotor diameter of each turbine is 112 m. Eight such turbines have been manufactured and tested.

Spain. With a wind power capacity of 3,337 MW, Spain ranks second in Europe after Germany in terms of new capacity. According to the Spanish Association of Renewable Energy Companies (APPA), in 2001 it commissioned 835 MW of wind power, i.e., there was a 33% increase. Spain had planned to receive 12% of its electricity from renewable sources by 2010. By 2006, wind turbines with a total capacity of 10,027 MW were already in operation (1,764 MW was commissioned in 2005). As wind energy grows rapidly, its development is starting to struggle with power transmission bottlenecks and grid connection regulations. The issue of the accumulation and redistribution of energy is becoming increasingly more urgent. The first practical steps have been taken to create the "WPP + hydrogen" complexes.

Denmark. The total number of new wind turbines commissioned in Denmark in 2001 was estimated to be at 117 MW, i.e., three times less than that in the previous year. By 2006, the problem had become even more acute. In contrast, in 2003, Denmark was the country with the largest share of electricity generated using wind power; in 2001, the share was more than 15%; in 2002, 27% of electricity consumed was provided by renewable sources, which significantly exceeded the country's targets, constituting 20% by 2003.

Long a leader in this field, Denmark had been facing the challenge of its wind power by the new conservative government. The country's leadership announced that it would cut subsidy programs for the development of renewable energy sources, including programs for wind energy. This side of the national budget was made public in January 2002. The cut put an end to 20 years of funding for international standardization, certification, and selection of optimal wind turbine types. Further announced steps include ending government support for offshore wind turbines and reducing environmental incentives for energy from existing wind turbines. This turn of government policy had hurt the Danish wind power industry, in particular calling into question its leadership in basic research and production of relatively cheap large wind turbines flooding the global market. Wind farm growth in Denmark stopped!

Great Britain. The increase in wind power capacity in the UK in 2001 was 64.6 MW, and in 2005 it was 446 MW. According to the British Wind Energy Association (BWEA), the total UK wind capacity in 2002 was 473.6 MW, generating 1.24 billion kilowatt-hours annually, which met 0.37% of the country's electricity needs. By the end of 2005, the total capacity of wind turbines had reached 1,353 MW, e.g., it approached 1% of the total capacity of the country's energy system. In the UK, the country with the largest wind energy potential in Europe, investment in wind energy remained modest. In 2002, about 200 MW was commissioned, and in 2005, that number rose to 446 MW, which was a record for the UK wind power industry.

Larger proposals are slated for longer periods. The Crown Estate, an owner of the seabed in UK territorial waters, approved the placement of 13 offshore wind turbines in September 2001. The total capacity of these 13 wind parks was expected to be between 1,000 and 1,500 MW. Engineering firm AMEC and atomic energy company British Energy, in December 2001, revealed their intention to build what was expected to be the world's largest wind farm. This was a project on the Isle of Lewis in the Hebrides. According to the London *Times*, the project had a total capacity of about 2,000 MW. British utility firm Powergen had announced that it was looking forward to building a wind farm which could have a capacity of about 500 MW, at the Thames estuary near London.

Powergen prepared project proposals in the range of 400 to 800 MW, which were accepted. The British government demanded that in 2002, electricity suppliers purchase at least 3% of their energy from renewable sources. By 2010, this figure was expected to be 10%. Offshore wind farms were slated to supply at least 20% of the required renewable energy generation. However, all these projects have not yet made England one of the five largest wind energy countries. In recent years in England, attention to the use of tidal energy has sharply increased. The developments by the Marine Currents Turbine (MCT) Company, which are very similar to modern wind turbines, have been adopted as a base element. The MCT offerings cover the entire coast of England (Fig. 1.5).

The practical reality of this project will be determined by the economic results that will be obtained with the implementation of a tidal power plant with a capacity of about 10 MW with 1.5 MW units (Fig. 1.6).

The first experience with a 300-kW unit (actually 240 kW – Fig. 1.3), which required £1.4 million from the government, did not inspire investors for economic reasons.

By the end of 2001, the future in the United States looked bright, and the American electrical industry was showing a huge and growing interest in wind power. But wind power was temporarily knocked out of the rut by another tilt in US government policy, as the federal wind power tax credit was about to expire at the end of December 2001 due to a congressional delay. The situation in this regard improved from 2002 to 2003 when, in March 2002, after months of uncertainty, the timing of the federal tax credit was extended to December 31, 2003. In the interim, however, the costs of the up-and-down cycles caused by the expiration of the tax credit

Fig. 1.5 Layout of damless tidal power plants in England.

Fig. 1.6 Anglesey Coastal 11 MW Tidal Power Plant Project with 1.5 MW MCT Turbines.

came to the fore dramatically as proposed projects worth about $3 billion were delayed. A record was established in 2001, when nearly 1,700 MW of new capacity at a cost of $1.7 billion was commissioned, more than double the previous record in 1999 when 732 MW units were commissioned. The 2002 downturn to 410 MW forced many states to turn to domestic sources of industry support. The total commissioned capacity in the USA at the end of 2002 was 4,671 MW; 26 states had utility-scale wind turbines. New facilities included the world's two largest wind farms, the 278 MW King Mountain wind farm in Texas and the 261 MW Stateline wind farm on the Oregon-Washington state border, and several other wind farms in excess of 100 MW. The massive success of wind turbines in Texas in 2001 was the result of a successful combination of the state's renewable investment standard and the wind power tax credit. The Texas Renewable Investment Standard was adopted in 1999 after a survey by utilities found widespread support for renewable energy use across the state from all types of consumers (including commercial and industrial consumers). This standard required state utilities to purchase new 400 MW renewable energy units by January 1, 2003. In fact, the 915 MW wind turbines were commissioned a year earlier, which was greater than double the standard requirements. Bids for the supply of new renewable energy facilities proved to be price competitive with other resources, and a voluntary agreement for oversize products followed. In March 2002, the federal wind power tax credit—an important factor in financing new wind turbines—was extended by two years, until December 31, 2003. This action by Congress and President Bush put wind power back on track so that it could significantly increase US wind power capacity over the next year. After 2003, the benefits were systematically extended. This restored impact was heavily promoted in the US, and required a strong and consistent support policy. With this support, wind power in the United States was expected to produce about 6% of the electricity needed by the entire country by 2020. Of course, the distribution of wind resources across the United States is uneven (Fig. 1.7).

The actual distribution of the commissioned capacities (Fig. 1.8) does not repeat the picture of wind resources—the policy of the administration of individual states, public activity, and investment opportunities of energy companies and individuals are of great importance.

The numbers indicate the total installed capacity of wind turbines (MW) in the states at the end of 2006.

As wind energy expands, it increasingly faces the challenge of gaining access to a wind power transmission system at a reasonable price so that it can be sold profitably in the wholesale electricity market. In the past, most wind turbines sold their products directly to local

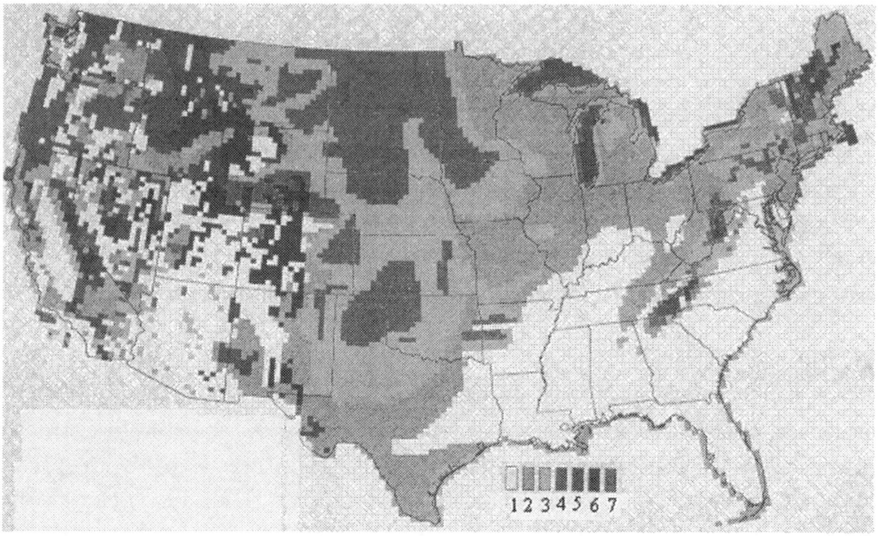

Fig. 1.7 Average annual wind flows over the United States at 50 m above the earth's surface. 1, 2, 3,... 7 – energy classes corresponding to energy flows (Watt/m^2) <200, 200 to 300, 300 to 400, 400 to 500, 500 to 600, 600 to 800, >800, which corresponds to average wind speeds (m/s) and Rayleigh distribution – <5.6, 5.6 to 6.4, 6.4 to 7.0, 7.0 to 7.5, 7.5 to 8.0, 8.0 to 8.8, >8.8.

utilities. But to restructure the "disconnectedness," the vertically integrated monopoly utility power industry requires utilities to make their networks available to all energy producers at a "free access" payment structure or tariff, which implies that power generators will accurately

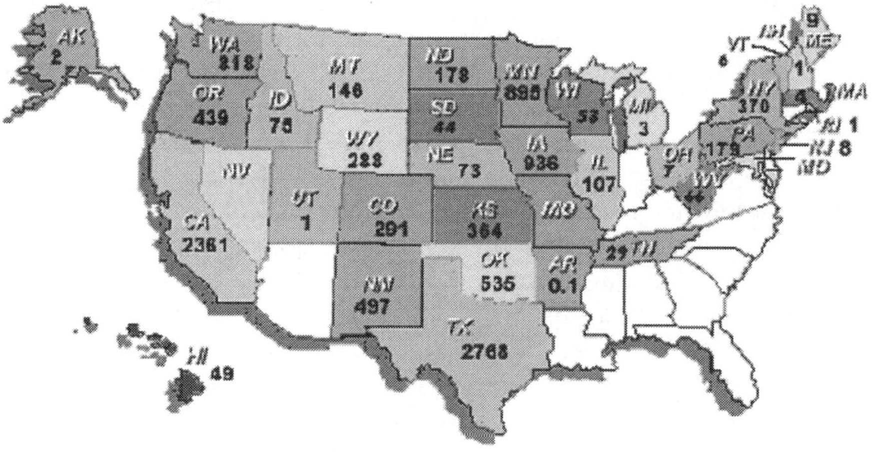

Fig. 1.8 Distribution of installed wind farm capacity among several US states.

schedule and monitor their networks and energy production in advance to fit this schedule.

These free access tariffs impose fines of 2.5 to 3.5 cents per kilowatt-hour on wind power generators for their use of the grids, in addition to the normal grid access charges. These penalties can double the wholesale price of wind power electricity. Fines are levied due to the variable nature of the wind and the fact that the wind farm cannot guarantee the supply of a certain amount of electricity at a given scheduled time and day. The fines are not based on the real costs of not delivering electricity; they are punitive fines to reinforce "good behavior" by large producers who can accurately control power generation.

As control over the interstate transmission system falls on the "Independent System Operators", the FEC is working to modernize the previous version of the regulatory tariff, and the impact of these penalties is also being reassessed. The New York and Texas Independent System Operators have specific rules for "volatile" resources such as wind power, exempting them from these penalties. The Mid-Atlantic States Independent System Operators have their own penalty-free market system that can accommodate intermittent wind farm production without incurring fines. An agreement was also recently reached to reduce fines for wind farms in California. However, fines are still a problem in important and windy regions such as the Pacific Northwest and Midwest. In the Northwest, for example, the Bonneville Power Administration (BPA) said it hopes to add 830-megawatt wind turbines to its system, but the development of those projects (worth $800 million) is being held back by fines.

Another problem for the industry is the chaos in the California electricity market and the failure of the Californian government to sign contracts with wind turbine developers to supply new capacity during the state's recent electricity crisis. What initially appeared to be a potential bonanza for resource-rich wind power in California and neighboring states, quickly turned into a roadblock as electricity prices skyrocketed and governments began hastily signing expensive long-term natural gas contracts, while refusing to buy electricity from cheaper but "unstable" producers such as wind farms.

The California Energy Commission has for some time followed a course of inviting and incentivizing renewable energy projects (including wind power) using utility funds. But the vast majority of proposed wind projects—with a total capacity of more than 1,500 MW—were never built. This was due to financial troubles in California's utilities and delays in new deliveries which emerged when California was implementing long-term contracts purchased at the height of the crisis. Thus only new wind power projects are progressing in this state.

Consumer interest in small-scale wind systems rose with concerns about high tariffs and diminishing street and window lighting during the California energy crisis in early 2001 and it led to significant sales even as the crisis later eased. Small wind power systems market, i.e., turbines with a capacity of less than 100 kilowatts are also expanding as a result of policies adopted in an increasing number of states. California and Illinois have established rebate programs that help reduce the high upfront cost of wind power systems. New York, New Jersey, Delaware, Wisconsin, and Rhode Island also followed suit. In California, the small wind power industry is also welcomed by a law introduced in 2001 that exempts local regulations unduly limiting the height of towers and structures. The Net Metering Law allowing the owner of a small wind farm or other renewable energy system to spin their electricity meter backward if the system is producing more energy than it is consuming has been passed in more than 30 states. This policy, along with simplified, standardized grid connection rules, reduces the possible cost and time spent by consumers installing a small hydro, wind, or photovoltaic system for their home or business.

Prospects for a nationwide incentive for small wind power systems have improved significantly with the introduction of a Bill (House of Representatives, 2322) by Congressman D.S. Watts (Oklahoma) to provide a federal 30% investment tax credit per unit. Similarly, the Twin Bill (Senate 1810) was introduced by Senator Richard Durbin (Illinois). Watts said the current value of small wind turbines is slowing down their sale, and a tax to lower the upfront cost of units will boost sales, help manufacturers increase output and thus further lower prices. In 2002, several deals took place, indicating that large energy corporations were increasingly interested in wind energy. Shell WindEnergy, a division of the Anglo-Dutch oil giant, now owns two wind farms in the United States. Entergy, a large US utility, recently bought a majority stake in an 80-megawatt wind farm in Iowa, while TXU (a Texas-based utility firm that is one of the largest buyers of wind power) has acquired property rights to two new wind farm projects in Spain. American Electric Power, a large utility firm, developed a 150-megawatt wind farm and acquired another one, both in Texas. PacifiCorp Power Marketing, a subsidiary of PacifiCorp, is acquiring and marketing power from three wind farms in the West, including the 261-megawatt Stateline project, and has announced plans to add significant wind power capacity to its equity portfolio over the next few years.

The US Department of Energy's wind energy program and a growing number of projects in the Great Plains are generating interest among farmers and other rural landowners in wind energy. This, in turn, is leading to increased political support for wind energy in the Midwest and West. The states of New York, Wisconsin, Michigan, and Pennsylvania are using public procurement of renewable electricity as a means of stimulating and supporting clean energy sources. The former governor of New York, George Pataki, once issued a directive to state bodies to purchase 10% clean energy by 2005, and to raise that number to 20% by 2010, while in Pennsylvania, the state purchase of clean energy in 2002 was 5% of the total capacity. The emerging research into the real cost of integrating wind energy into the grid, as well as a number of other aspects of wind energy having not yet been taken into account, makes some authors doubt the effectiveness of this direction. The criticism of the privileges provided continues unabated. In view of this, the 2002 balance in the United States turned out to be relatively modest, but the impact that was obvious in 2001 was later restored by the extension of the tax credit, which was immediately reflected in a sharp increase in the commissioning of wind power capacity (Fig. 1.9), although the total contribution to the country's energy balance had not reached 1%.

The average growth rate of wind energy over 5 years was significantly lower than the maximum reached, but the growth rate of 2,431 and 2,454 MW in 2005 and 2006, respectively, suggests that the power balance of "supporters and opponents" of wind energy in the United States was still on the side of traditional wind energy. The influential *Washington Times* on July 27, 2006, published an article titled "Energetic Turn to Wind Power", which cites President George W. Bush's desire to bring energy consumption from wind farms to 20% of the country's total demand. With respect to the real wind regime in the atmosphere surface layers in the US, this would mean that the installed capacity of wind turbines would reach 60 to 80% of the country's capacity balance. Such an intention, if it is based only on generally accepted technologies, seemed quite implausible. Yet, if we focus on the qualitatively new directions discussed in this book, then together with the energy of a new type of tidal power plants and the use of hydrogen energy as a universal energy accumulator and distributor, the outlook proposed by George W. Bush would not only be plausible, but would also be economically viable.

Meanwhile, the density of the wind energy flux at those heights of the surface layer that are now technologically accessible remains relatively small, reaching economically effective values in small areas located mainly in mountainous regions or in coastal zones of seas. There are also a number of fundamental negative features of wind energy, which begin to manifest themselves as the scale of its use increases. First of all, it is obvious that separate wind turbines and wind farms located in one place do not provide

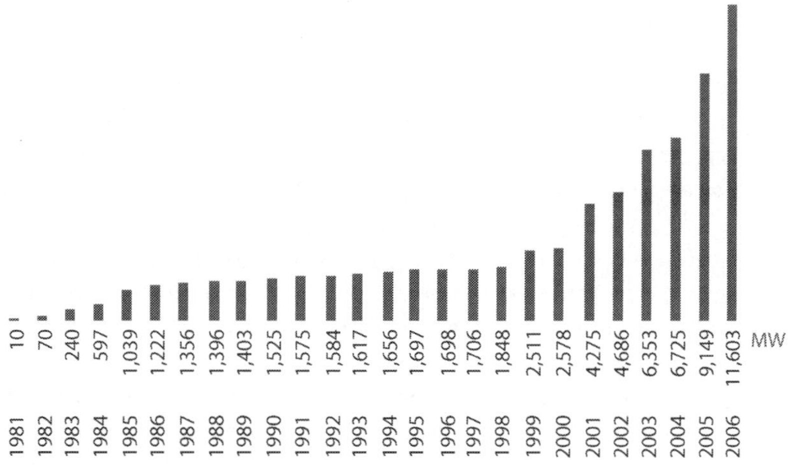

Fig. 1.9 Total U.S. Wind Farm Capacity at the End of the Year.

guaranteed power. They are required to be built in specific areas that span a considerable size, and can significantly change the local wind regime, along with generating significant acoustic radiation. Direct economic assessments carried out at a relatively high utilization of the installed capacity of wind farms (about 30%) indicate the following increase in the cost of wind farm energy (EUR/MWh):

- integration costs 3 to 4
- balancing and redundancy costs (when the wind power plant capacity is less than 10% of the system capacity) 1 to 2
- modernization of electrical networks 0.1 to 4.7
- operating costs 12 to 25

The actual rise in price should be much higher, since the relative utilization of the installed capacity in many cases is to be significantly less than 30%. Wind farms occupy specific areas—according to the US experience, from 4 to 32 hectares per 1 MW, which has not yet been taken into account when calculating the cost involved. Moreover, changes to the landscape introduced by wind farms (Fig. 1.10) are not welcomed by the population. There is no reliable evidence of the environmental safety of modern large wind turbines, the ends of the blades of which move at a speed exceeding 100 m/s. Such speed is destructive for the living inhabitants of the atmosphere.

In developed countries with a high population density, the owners of wind farms are forced to remove restrictions imposed on the "exclusion zones for wind farms". In Japan, Germany, and France, wind turbines are located right by the roadsides, which, of course, does not contribute to traffic safety. More attention is being paid to the placement of wind turbines in the coastal zone of the seas. Thus far, structures based at the bottom of the sea were used (Denmark, Germany), which caused an additional rise in cost and was not always possible, for example, due to a sharp increase in sea depth. In 2007, a 3 MW demonstration of a floating wind turbine with a rotor of about 90 m in diameter was launched in Norway (the Hydro Company, project Hywind: Fig. 1.11). The company intended to increase the diameter of the three-bladed rotor to 120 m, and the installed capacity of the unit, to 5 MW. For a demonstration project, the company needed $22.8 million.

The world's largest wind turbine Enercon E-126 was installed in 2007 in Emden, Germany. The height of the base column is **135** meters, and the rotor diameter is 127 meters. The maximum power is 7.58 MW (a design nominal of 6 MW), n = 5 to 11.7 rpm, and the weight of the column is 2,800 tons. U_{min} =3 m/s, U_{nom} = 9m/s. The installation foundation weight is 2,500 tons; the weight of the bearing tower is 2,800 tons; the generator nacelle has a weight of 128 tons, and the weight of an electric generator is 220 tons; the weight of the rotor together with the blades is 364 tons. The total weight of the wind turbine is about 6,000 tons. The largest HPP in the world with a capacity of almost 7.9 GW is the Chinese "Gancy" projected to produce 20 GW by 2020. In 2011, the Indian HPP "Muppandal" was commissioned, the installed capacity of which was 1.5 GW. The world's third power plant of 1 GW is the Indian Jaisalmer Wind

Fig. 1.10 One of the first large wind farms in the United States is Tehachapi Pass, California. The landscape is clearly depressing. There are reports that snakes have left this area.

Fig. 1.11 A 3 MW floating wind turbine with a rotor diameter of 90 m. Norway, 2007.

Park, operating since 2001. Source: https://energo.house/veter/krupnejshie-ehlektrostancii.html.

In recent years, wind energy has continued to develop. However, the growth rate of world energy is declining: if in 2015 a record 62.6 GW was commissioned, then in 2016 the indicators declined to 54.6 GW. In 2017, 52.57 GW of new capacities were commissioned around the world. By the end of 2017, the total installed capacity of wind power plants in the world reached 539.58 GW. The modern leader in wind energy is China. According to Windpower Intelligence, the installed capacity of the wind farms in China is 171.8 GW. From a ten-year perspective, China intends to commission more than 20 GW annually until 2027.

In Europe, 16.8 GW was added in 2017, of which 15.7 GW fell on the countries of the European Union. Germany and Great Britain are leading with 6.6 GW and 4.3 GW each, respectively.

In North America, growth rates have been falling in recent years. In the United States, in 2017, the increase was only 7 GW, and only to rise to 7.8 GW in total throughout the continent.

Among the Latin American nations and the Caribbean, Brazil excels with the introduction of new capacities, adding 2 GW out of the total of 2.6 GW.

In the combined African and Pacific region, only two countries showed significant growth: South Africa with 621 MW, and Australia with 245 MW.

At the end of 2017, the top three leaders in terms of total wind power capacity in the world were China (188.2 GW, 35%), the USA (89 GW, 17%) and Germany (56.1 GW, 10%). The share of the top ten countries in the ranking in the global wind energy balance is 85%.

Despite the obvious successes, a comprehensive analysis of all aspects of the wind energy direction, focused on the use of traditional technological solutions, causes very restrained optimism[1]. In fact, the observed rapid growth of wind farms is largely a consequence of political decisions and financial (budget) benefits indirectly paid by the entire population.

This applies almost exclusively to wind farms with high-speed wind turbines having a traditional layout—a horizontal axis of rotation and a three-bladed rotor on a high tower (HAWT, in American terminology). At the beginning of the 21st century, 10 companies provided 95% of sales of wind energy equipment (Table 1.3).

Vestas is the leader in the supply of wind turbines (Fig. 1.12) with a rotor diameter of 80 m, a steel tower height of 60, 67, 78, 100 m, a power of 1.8 MW at a 15 m/s wind speed (power output starts at 4 m/s), a rotor weight of 34 t, a nacelle weight (excluding rotor) of 61 t, and a total weight above the foundation for the indicated tower heights is 205, 225, 265, and 315 tons. Thus, the specific consumption of materials is from 102 to 158 kg/kW.

---

[1] M. Schwartz, March, 2004,"Wind — Facts or blowing hot air?" The Virginia Land Rights Coalition POB 85, McDowell, Virginia FOC 24458
Glenn Schleede, April 14, 2005, "Big money" discovers the huge tax breaks and subsidies for wind energy while taxpayers and electric customers pick up.

Table 1.3

#	Manufacturer	Country	Share of sales, %
1	Vestas	Denmark	21.6
2	Enercon	Germany	17.9
3	Neg Micon	Denmark	13.9
4	Gamesa	Spain	11.5
5	General Electric	USA	8.6
6	Siemens	Denmark	6.8
7	Nordex	Germany	6.8
8	Made	Spain	3.3
9	Repower	Germany	3.0
10	Ecotecnia	Spain	1.6
11	Other		5.0

The largest machines from NEG Micon have approximately the same performance; the rotor diameter of 60 m, 2 MW power capacity. For areas with a relatively weak wind, the Enercon system units are good with the rotor diameter of 70 m, tower height of 65, 86 m, and the power capacity of 1.8 MW at a wind of 12 m/s (power output starts at 2.5 m/s).

The largest machines are offered by Nordex. The diameter of the projected rotor is 115 m, the tower height is 85 m, the power is 5 MW. As can be seen, for almost 70 years of development of wind energy, the maximum capacity of a single wind turbine with a traditional layout has not increased. According to Boeing's estimates, the length of cantilever blades made of modern materials cannot exceed 60 m and, accordingly, the effective maximum power of a wind turbine with a traditional layout cannot exceed 7.3 MW.

All large machines under construction are oriented to work in a powerful electrical network having a similar layout (Fig. 1.12), although they differ in some details.

The common features that primarily appear are the use of three-bladed rotors with a rotational speed of up to 30 rpm and swivel (P) or fixed (S) blades oriented downwind using a special electromechanical drive, with the use of planetary multipliers or low-speed generators (Enercon) and the

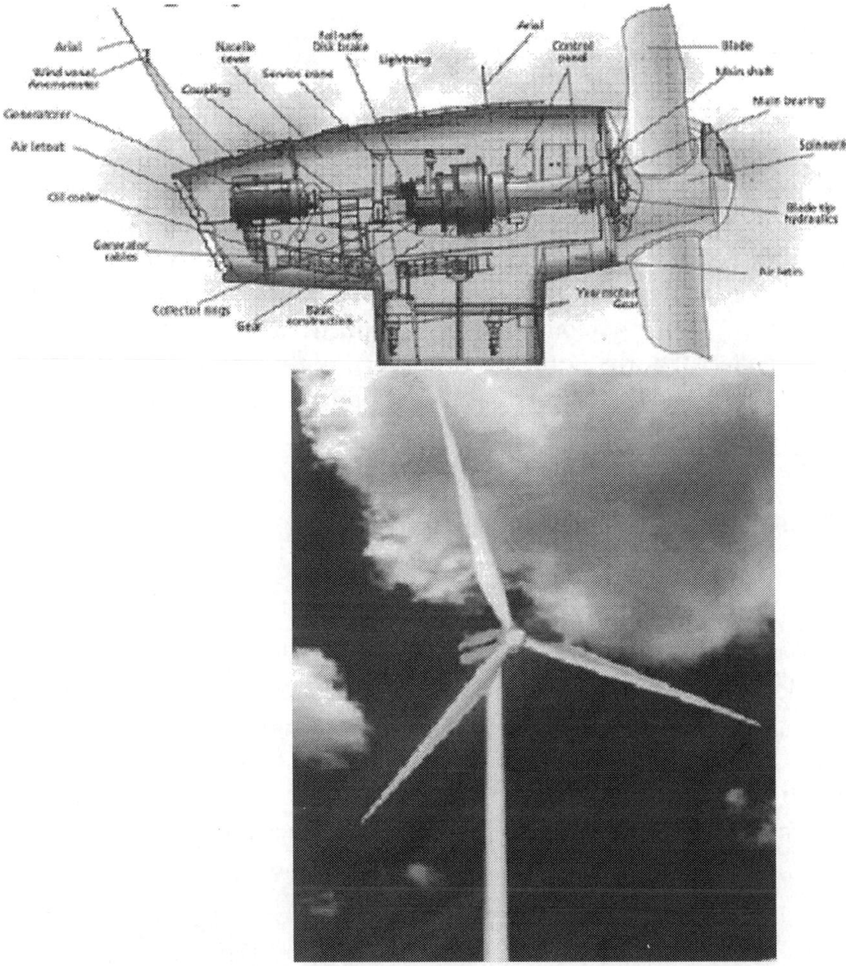

Fig. 1.12 A Vestas modern serial wind turbine. Power capacity: 1800 kW.

use of two different induction generators in some circuits with electronic speed control. Optimal blade control increases energy production by 15%, and when combined with speed optimization, the production at slight winds can be increased by about 80%.

Depending on the average wind speed, the average power of a wind turbine with fixed (optimized) blades and two generators with different nominal rotation speeds can be from 17 to 65% of the nominal power of the unit, produced at the wind nominal speed of U_n = 11 to 12 m/s.

Average wind speed, m/s:	5	6	7	8	9
Relative Average power, %	17 to 19	28 to 31	37 to 41	48 to 53	54 to 65

Real commercial wind turbines have significantly lower power generation rates. This is primarily due to the choice of the design wind speed U_n corresponding to the rated power. Fig. 1.13 shows the relative average power of wind turbines of different companies with almost the same diameters of three-bladed rotors (43 to 44 m) and the same rated power of 600 kW, achieved, however, at different wind speeds. At an average wind speed of 7 m/s, the best Enercon machine demonstrates the best performance due to the fact that the nominal wind speed of this machine is minimum – 12 m/s. In Fig. 1.13, the equation of the upper envelope (Enercon units):

$$t = 0.0760\ (U - 2.5),\ U > 2.5\ \text{m/s}.$$

Lower envelope equation (Micon units)

$$t = 0.0476\ (U - 2.5),\ U > 2.5\ \text{m/s}.$$

The largest wind turbines with a traditional layout, built in Russia, in Kalmykia ("Raduga" Scientific Production Association, Tushinskyi Machine Building Plant) have a capacity of 1,000 kW at a wind speed of about 12 m/s.

The rotor diameter is 48 m, the rotational speed is variable (from 21 to 42 rpm), the tower height is 36 m, and the weight above the foundation is 130 tons. The average annual capacity of this unit ranges from 29% (Crimea) to

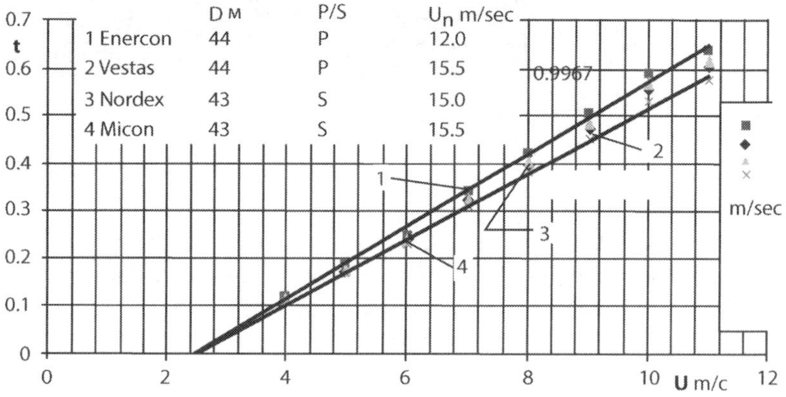

Fig. 1.13 Relative time of using the installed capacity of the wind turbine of 600 kW, depending on the average annual wind speed at the unit axis. The distribution of the probability of wind speeds, normalized by the average speed, is taken for the conditions of Kamchatka to be the same for different average speeds.

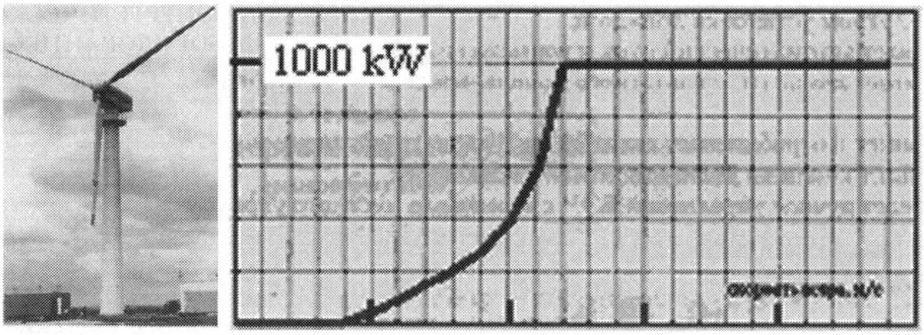

Fig. 1.14 A TMZ serial wind turbine, Moscow ("Raduga" Scientific Production Association).

56% (Cape Lopatka in Kamchatka) of the nominal value. These machines, in terms of their technical indicators (Fig. 1.14) and cost ($1,000/kW for single orders and up to $700/kW for serial production) could successfully compete in the world market.

However, this did not happen, as Russia was at the very beginning of its development in 2018.

In January 2018, the most powerful Russian project was launched the **"Fortum"VES-1 Ulyanovsk wind farm** with a capacity of 35 MW and a cost of about 65 million euros, built over the year in the Ulyanovsk region and managed by the Russian company Fortum. The wind field is made up of 14 wind turbines manufactured by the Chinese company DongFang with a capacity of 2.5 MW each. The level of localization is only 28%, and all the equipment is Chinese, following the approval of the project in 2015. According to the results of the first quarter, the Fortum wind farm produced 21.61 GWh. It was expected that the annual sales of electricity generated at UVES-1 would be 85 GWh.

In the development of the project, the government of the Ulyanovsk Region planned to create several wind farms with a total capacity of 250 MW within two years, and increase the total capacity to 600 MW within two years, to be followed up with a rise to 1 GW by 2030.

Zelenogradskaya wind power plant located near the village of Kulikovo, Zelenogradskyi District, Kaliningrad Region: The installed capacity of the WPP is 5.1 MW. The Zelenogradskaya wind farm has one Wind World 4200/600 wind generator installed with a capacity of 0.6 MW and 20 Vestas V27/225 wind generators with a capacity of 0.225 MW each. All wind generators were purchased secondhand; due to problems with spare parts for obsolete equipment, most of the generators now stay idle.

Wind farm Tyupkildy is located near the village of Tyupkildy, Tuymazinskyi district of the Republic of Bashkortostan. The installed electric capacity is 1.65 MW. It was commissioned in 2001. The equipment includes three ET-550 German-made wind turbines, 550 kW each. The fourth generator burned out in February 2017.

Kalmyk wind farm was commissioned on December 5, 2009. The installed capacity is 1 MW.

Wind farm (village Tamar-Utkul) is located in the Orenburg region. The installed capacity of the wind farm is 0.925 MW. Commissioned in 2013.

Wind farm in OrskCity — the installed capacity of the wind farm is 0.4 MW. It was commissioned in 2015.

Wind farm LLC, AltEnergo, is located in the Belgorod region. The installed capacity of the wind farm is 0.1 MW. It was commissioned by AltEnergo LLC on August 1, 2010.

The Anadyr wind farm is located at the Observatsii Cape of the Anadyr district of the Chukotka Autonomous Region of the Russian Federation. It provides electricity to the population of the settlements in Shakhtersky, Ugolnye Kopi, and to the Ugolnyi airport. The installed electric capacity is 2.5 MW. Until 2006, the power plant generated 3 million kWh; in subsequent years, the supply of electricity significantly declined. It was also planned to build 17 similar wind farms in the district, united into a single network, but these plans remained unfulfilled.

There are seven wind farms operating on the territory of Crimea:

- **Ostankino wind farm** with an installed capacity of 25 MW;
- **Tarkhankutskaya wind farm** with an installed capacity of 22,45 MW;
- **Sakskaya wind farm** with an installed capacity of 20.82 MW;
- **Presnovodnenskaya** wind farm with an installed capacity of 7.39 MW;
- **Donuzlavskaya wind farm** with an installed capacity of 6,78 MW;
- **Sudakskaya wind farm with an** installed capacity of 3,76 MW;
- **Eastern-Crimean** wind farm with an installed capacity of 2,81 MW.

Network wind turbines with a capacity of up to 1 MW are produced by many companies in Denmark, Germany, Spain, Italy, the USA, Canada, and Japan.

With an increase in the unit capacity of wind turbines and the creation of energy systems on their basis, unit costs are reduced several times. Thus, the Fuhrlander company (Germany) offered a series of similar traditional network machines with induction generators with a capacity of 30 to 1,000 kW with the following unit costs (ex-works), and designed wind speeds at the unit axis:

Rated power, kW1002508001,000
Design wind speed, m/s12151413
Per unit cost, $/kW (2000)1,200920837790

The prices quoted are typical for Europe in previous years. For example, 250 kW NEDWIND machines (Denmark) were offered in 2000 for $237,750 ($950/kW), and larger 500 kW machines — for $446,800 ($894/kW).

According to negotiations held in the summer of 2000, 750 kW network machines from the NEG Micon international company, with a fairly large order, could be purchased from the manufacturer at a price of about $700 per kW. A more complete price list for the company's machines had the following form in 2000:

TypeNM600/43NM750/48NM900/52NM1000/60NM1500/64
Power, kW600/150750/200900/2001,000/2501,500/300
Rotor diameter, m4348526064
Tower height, m4545444560
Factory price, $451,332551,628576,702877,5901,128,330

Specific cost, $/kW752735641878752. These machines have a high degree of assembly readiness which allows them to be erected in a short time. So, at the construction of the Le Nordais wind station in Canada (Quebec) in 1999, 57 units of 750 kW each were installed and commissioned in 6 months, which brought the total to 10 units per month! A typical energy characteristic of a 900-kW machine is shown in Fig. 1.15.

Taking into account the construction, installation, and transportation costs by the coefficient 2 (?), we find that with a 20-year service life, the cost of energy from these machines in an area with a wind speed of 7, 8 or 9 m/s will be 1.95, 1.43 or 1.23 cents per kWh, respectively. Of course, the quality of energy and, accordingly, additional costs for duplicating capacities and power lines, operating costs for maintaining complex mechanical systems, as in all machines of this type, significantly increase the energy cost.

Low-power machines designed for the individual consumer with power reserve on batteries or on the hydrogen cycle come out significantly more expensive as per unit of power capacity and output.

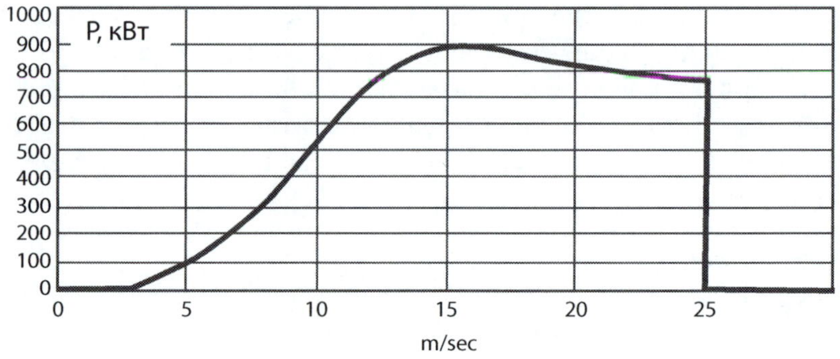

Fig. 1.15 Energy characteristic of the NEG Micon wind turbine. The rotor diameter is 52 m. A 900/200 kW induction generator, 4/6 poles, planetary gear with a gear ratio of 67.5 (for 50 Hz).

For example, one of the best wind power complexes for areas with medium or weak wind and a guaranteed power capacity of 2 to 5 kW, serially produced in Russia, has a price of about 9,000 euros. The rated power of the wind turbine in this complex with a low-speed contactless generator reaches 2 kW at a wind of about 8 m/s (this is the best indicator in the world), and a high guaranteed power is achieved by using capacious batteries and a powerful inverter. Another series of simpler, inexpensive wind turbines with a capacity of up to 1 kW (with a three-bladed rotor 1.5 or 2.2 m in diameter) has been developed and manufactured by the Russian enterprise Elektropribor (St. Petersburg).

This unit is peculiar becasue of the lightweight design of the blades (aluminum sheet with foamed polymer filler). The original scheme for avoiding stormy winds was to have the unit axis turn into a vertical position under strong winds, which was a special design that included a contactless generator with permanent magnets on the generator rotor, and an automated control system. The rated power is achieved, as usual, at relatively high wind speeds of 12 to 13 m/s (Fig. 1.16); however, the mass of the unit complete with the tower, excluding batteries, is only about 50 kg.

A series of 0.2 to 5 kW single-blade wind turbines with an estimated wind speed of 8 to 9 m/s is offered by the Russian company VINDEK at a price of about $3,000/kW.

These units with 0.9 to 2.5 m long blades and a "tail" orienting the rotation axis in the wind, operate on a 12, 24, or 48 VDC battery. The cost of a 6 to 8 m mast, fasteners, and control system is included in the price. Batteries, inverters, and phase matching systems at the inverter output are not included in the specification. Just like all machines of this type, these wind turbines require constant monitoring—in a turbulent wind, they can

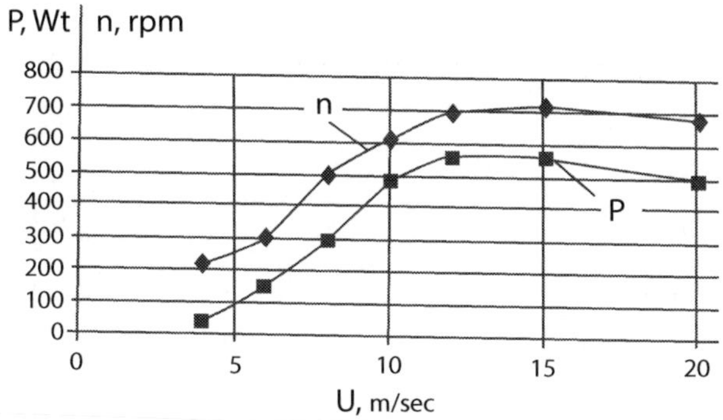

Fig. 1.16 Energy characteristics of the wind turbine of the "Electropribor" Central Research Institute RF.

be twisted for several turns around the vertical axis thus causing a cable break. A large list of manufacturers of small wind turbines up to 100 kW can be found on the websites of national wind energy associations, in particular on the AWEA website.

The modern layout of wind turbines, which now can already be called traditional, was the result of a strict selection from numerous layouts seeming promising. Nevertheless, the economic indicators of these "best" machines are such that the entire industry is tightly associated with state support (patronage, subsidies, benefits, etc.), and where this support is weakening (USA, Denmark, possibly Germany) or has not yet emerged (Russia), wind energy is losing investors. Wind power plants are economically viable without external subsidies and are used only in extreme conditions when there are no other sources of energy available, or when they appear to be very expensive. In these cases, for example, the WTIC company (USA) offers the "wind turbine + solar battery + diesel" complexes, in which a 10 to 17.5 kW traditional three-bladed wind turbine with plastic blades (Fig. 1.17) with a 24 m tower has a price from $23,800 to $26,100.

According to the authors, orthogonal wind turbines in which the rotation axis is perpendicular to the direction of the wind, have more promising prospects.

Their operation does not depend on the direction of the wind, and a significant environmental advantage lies in the reduction of acoustic radiation in those layouts in which all points of the rotor blades move at the same speed. Of particular interest in the middle of the 20th century were the aggregates designed according to the Darrieus scheme (Fig. 1.18).

The energy efficiency of optimized orthogonal units can be no lower than that of traditional ones. However, in the form in which the orthogonal

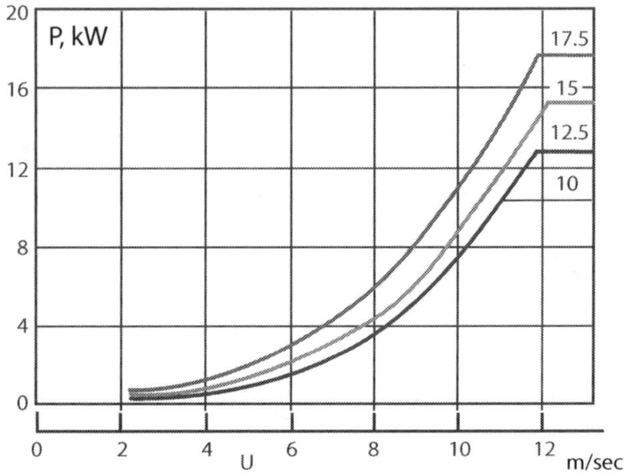

Fig. 1.17 Energy characteristics of a WTIC wind turbine (USA).

Fig. 1.18 Orthogonal wind turbines designed according to the Darrieus scheme (1931). On the left is a serial 500 kW unit (Canada); on the right is one of the first experimental 40 kW machines (USA).

units were initially tested and constructed everywhere, they had an important drawback—pulsating torque on the generator shaft and pulsating forces acting on the blades and the whole unit.

This circumstance was known to the designers and was taken into account in the projects. However, the experience of testing and operating such machines showed that this accounting was not sufficiently complete and comprehensive.

The largest wind turbines were designed and constructed precisely in vertical-axial design—a 4 MW wind turbine was built in Canada (Fig. 1.19).

In England, from 1980 to 1985, intensive research was carried out on an orthogonal wind turbine with variable geometry. Calculations and tests of models and full-scale samples have shown that the energy efficiency of such a layout can exceed 40% (Fig. 1.20).

In the USSR, the first series of orthogonal wind turbines with blades from helicopters and induction generators was manufactured and tested from 1981 to 1985 (Fig. 1.21).

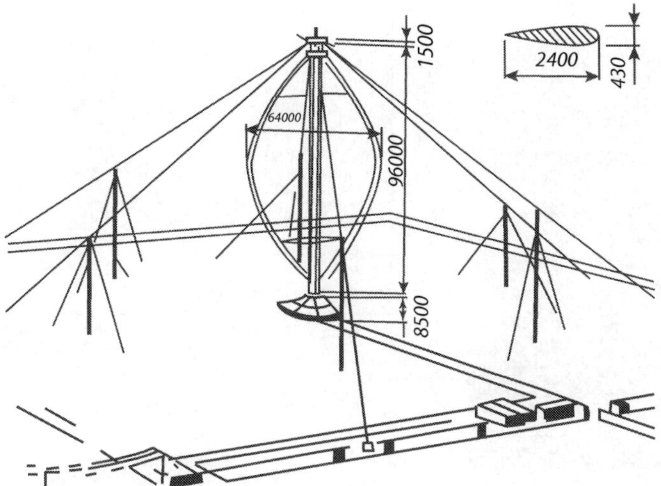

Fig. 1.19 A 3.8 MW EOLE orthogonal wind turbine at Cap Chat cape on the southern shore of St. Lawrence (Northeast of Montreal), Canada, 1981 to 1987 Hydro Quebec Project.

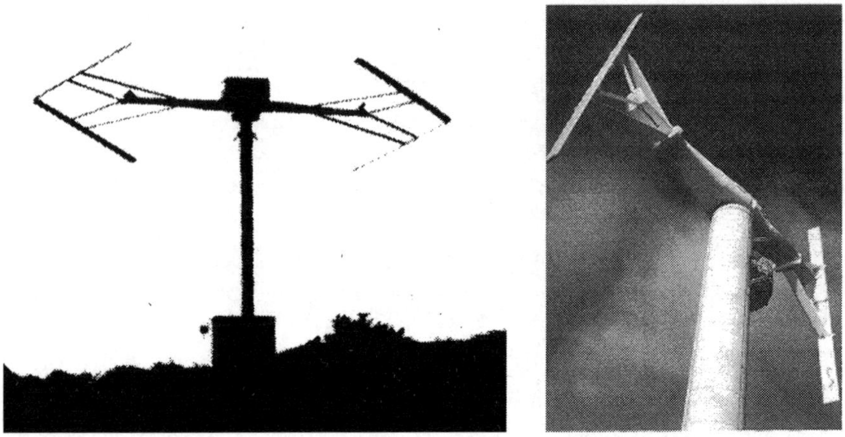

Fig. 1.20 Wind turbines 6 m (left) and 25 m (right) in diameter with variable rotor geometry of the P.J. Musgrove design. England, 1982. The power capacity of the large unit is 100 kW. England, 1982.

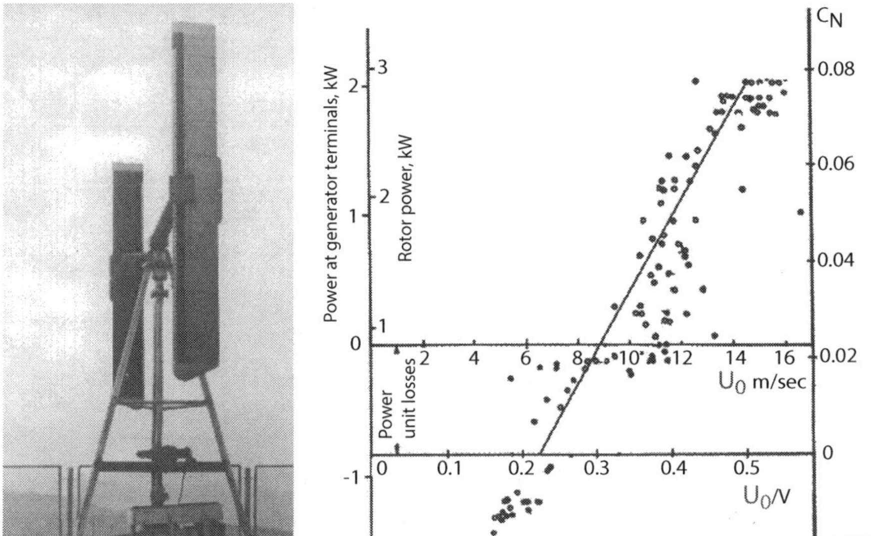

Fig. 1.21 Wind turbine in the village of Dubki. The rotor diameter is 4 m. NACA 23012 profile (Blades of Mi-2 helicopter), a chord of 0.4 m, $\sigma = 0.2$. The flat side of the blade is directed towards the rotation axis. The rotation speed is 140 rpm. Field test results. B_0=0.228, B=0.27.

Fig. 1.22 A 16-kW wind turbine, the village of Dubki (Chirkeyskaya HPP), 1981.

Experiments and calculations have shown that units with a similar arrangement can have a high power[2]; however, it is advisable to use them as part of energy networks. In the same period, a project was developed, models were tested and a prototype of a 16-kW serial wind turbine was built near the village of Dubki in Dagestan (Fig. 1.22)[3].

Larger 130 kW and 1,000 kW wind turbines designed with rotors made of wings from the Yak-40 aircraft, were built in Tajikistan (Fig. 1.23) and Kamchatka (Fig. 1.24).

---

[2] Malyshev, N.A., Lyatkher, V.M. "High power wind turbines", journal Hydraulic Engineering, 1983, No. 12, pp. 38-44

[3] Malyshev, N.A. and others "Wind turbine..," US patent 4281965, USSR patents 1242637, 1257276

Fig. 1.23 A 130-kW orthogonal wind turbine at the Chormozak pass, Tajikistan, 1988.

Fig. 1.24 A 1,000-kW wind turbine in the village of Beringovskyi (Magadan Region), 1991.

Orthogonal units in their original form and in the simplest industrial designs were not widely used. However, newly created design solutions along with the accumulated data of laboratory and field tests, which will lead to the optimization of rotors, may reasonably restore the interest in this direction. In parallel, experiments were started with models and full-scale samples of orthogonal hydraulic turbines for damless hydroelectric power plants[4] (Fig. 1.25) including those based on tidal currents[5].

Orthogonal hydraulic units in the pressure version were practically implemented at the Kislogubskaya tidal power plant in the Russian Federation[6]. Modern solutions (Fig. 1.26) are focused on application in free rivers, canals, and straits (damless – item 6). The turbines can be installed at the bottom or fixed under pontoons (Fig. 1.26, right).

Free-flow turbines of the proposed orthogonal type, due to the cross-sectional shape, have an advantage over traditional hydroelectric power stations, especially for flat rivers and river sections without an ice cover.

For river sections that do not allow the creation of dams and reservoirs, underwater extensive grounds offer a series of standardized environmentally friendly generating units. These are fixed directly in line with the bottom of the river or placed under the pontoon to be installed along the stretch of the river, where there is maximum velocity and flow depth and a consumer of electricity.

The feature of the orthogonal turbine is that the turbine axis of rotation is perpendicular to the direction of flow, and the speed of movement of the blades is 2 to 3 times more than the flow rate at the mouth of the turbine. The blades in these turbines have the profile of an airplane wing and use pull force when the flow occurs around this profile. The most effective scheme in the streams of great depths at speeds of 2-3 m/s may be mechanically trimmed using the tiered single-vane turbines (V. Lyatkher, US Patent 8007235 B1) and two-bladed turbine with two counter-rotating generators (V. Lyatkher, US Patent 7741729 B1) shown in Figs. 1.27, 1.28.

---

[4] V. M. Lyatkher, I. I. Ivanov, S. M. Skosareva. Experimental studies of orthogonal aggregates for using the energy of currents, Hydrotechnical Construction Journal, 1986, No. 11, pp. 33 to 37

[5] V.M. Lyatkher. Environmentally friendly energy conversion of air, river, and ocean currents, *Hydraulic Engineering Journal*, 1989, No. 8, p.

[6] Istorik, B.L., Usachev, I.N., Shpolyanskyi, Yu.B. New orthogonal hydroelectric unit for the tidal power plant at the Kislogubskaya TPP, *Hydraulic Engineering Journal*, 2007, No. 12, pp. 38 to 44 *Vertical–axial two–rotor wind power units Bidarrieus-1:* Yershin, Sh.,Yershina, A.K., Ydyryssova, A.A. *Eurasian Physical Technical Journal*, 2017, vol. 13, No.4(28) – pp.108-112., p-ISSN 1811-1165, e-ISSN2413-2179.

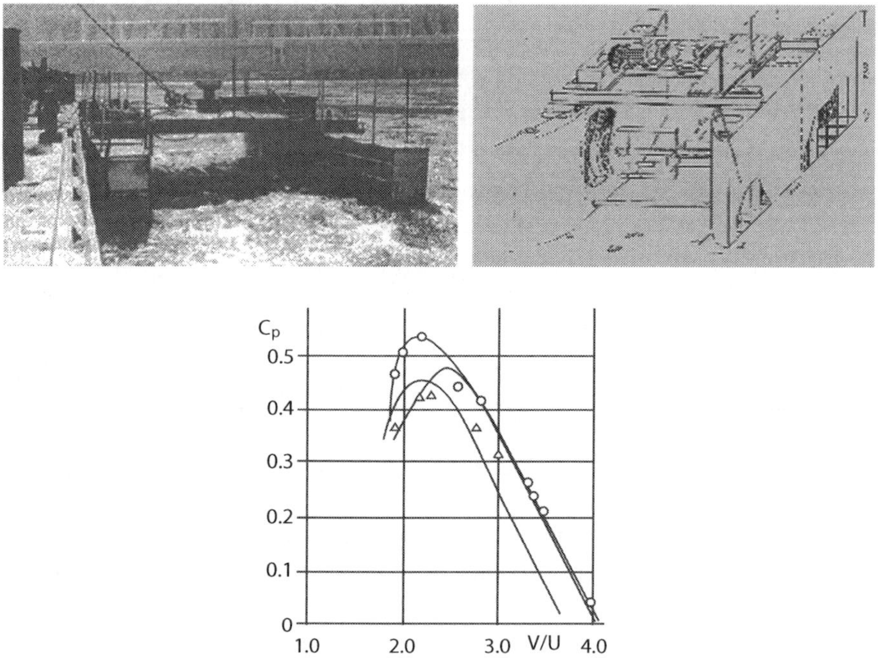

Fig. 1.25 Turbine with straight blades in the downstream of the Krasnoyarsk HPP (1988). The turbine diameter is 1m, length — 2.45m, chord — 0.15m, power capacity — 11 kW. The flow speeds vary from 1.79 to 2.14 m/s.

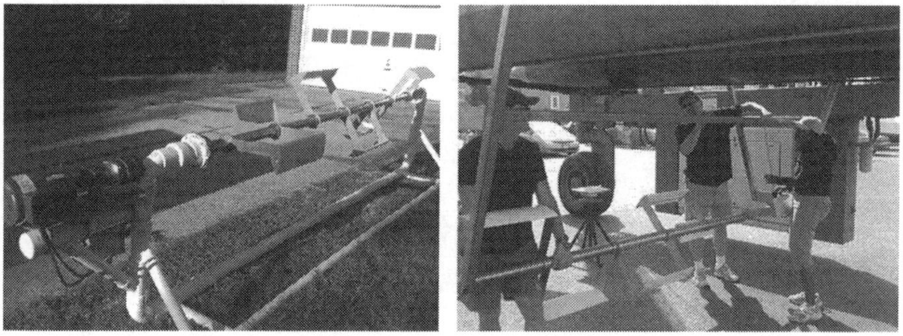

Fig. 1.26 Single-blade 6-stage turbine (US Patent 8007235). A 160 mm chord, one tier is 0.45 m long. The water flow rate is U = 0.5 to 2 m/s, the optimum rotating speed is n_{optim} = 103 U, rpm. The maximum power is P_{max} = 617.2 U^3 (W), The accelerating rotating speed (no load) is n_{max} = 160 U, rpm. (2010, USA).

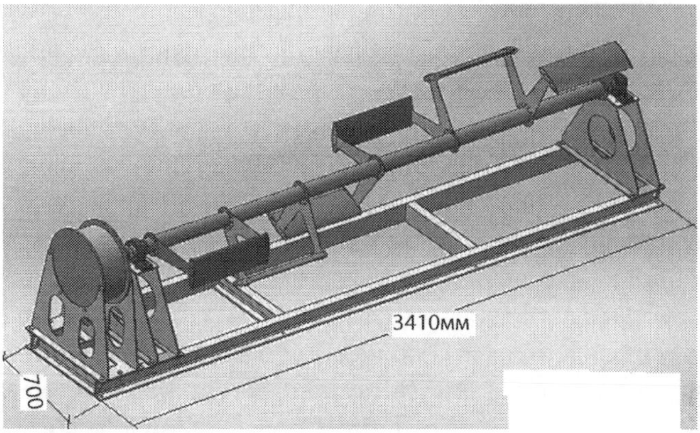

Fig. 1.27 Optimized single-blade turbine.

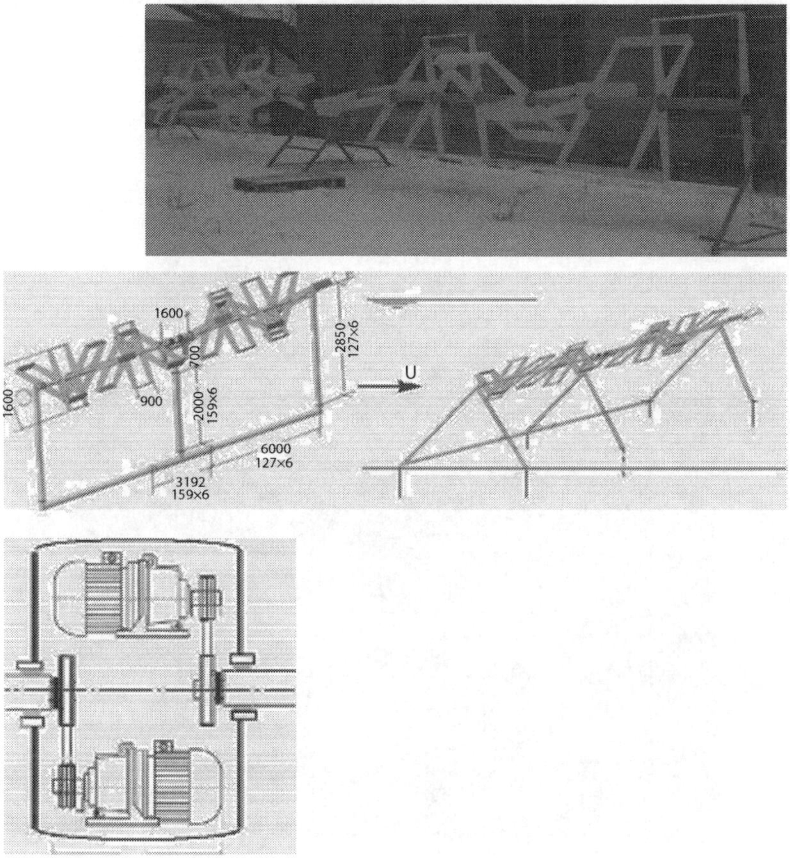

Fig. 1.28 Optimized two-blades turbine.

Fig. 1.25 features turbines that have been tested in the laboratory, and also those that have been tested in nature. The turbine, together with the generator, can be installed at the bottom of the river. The fixing of the turbine under a pontoon (Fig. 1.26) or inside of it makes the system agile. Another way could be to place the turbine in a hanging position inside a stationary or mobile variant (Fig. 1.29).

Turbines with a fixed shape and set dimensions of the blades (there is now ready production of blades with a chord of 160 or 180 mm, the length of one blade to 3M, usually 900 mm), are different than those with one blade, as the blades in each differ in diameter (radius R of the track of the blades). The energy characteristics of turbines of different diameters in a deep and wide stream of water at a rate of 1 and 2 m/s are shown in Fig. 1.30.

Turbines of this type are applicable to flows with a speed of from 0.5 to 3 m/s. The frequency of rotation is proportional to the speed and power of the flow velocity in the cube. Machines are labeled with the length of turbine (L=2.7 or 5.4 m)

the rotors diameter 2R= 0.64 m,
1.28 m
1.92 m
and the estimated flow velocity U= 1 +/- 0.25 m/s;
1.5 +/- 0.25 m/s;
2+/-0.5 m/s;
3+/- 0.5 m/s.

Fig. 1.29 Turbine lowered into the water discharge channel is fixed in a rigid frame.

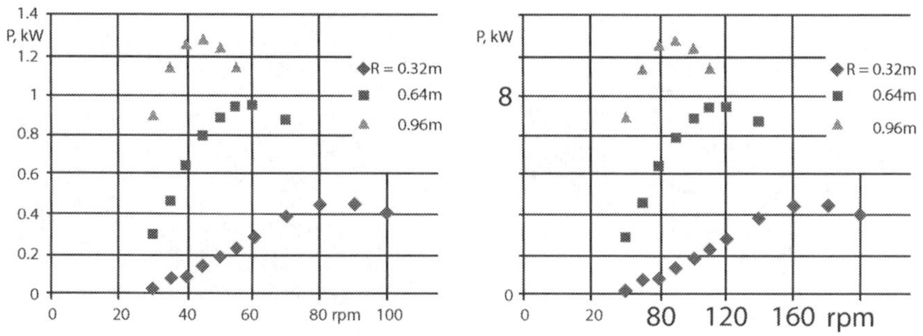

Fig. 1.30 Capacity of turbine with L=5.4 m. with different R for water speed U=1 m/s (left) and U=2 m/s (right).

Depending on the design, speed of the flow changes the electrical Assembly and the attachment system to the bottom or to the pontoon (Table 1.4).

In that case, if it can be used as a significant part of the flow in the cross-section (20%), hydropower capacity can be considerably improved by increasing the chord length, the number of blades, and the solidity of the turbine (ib/D).

At a constant flow through the turbine, power and the pressure in the turbine increases in proportion to the shading of the turbine. By increasing the number of blades in each tier, balancing of the turbine can be achieved with a smaller number of tiers.

For example, if the depth of flow is 1.5 m, width is 20m and flow rate of flow is 2 m/s (a common occurrence for smaller rivers of Nepal in the foothills), the recommended turbine of use can be a three-tiered turbine with a length of 10.8 m with a diameter of 1.28 m with 2 or 3 blades in each tier. This turbine would have maximum power 42кВт (Fig. 1.31).

Table 1.4 Maximum power (kW)/optimal speed (rpm) six-tier a single blade turbine with a length of 5.4 m with a diameter D, m.

U, m/s	**1**	**1.5**	**2**	**2.5**	**3**
D, m					
0.64	**0.4**/90	**1.4**/130	3.5/180	7/225	**11**/260_
1.28	**0.9**/60	3/90	7/120	14/150	**24**/180_
1.92	**1.2**/50	4/75	**9.5**/100	**21**/125	**32**/100_

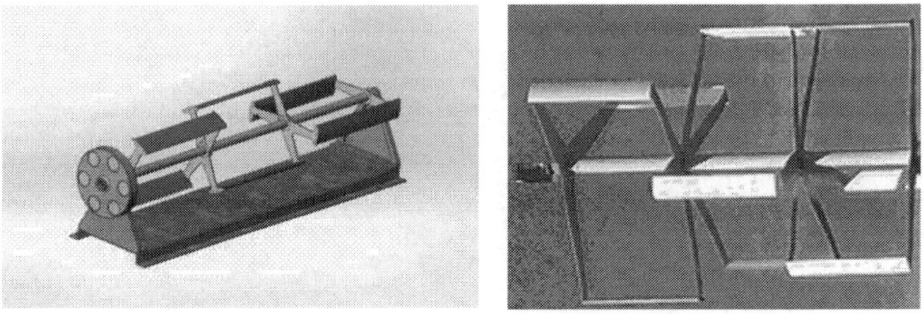

Fig. 1.31 Three-tiered turbine with two (left) or three (right) lobes in each tier.

Fig. 1.32 Spiral-blade turbine.

Fig. 1.32 shows a weak-willed turbine with spiral blades that is currently being developed. Such turbines are touted to cause minimum impact on aquatic life.

The turbine may be connected to a power source that is 220(127) V, 50 (60)Hz (option 1), or work on the individual consumer (option 2).

Radical increase in the effectiveness of use of flow energy from wind and tidal waves is proposed by using orthogonal turbines (Chapter 6). This increase proves to be possible due to the new principle of designing tidal power plants (TPP) without the head front – without the dams, power-house, by-wash and other hydraulic structures[7]. The proposed approach excludes the necessity of fulfilling the construction and installation works at the site of TPP use, thereby lowering the expenditure per unit capacity and power production. Above all, this would shorten the time of setting the units into operation, and the time of return on investment. At the same time, it would make it possible to improve the quality of the obtained energy, ensuring the issuing of the guaranteed power according to the user schedule.

[7] Лятхер В.М. Оптимальное использование энергии приливов, ДАН РФб 2006, т.411, №4, стр.476-479

The realization of the proposed idea can significantly change the structure and balance of Russian and world energy. The deployment of power stations of the proposed type does not require large lump-sum capital investments or long durations; neither does it create ecological problems; this work with minimum risks can be begun immediately.

High efficiency may also be achieved by using the waves and winds in Bangladesh. Figs. 1.33 and 1.34 show information about winds and waves in this area.

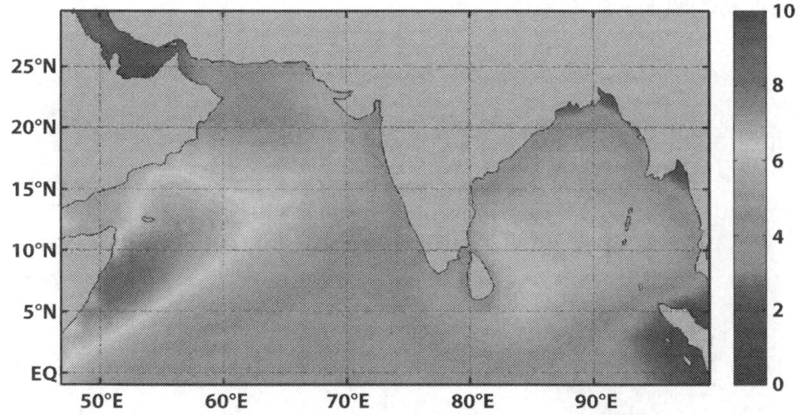

Fig. 1.33 Field of average driving wind speed over the Northern Indian ocean for 2009-2015 according to data NCEP/NCAR (m/s).

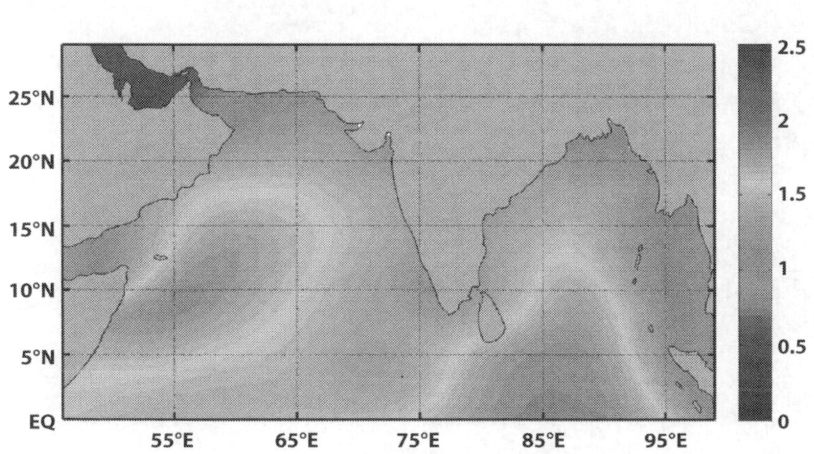

Fig. 1.34 Average height of significant waves in the Northern Indian ocean for 2009-2015 according to AVISO data, (m).

Energy complexes in the Bay of Bengal may be highly efficient. It was proposed to locate a floating energy complex in the territorial zone of Bangladesh in the Bay of Bengal, including

1) Receiving point of energy from a high-altitude power plant on a jet stream of atmosphere with a capacity of 10-50 MW;
2) An above-water wind farm with a capacity of 1-10 MW;
3) Wave power plant with a capacity of 0.5-1 MW;
4) Set of equipment for the production and storage of liquid hydrogen with a capacity corresponding to the selected power of energy sources.

The first step may be made on the island of St. Martins (Fig. 1.35).

The island is located on a shallow continental shelf with a maximum depth of **25m**. The maximum depth of the coast of Narikel Jinjira is only **10m**. The main shoreline habitats are sandy beaches, dunes, scattered rocks, and coral boulders. Some boulders are also found on the interior of the island. The shallow water marine habitats include rocky and sandy intertidal, intertidal rock pools, offshore lagoons, rocky and sandy sub tidal and offshore soft bottom habitats.

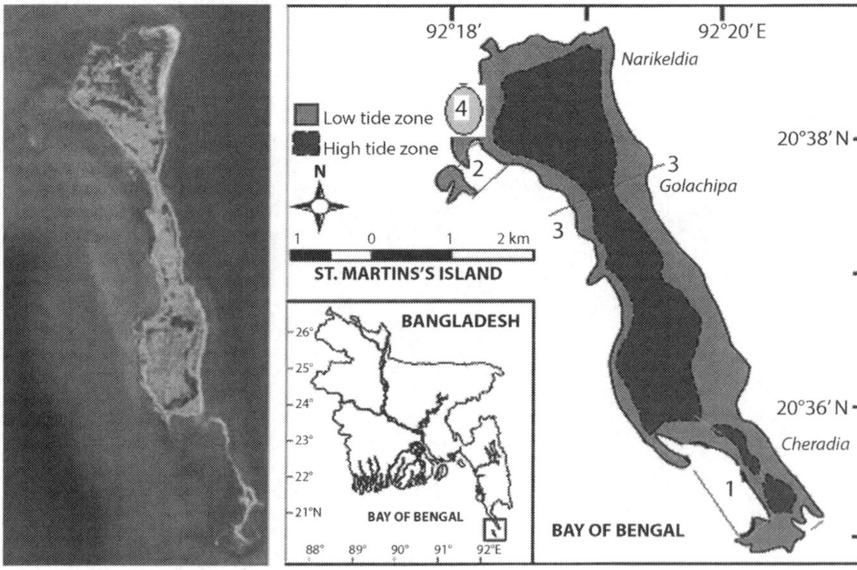

Fig. 1.35 Plan of St Martin island. Red lines – locations power systems. 1-area 1.06 km^2, TPP 173 A^2(m) kW, P_{max} = **2.2 MW;** 2 – area 0.28 km^2, TPP 46.2 A^2(m) kW, P_{max} = **0.6 MW;** 3-3 - line of pressure pipe of Hydro Power Plant. 4-wind-power system.

The tides affecting Narikel Jinjira are semi-diurnal (i.e., two high tides and two low tides per day). The mean tidal range at Shahpuri Island (about 9 km northeast from Narikel Jinjira) in the Naaf estuary is **1.874m**.

The mean low spring tidal range is	**0.19m**
The mean high water spring tidal range is	**3.56m**
The averaged tidal range is	**0.3-0.6m**
The highest astronomical tidal range is	**4.lm**
Winds are the Easterly Trades generally at	**5 -10 m/s**
In the hurricane season during 4-9 days	**12- 24 m/s**
During 1-2 days	**up to 60 m/s**
At the wind action waves is dominating above	**2.4-2.7 m.**

In general, the plan of building the renewal power system in Bangladesh should include work in all four directions, in addition to new works with the accumulation of power.

1) High-altitude power plant on a jet stream with a capacity of 20-50 MW
A high-altitude wind turbine for using the energy of jet streams of the atmosphere can be [Lyatkher V. M., Lyatkher A.V. High-Altitude wind power plant (variants), RF Patent 2686538, priority 13.08.2018, BI publication No. 13,29.04.2019] the closest match to this proposal. A modified version of this unit scheme is proposed for use in this new design of a high-altitude wind power plant. The proposed design contains the following new elements that increase its efficiency. The structure contains two rings that are the rotors of a two-way linear motor generator, the stator of which is fixed to an openwork metal structure that is supported by racks that pass through the cables when the structure is lifted from the ground to the zone of exploitation (in the zone of jet flows). The ring rotors above and below the generator on a magnetic suspension carry the working blades of the aerodynamic profile, moving in opposite directions (Fig. 1.36).

Counter rotation of the rotors compensates for the torques acting on the system and increases the efficiency of the linear motor generator. The entire system "floats" in the zone of jet streams of the atmosphere. The cross-section of the support ring is shown in Fig. 1.37 for two sections of the ring passing through the blade (lower Fig. 1.37) and between the blades (upper Fig. 1.37) of the unit.

The stator of the linear generator can be made of separate blocks, between which there are supports that perceive horizontal forces. Part of the blades are fixed to the crossbars that allow the position of the blades to be changed. The working position of the blades is almost vertical. When

Fig. 1.36 General view of the power unit for high-altitude jet flow in the assembled state (before lifting).

the system is raised or lowered, the blades are moved to a position close to the horizontal position (dotted line in Fig. 1.37). The generator operates in engine mode, providing an action on the unit of the lifting force that ensures a smooth rise or fall of the system. The blades are designed in a manner that allows their center of mass to coincide with the axis of rotation of the support hinges, and are located at an adjustable distance slightly closer to the tip of the blade than its aerodynamic center. This ensures that the blade turns (within an acceptable, adjustable angle) with the toe towards the incoming flow. In large-sized turbines, increasing the pulling forces on the blades and the turbine torque can be achieved by using flaps that turn in the desired direction to the optimal angle under the action of forces from the flow side on all sections of the blade route. It is known that the center of the force acting on the flap is located at a distance of 0.38-0.40 chord from the tip of the flap. By placing the axis of rotation of the flap to have it pass through its center of inertia, such as placing it in the middle of the flap, it is ensured that the flap turns in different directions from the neutral on the front and rear sections of the route to increase the pulling force of the blade in both sections. The angle of rotation of the flap is limited by the controls placed in the traverses (Fig.1.38).

Circuit blades are made in the form of an elongated deformed ellipse, the major axis of which is in the form of an arc, the concave side of which is oriented against the predominant flow direction with velocity U, and the distance between the front and rear sections of the route are much greater in height (length) (Fig. 1.4 on the left). The front and back sections of the blade route may be in the form of parallel straight lines (Fig. 1.39 right).

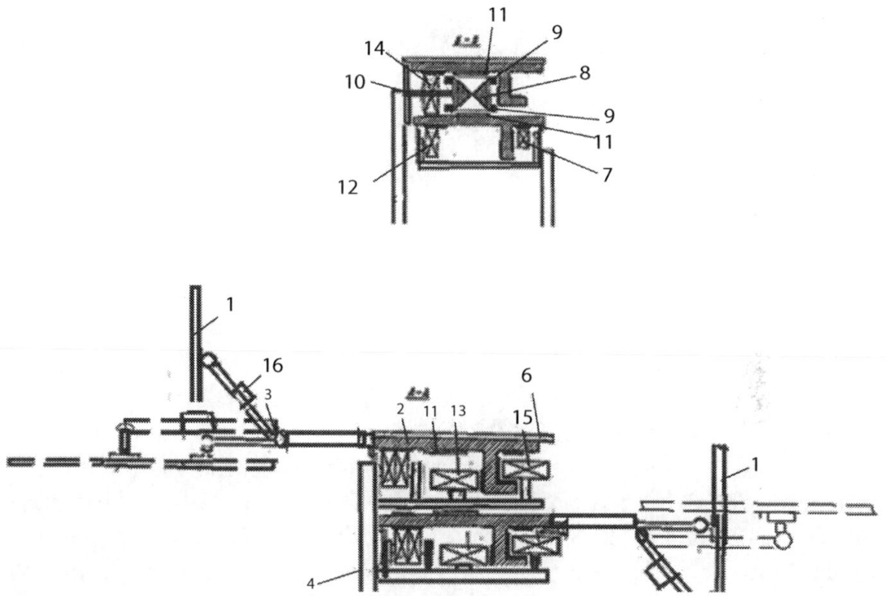

Fig. 1.37 Cross-section of the rotor. Support 14 fixes the mutual position of the rotors moving in opposite directions, and 15 holds a radial thrust bearing. The dotted line shows the location of the blades when lifting and maneuvering the unit. 1 is the blade in the working position, and 2 is the support ring of the rotor of the linear motor-generator, with a rigid carrier element of the primary moveable ring rotor; 3 - bearing blade Assembly, the locking of the horizontal beam and allowing rotation of the inclined traverses; 4 - contact design with the pillars of the earth (pontoons) or using wires control; 5 - power body carrying the stationary ring of the turbine; the 6 - element telescopic contact rigid elements of the main ring; 7, 12 - support, perceiving a vertical exposure of the turbine rotor; 8, 9, 10 - elements of the stator of a linear motor generator that includes a core 8 in a single design with a carrier truss 10 and a winding 8; 11 - a short-circuited rotor; 13 - supports that receive horizontal forces from the turbine rotor; 16 - hydraulic cylinder that provides extension (shortening) of the traverse and rotation of the blade in the vertical plane.

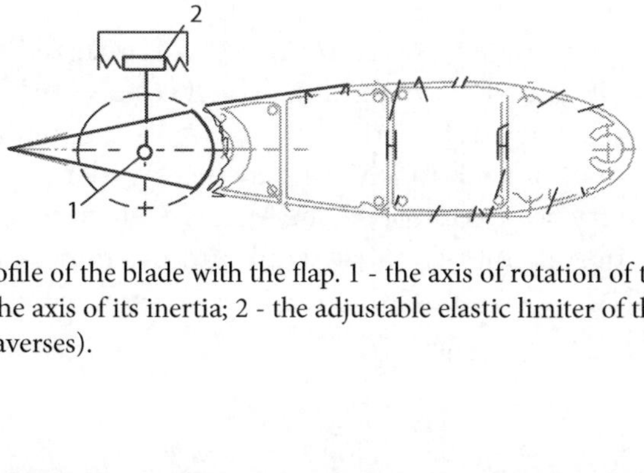

Fig. 1.38 The profile of the blade with the flap. 1 - the axis of rotation of the flap coinciding with the axis of its inertia; 2 - the adjustable elastic limiter of the flap rotation (located in the traverses).

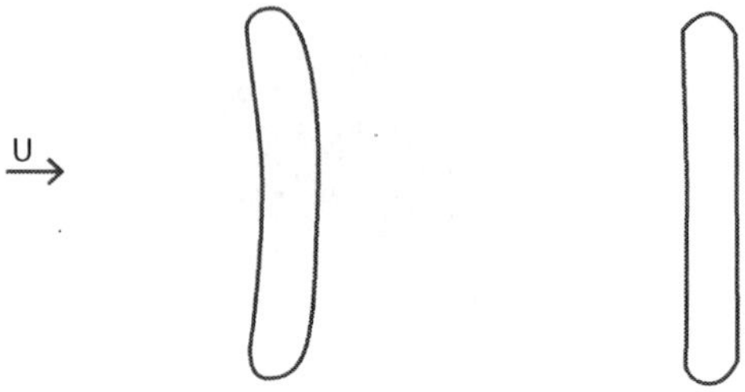

Fig.1.39 Plan of the blade's way.

It is advisable to accept the first **HIGH JET POWER** experimental plant capacity of not less than 20 MW, as a considerable share of the plant development and manufacture self-cost would fall on the gas-filled cable. Furthermore, as the plant capacity doubles, the cable self-cost increases by not more than 20-30%, as most of the cable operation systems do not depend on the plant capacity. The first experimental model of 20 MW capacity would cost approximately 1500÷ **2000 US dollars/kW+ cable**, including all the design and research expenses and costs of the equipment manufacture and plant construction, installation and commissioning. Taking into account the economic and ecological prospects of the proposed energy source, the above-mentioned cost of the first model should not be considered excessive.

2) On ground or sea wind farm with a capacity of 10-20 MW

Power wind multi-blade types may be substantially greater than that of traditional execution units, and specific capital investment and operational costs are lower. According to the draft made by the author in 2002, when the wind flow is 15-16 m/sec (beginning of distribution of power in the wind 5.5 m/s) multi-blade wind turbines with a rated capacity of 10-20MW (Fig. 1.40) must have a mass above the base, about 800 tons (80 kg/kWh), and have relative average power that ranges from 19 to 43%, with average annual wind speeds of 5.5 to 8.5 m/s.

The maximum installed power of 20 MW is realized under normal atmosphere and intense vertical mixing, which is a lower value of power in the presence of a stable stratification that prevents the mixing of air vertically. Energy output of such machines may be considerably increased due to the jet boundary layer control on the blades or devices of rotary blades. The central support pylon of similar wind turbines may be on the ground or on the water attached to an underwater pontoon, having excess buoyancy (Fig. 1.41).

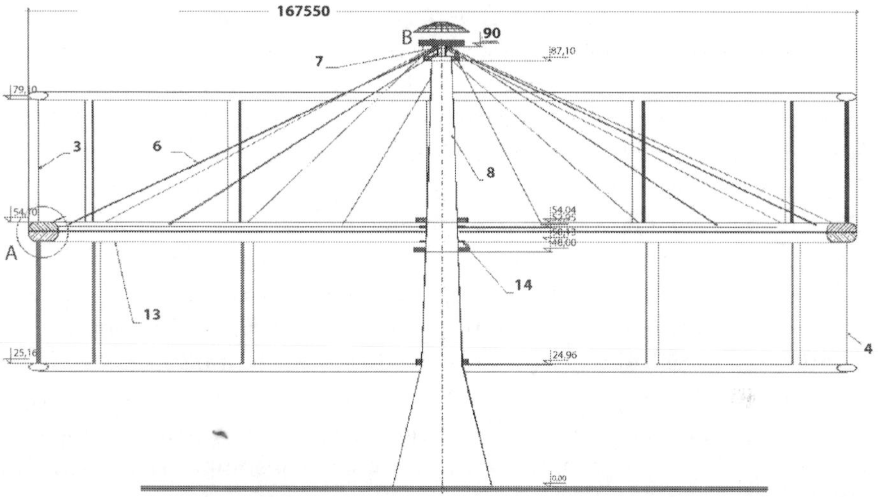

Fig. 1.40 Two-tier multi-blade wind turbine with counter-movement of the rotors. 10-20 MW.

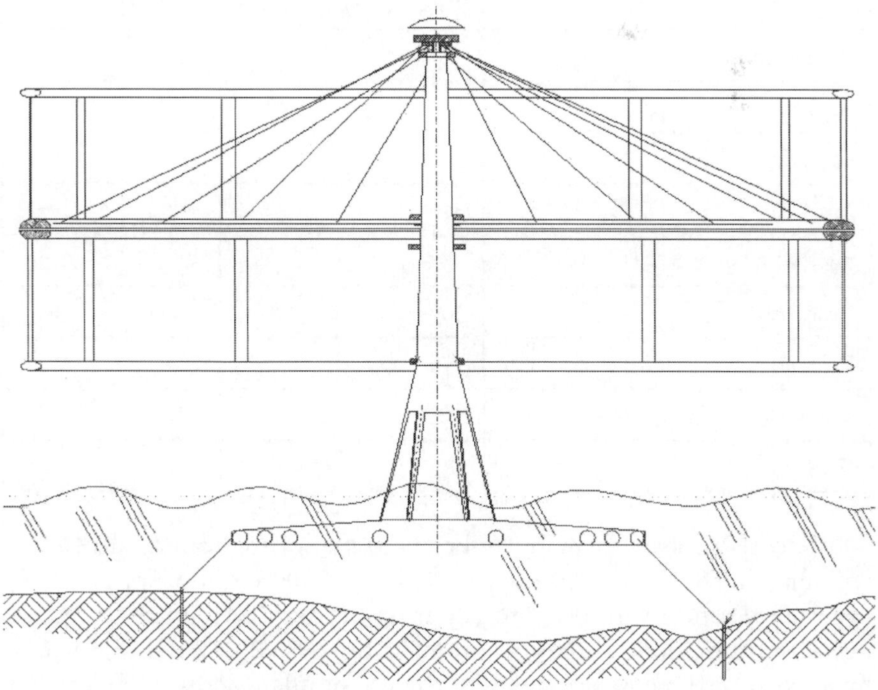

Fig. 1.41 The wind turbine is going on a pontoon in the dry dock and a float is delivered to the place of work.

With the help of cables and systems of piles, the pontoons are fixed in space, and the whole system is able to withstand hurricane-force winds and storm surges. The corresponding calculations performed for plants with a capacity of 10-20 MW and water depth of up to 20m (conditions of a wind farm in New York), show that the total costs of the installation during its serial production at the Tushino Engineering Plant did not exceed 1,800-1,900 dollars per kilowatt, and the number of hours of use of installed capacity in the New York area, where average wind speeds exceed 8.5 m/sec, were about 3,800 hours. The proposed wind turbines of the big capacity have significantly fewer material than the best machines of the traditional line up. The table presents a comparison of specific indicators of a new machine with a capacity of 10-20MW wind turbine with Vestas-1800. The offered machine is relatively easy and, more importantly, is much simpler – it requires relatively less machine work and less consumption of expensive materials.

Comparative table of consumption of material units Vestas-1800 and MLP-10-20.

	VESTAS 1800	**MLP-10000**
Blades, tons	34	22.8
Kg/kW	18.9	2.28
Generator +		
tons	61	90
Kg/kW	33.9	9
Base + tons	200	758
Kg/kW	110	75.8
Just above the base, Tons	315	871
Kg/kW	175	87.1

The cost involved in manufacturing 10 sets of the proposed turbines amounts to USD1 9,000,000/set, i.e., at least 1,900 U.S. dollars per 1 kW of installed capacity. In the future, during further development of higher power consumption, the cost of products can be reduced which, in combination with the new principles of the use of tidal energy, the energy of the rivers in the lower reaches can qualitatively change the structure of the energy balance of many countries. For example, according to the project

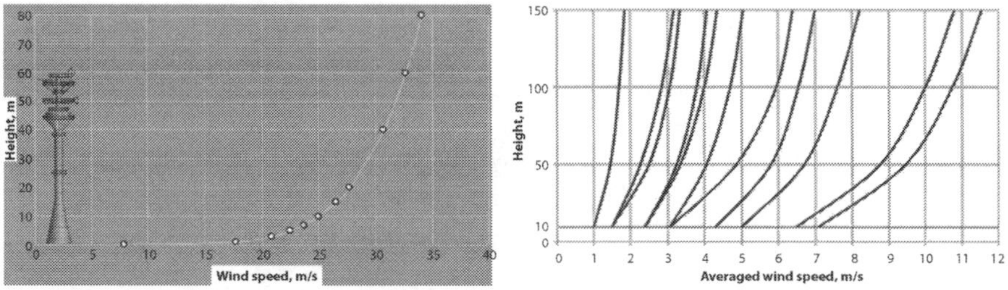

Fig. 1.42 Wind speed (m/s) on different height (m).

executed by the author, the windmill MLP with rated power of 10 MW has a mass of 748 tons (75 kg/kW) and the relative average capacity of 19% to 43% at mid-annual wind speed from 5.5 to 8.5 m/s. Average wind speed in this area is 6 m/sec (Fig. 1.33). The proposed location of a floating complex of wind and wave power plants will have coordinates 15^0N, 90^0E.

At a height of 20-80 m, corresponding to the position of the working blades of the unit (Fig. 1.40) the wind speed can be 4-5 times higher than the driving speed (Fig. 1.42), i.e., about 15 m/s can be expected.

If, for more cautious estimates, we take the average annual wind speed at the average station mark of 8.5 m/s, then the output of a station with an installed capacity of 10 MW will be at least 40,000 MWh/year, which corresponds to an average capacity of 4.6 MW. Total costs of the installation in the case of its serial production at the Russian Engineering Plant will not exceed 1800 dollars for kW. For the first instance, the cost may be two times higher **($3÷3,600.00/kW)**.

3) Wave power plant with a capacity of 0.5-1 MW

The height of the waves in the planned location of the complex is about 1.5-2 m (Fig. 1.34).

It is proposed to convert the energy of wind waves using our horizontal orthogonal turbines, buried below the sea level to a depth of about the average length of wind waves, and connected to a float on the sea surface. The vertical movements of the float follow the wave vibrations of the water surface and provide the corresponding relative velocity of the turbine flow (Fig. 1.43), which is almost equal to the vertical component of the float speed.

The weight of the turbine in the water must be sufficient to ensure the tension of the cables connecting it to the float in all phases of oscillation holds. The amplitude of fluctuations in the speed of currents caused by a wind wave decreases exponentially with depth – at a depth equal to the

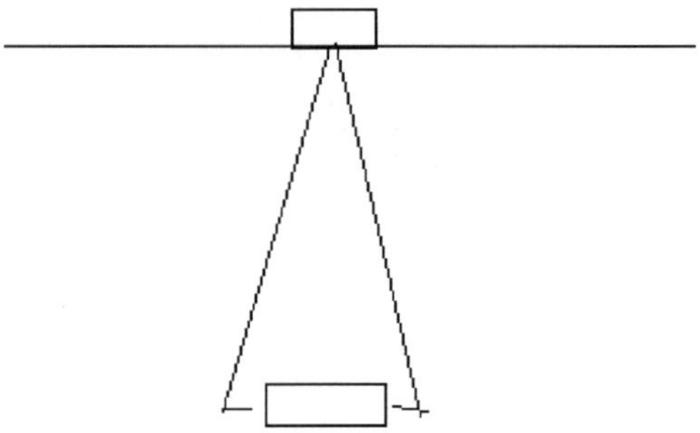

Fig. 1.43 Wave power unit with an underwater orthogonal turbine.

wave length, the amplitude of the water speed decreases by 500 times – and the turbine will make vertical movements in still water. A favorable feature of orthogonal turbines is that their operation does not depend on the direction of flow. Therefore, when the float with the turbine moves in opposite directions (up and down), the turbine will rotate in the same direction and produce power proportional to the speed of the float. The dynamics of the system is subject to special research. In the first approximation, the amplitude of the turbine speed is equal to the amplitude of the vertical component of the water surface speed: $U = \frac{1}{2}\omega h$, where h is the wave height, and $\omega = 2\pi/T$ is the circular frequency of water surface vibrations with a period of T. For the selected position of the complex with our turbine at h=2m, according to Fig. 1.44, T=3sec, ω=2.1 1 / s, U=2.1 m/s.

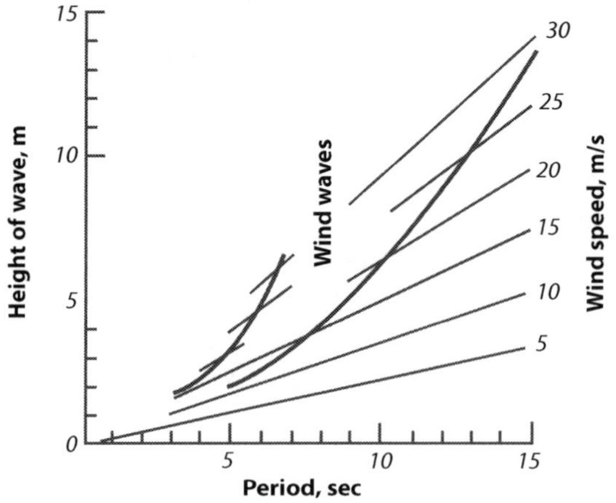

Fig. 1.44 Wave heights (m) as function of period T (sec) for different wind speed W (m/s).

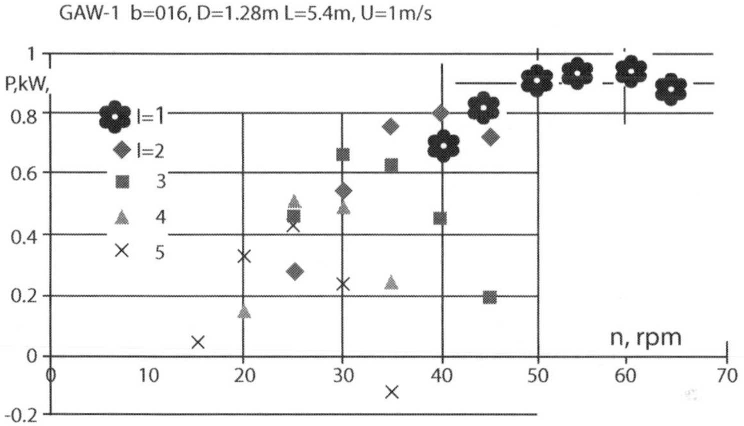

Fig. 1.45 Power of straight blade hydro turbines with length L=5.4m, Blades GAW-1, b=0.16m, α=0÷20. D=1.28 m, Water speed 1 m/s.

Fig. 1.45 shows the power of an orthogonal turbine with a diameter of 1.28 m and a length of 5.4 m with a different number of blades at a flow rate of 1 m/s.

At a flow rate of 2 m/s, the rotation speed is two times higher, and the power is eight times higher than the values shown in the Figure, i.e., it will reach almost 8kW (about 1.5 kW per linear meter of the turbine length). If such turbines are placed around a circle with a diameter of 160m (similar to the windmill in the proposed complex), the installed capacity of the wave complex will be Pwave = 3. 14x160x1. 5= 754 kW. The cost of manufacturing turbines in mass production will be approximately 1500÷2000 USD / kW; for the first samples, this would be 1.5÷2 times more expensive (it means **$3,000.00/kW**).

4) Set of equipment for the production and storage of liquid hydrogen with a performance corresponding to the selected power of energy sources

The universal method of hydrogen production is to obtain hydrogen from water using plasma-chemical and electrochemical technologies that require energy costs of about 4 kWh/m^3 H_2 (45 kWh/kg) using the most efficient modern plasma-chemical non-equilibrium and electrochemical systems. In modern factory conditions of large-scale production, energy costs are higher. For example, at a plant with a capacity of 3 tons of hydrogen gas per hour, built by Brown Boveri in Egypt on the basis of the Aswan HPP, the actual energy consumption per 1 kg of normal hydrogen is 54÷56 kWh.

To liquefy hydrogen from the gaseous state, it is theoretically necessary to have 3.3÷4 kWh/kg, in practice and, in good production, up to 11 kWh/kg.

With small production volumes, energy costs are significantly higher. For example, at a plant near Moscow that has a capacity for liquid hydrogen production of up to 900 tons/year (up to 160kg / hour), the energy consumption for producing hydrogen gas is 89.5 kWh/kg, and for liquefying it, an additional 53.3 kWh/kg (total142.8 kWh/kg) is needed. More modern equipment manufactured by Hydrogen Systems N. V. (Belgium) under its IMET patent (Fig. 1.46), allows energy consumption to be reduced to 4.6÷4.8 kWh/m^3(50.5÷52.7 kWh/kg) at a pressure of hydrogen gas at the outlet from 3.5 to 10 bar.

The largest products of this company IMET-60 have a capacity of 5.4 kg/h(47 total weight of 5.2 t (including an electrolyzer of 2.6 t). This product with the maximum specific energy consumption requires minimal capital investment – its cost in 2003 was about 404,000 euros. There is information that similar and larger products can be offered by NTC HYDROGEN (Protvino, Moscow region). The advantage of membrane technologies is the absence of an alkaline electrolyte, where complete separation of the resulting hydrogen and oxygen (there is no danger of formation of a rattling mixture), a high degree of purity of hydrogen (99.9-99.995%), environmental cleanliness, relatively low energy consumption, and high reliability can be achieved. Traditional electrolyzers with a capacity for normal hydrogen from 4 to 41 cubic meters per hour are offered, for example, by ZAO TKH (Yekaterinburg). The weight of the electrolyzer with complete technological equipment is from 10578 to 27335 kg, plus control panels (1284 kg)

Fig.1.46 Imet complex for economical production of pure hydrogen using membrane technology (Belgium, 2002).

and 3 receivers of 10 cubic meters (2295 kg). Energy consumption in this case ranges from 24 to 215 kW (6.0-5.2 kWh per cubic meter of normal hydrogen). The cost of such a complex in the minimum set for hydrogen performance of 4 m^3/hour is $79358. New ideas that promise to reduce energy consumption by an order of magnitude when producing hydrogen were put forward by a group of authors and owners of Russian patent 2224051 with priority from 17.02.03. The idea was to use centrifugal forces to bundle hydrogen and oxygen ions. The experiments demonstrated by the authors (Mogilevsky I., Surikov A., Ovsyannikov E. M.) indicated the possibility of reducing energy consumption to 0.3 kWh/m^3(3.3 kWh/kg). The degree of purity of hydrogen, however, was not regulated.

Hydrogen liquefiers are produced by the Swiss company Linde and the Russian companies Cryogenmash and NPO Heliymash. The Russian OV-2500 liquefier with a liquid hydrogen capacity of 2.5 m^3/hour (178 kg/hour or 1560 t/year), together with two tanks with a capacity of 250 m^3 each, can cost about 19 million euros. Lowering the energy price of hydrogen production remains a fundamental task. This problem requires a wide range of diverse studies, including fundamental ones. It is also important to reduce the cost of electrolyzers and liquefiers by using new design solutions. If, however, we focus on the best results achieved and take a cautious estimate of the energy costs to produce hydrogen gas of 53 kWh/kg, and for 65 kWh/kg of liquid hydrogen (in general) then only 10% of the annual output of the Sayano-Shushenskaya HPP (this is an average of 2.35 billion cubic meters) would be used, and it would be possible to produce 36.2 thousand tons of liquid hydrogen or 44.3 thousand tons of gaseous hydrogen per year. To get this amount of liquid hydrogen, 770 I MET-60 devices and 24 OV-2500 liquefiers would need to be used. The total cost of this equipment, together with tanks for short-term storage of products, would be 770x0. 404 + 24x19 = 311 + 456 = 767 million euros. If the cost of energy on HPP tires is considered to be 0.01 euros per kWh, then payments for hydrogen would amount to 23.5 million euros per year, and the cost of liquid hydrogen for 20 years would not exceed 1.7 euros per kg. The cost of liquid hydrogen in 1980 was 5.6 dollars per kg.

Tidal power systems with orthogonal turbines may be much more efficient than the traditional ones.

Sihwa Lake Tidal Power Station is the world's largest traditional installation, with a total power output capacity of 254 MW, surpassing the 240 MW Rance Tidal Power Station which was the world's largest for 45 years prior to 2011. It is operated by the Korean Water Resource Corporation. The tidal barrage made use of a seawall constructed in 1994 for flood mitigation and agricultural purposes. Ten 25.4 MW submerged bulb turbines were driven in an unpumped flood generation scheme; power was generated on tidal inflows only and the outflow was sluiced away. This slightly unconventional and relatively inefficient approach had been chosen to balance a complex mix of existing land use, water use, conservation, environmental and power generation considerations. The tidal power station provided indirect environmental benefits as well as renewable energy generation. After the seawall was built, pollution built up in the newly created Sihwa Lake reservoir, making its water useless for agriculture. In 2004, seawater was reintroduced in the hope of flushing out contamination; inflows from the tidal barrage were envisaged as a complementary permanent solution. Cost of the project was met by the South Korean Government, totalling 313.5 billion won (£180 million, or $US293 million, roughly $1m per MW).[Mean operating tidal range was 5.6 m (18 ft), with a spring tidal range of 7.8 m (26 ft). The working basin area was originally intended to be 43 km^2 (17 sq mi)[5] although this has been reduced by land reclamation and freshwater dykes to 30 km^2 (12 sq mi) and is likely to be reduced further.

It may be proposed to consider options for the integrated development of renewable energy in the country with the small power incomes in two directions:

1. The provision of energy from wind and hydro units with a capacity of 1-10 kW (100 kW) for individual (mostly rural) consumers;
2. The creation of a network of power plants with a capacity of 100 MW with the delivery of energy to a public or private electrical network according to the schedule defined by the mode of the energy source (wind, river currents, the intensity of tidal).

The first line of action may be to buy wind turbines manufactured by companies in the USA, Europe, Russia, and by various companies in China and other countries.

A negative feature of these machines is that they provide power only when there is a relatively strong wind. This book proposes a new development that increases efficiency in complex system optimization of the wind turbine and energy storage. This would allow a stand-alone consumer to fully provide itself with energy. This proposed design for an experienced industrial complex can be created by setting a price for each serial of the unit. Further serial production could be organized with the help of our documentation in any country to supply the separate component elements.

The second line of action can be to modernize large wind turbines through the use of the proposed solution. The proposed solution may be sent to the chosen company from among the ones in the countries mentioned in the first line of action. The total initial cost for the creation of an experimental-industrial model of turbine with double the installed capacity may be correctly estimated from this initiative.

The manner of composition of units laid out in the first line of action may provide some level of protection from biological contamination to the water. The proposed solution can have some of the contaminants withdrawn, to later have these processed into fertilizer for use in agriculture.

The floating dam hydroelectric turbines favor the windmills' stability and power delivery; however, because that is dependent on the strength of the current/flow of the water in the rivers, the cost of an installed kilowatt hydropower station is usually not above $4,000 per kW. The average power will be close to what is estimated. Hydropower can be supplied in sets, and it can be arranged to have the major components mentioned in the proposed documentation manufactured locally.

This book offers variants of the many methods of using the energy harnessed from streams and sea (tidal) currents, wind and wind waves in any condition with the highest efficiency. All the methods proposed take into consideration the originality of the constructive circuit so as to ensure that the best possible outcome can be generated against the set conditions.

Three general types of windmill installations

- high-power (5-15 MW in one unit) for use in powerful electric networks on land or under the sea at mid-annual speed of a wind, the greater of 6 m/s - types WMB and WCOL;
- average capacity (20 - 500 kw in one unit) for use in complexes for clearing or (and) desalinating water to provide individual power supply, in particular, for installation on roofs of inhabited or industrial buildings especially in remote places, for example, in mountainous districts, and in areas with severe climates at average and high speeds of a wind - types WHEL and WST;

- average capacity for guaranteed power supply at high variability of winds, and concerning low mid-annual speeds of a wind - Type WAI.

Three general types of hydropower installations for use in harnessing energy of currents

- high power (from 1.5 up to 75 MW) for use of energy of currents with great speeds (2 - 4 m/s) and large depths - type HMB;
- the same at small depths - type HORT;
- universal for transformation of energy of streams where speed varies in a wide range - from 0.1 m/s without restriction of the top limit of speed, without restrictions on width and depth of a stream - type Hydro Impact. Capacity of the unit is determined by dimensions of a stream and requirements of the c ustomer.

Wave power installations for use of energy of wind waves in a coastal zone are offered with accommodation of the electro power equipment on the coast with usual (type Wave) and raised power output (type Wave & Wind). Offered installations work at any parameters of waves. Capacity of separate installation or a complex is defined by requirements of the local conditions. ***The description of each type of power system, providing full ecological cleanliness and the lowest cost of energy, goes under the first requirement of the customer and local conditions.***

2

Conversion of the Energy of Currents

"Being first isn't always what counts...
Being best is more important."

Mitchell Wilson, *My Brother, My Enemy, 1955*

Systems converting the energy of currents can be of mechanical action in combination with electrical, pumping, frictional, or thermal units, as well as non-mechanical action, using, for example, the effect of ionization and electrical conductivity of a stream passed through an electric or magnetic field. All mechanical systems called power plants (PPs) use the work of the forces arising from the flow around the moving elements of the power plants structurally united into the power rotors (PRs) of these power plants. PRs' working elements move along closed paths sweeping some surfaces, the axis of symmetry of which can be parallel (collinear) or perpendicular (orthogonal) to the flow. Accordingly, we distinguish the two groups of PPs:

Collinear, in which the axis of symmetry is approximately parallel to the flow, and
Orthogonal, in which the axis of symmetry lies in a plane orthogonal to the direction of flow.

The PR elements can move slower or faster than the flow on the approach to the power plant. Accordingly, they are called low-speed (active) or high-speed (jet) machines. Collinear, high-speed units (Fig. 1[1], 0.1,0.2,0.10,0.12,0.14) are currently the most widespread ones.

Orthogonal aggregates can have a horizontal or vertical axis (Figure 2.0.18-0.26).

[1]The numbering of figures and formulas in each section is different.

Victor M. Lyatkher and Ziaur Rahman. Wave, Wind and Current Power Generation, (59–94) © 2022 Scrivener Publishing LLC

Fig. 2.1 Modern collinear wind turbine. The axis of rotation (1) is parallel to the flow. The blades (2) move in a plane perpendicular to the flow.

Fig. 2.2 Orthogonal high-speed units with an axis of rotation (1) perpendicular to the direction of flow. The blades (2) move along circular paths. On the left is a horizontal hydraulic unit before being placed in the stream, on the right isa verticalwind unit.

Comparison of various types of power units, even within one class, causes certain difficulties, since it is necessary to take into account several criteria measured in different units—specific (per unit of capacity and output) material consumption, quality of generated energy, degree of simplicity and factory readiness, convenience and reliability in operation, labor intensity of construction and operation, and environmental friendliness. The combination of these criteria determines the multidimensional characteristic of the quality of a PP. Comparison of the quality of power plants is especially difficult for machines of different classes. Therefore, the choice of promising schemes is largely arbitrary. Changes in the value (social significance) of certain coordinates in the quality of the space can also change the conclusions. However, economic, energy, and environmental indicators which are now accepted as the main ones, presumably will retain their importance for yet many years.

Due to the fact that, as the velocities of bodies approach the speed of sound, the wave of aerodynamic resistance and acoustic radiation sharply increases, and the elements of wind turbines tend to be designed so that they are flown around at speeds significantly lower than the speed of sound (with low Mach numbers). In this case, air can be considered practically incompressible, or compressible according to any law (adiabatic or isothermal), and all the results of aerodynamic calculations—suitable for recalculation for the conditions of motion of an incompressible fluid with any other density. In particular, all the results can be directly used to analyze river or ocean power plants with similar layouts. It should be noted that, for calm water flows with low Froude numbers, the role of the free surface of the flow is, in a sense, close to that of a solid smooth wall[2]. Therefore, the analogy, for example, between ground-based wind and river (oceanic) installations, can be twofold—when the units are located near the water surface or at the bottom of a reservoir. Of course, the designs of assemblies, materials used, strength and technical and economic assessments for power units in air and water can be significantly different.

In any scheme of power take-off from the flow, it makes sense in the first approximation to replace the unit carrying out this process with some hypothetical flat permeable figure, the contours of which are the projections of the boundaries of the body swept by the rotor onto the figure plane perpendicular to the flow direction on the approach to the unit (Fig. 2.3).

---

[2]Lyatkher, V.M., Prudovskiy, A.M. Hydraulic Modeling, M., *Energoatomizdat*, 1984, 392 p.

Lyatkher, V.M., Prudovskiy, A.M. *Studies of Open Flows on Pressure Models*, Moscow, Energiya, 1971.

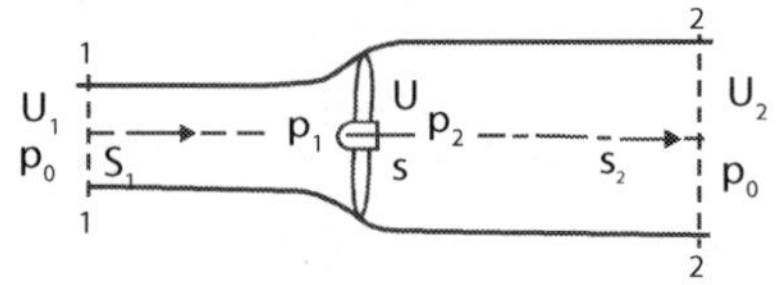

Fig. 2.3 Flow diagram in the turbine zone.

In the case of traditional, collinear aggregates, such a figure will be a circle or a ring; in the case of orthogonal aggregates, it can be a strip (as in the considered model), a rectangle, a trapezoid, an ellipsoidal figure (section of a Darrieus unit), or a triangle. Whatever this figure may be, at a distance in front of it, the flow rate and pressure remain unperturbed:

$$U_1 \text{ and } p_0 \tag{2.1}$$

and at some distance from it downstream, where the pressure distribution across the flow levels equalizes and returns to the initial value p_0, the velocity inside the stream tube resting on the selected figure has a slightly lower value of

$$U_2 = (U_1 - u).$$

If the flow velocity through the selected figure is denoted by U, and the pressure from the front and back sides of the figure by p_1 and p_2, respectively, then the Bernoulli equations for the front and rear sections of the stream tube will have the following form:

$$p_1 + \rho U^2/2 = p_0 + \rho U_1^2/2 \tag{2.2}$$

$$p_2 + \rho U^2/2 = p_0 + \rho(U_1 - u)^2/2 \tag{2.3}$$

If the flow rate and pressure on the surface of the selected current tube are taken to be constant, then the equation for changing the momentum for the selected volume has the following form:

$$Q\rho u = \Delta p S \tag{2.4}$$

$$Q = U\,S \tag{2.5}$$

– the flow rate of the medium moving in the considered stream tube, S, is the surface area that simulates the turbine,

$\Delta p = p_1 - p_2$ the pressure drop on the selected flat figure replacing the power unit is calculated by subtracting equality (2.3) from equality (2.2) taking into account (2.4) and (2.5):

$$\Delta p = \rho u(U_1 - u/2) = \rho 2(U_1 - U)U \tag{2.6}$$

From equations (2.2) to (2.5) we find:

$$U = U_1 - u/2 \quad \text{or} \quad u = 2(U_1 - U) \tag{2.7}$$

The power lost by the flow and transferred to the turbine is

$$P = U\Delta p\, S = \rho u(U_1 - u/2)^2\, S = \rho 2(U_1 - U)U^2\, S \tag{2.8}$$

The found solution (2.6), (2.8) represented in the form of the drag factor of

$$C_D = 2\Delta p / \rho U_1^2 \tag{2.9}$$

and the efficiency factor of the unit

$$C_P = 2\Delta p / \rho U_1^3 \tag{2.10}$$

turned out to depend on only one parameter which is the relative flow velocity in the unit zone, U/U_1.

$$C_D = 4\,(1 - U/U_1)U/\,U_1 \tag{2.11}$$

$$C_P = 4\,(1 - U/U_1)(U/\,U_1)^2 \tag{2.12}$$

The maximum value of the power, P takes place at $U/U_1 = u/U_1 = 2/3$ and is

$$P_{max} = 16/27\; S\,\rho U_1^3 / 2\,, \qquad (C_P)_{max} = 16/27 = 0.593 \tag{2.13}$$

The obtained relations (2.6), (2.8), (2.13) are usually associated with the works of N. Ye. Zhukovsky (1912) and A. Betz (1920), giving these relations the meaning of certain limiting laws[3].

---

[3]B.B. Kazhinsky, *Wind power plants*. GIz., M-L, 1928, 311 p. D. Le Gourieres, *Wind Power Plants*, Pergamon Press, 1982, 285p. E.E. Shpilrain, *Renewable Energy Sources and Their Prospects for Russia*, coll. *Energy of Russia. Problems and Prospects*, Moscow, "Nauka", 2006, pp. 284–292.

In fact, this is not the case. The original equations (2.2)–(2.5) are not exact and reflect only some model of the phenomenon. In particular, equation (2.3) is certainly not valid as $U \rightarrow 0$.

In this real limiting case (impermeable aggregate), the pressure p_2 behind the aggregate will certainly be less than the pressure p_0 at a distance from the aggregate, and not more than it is, as follows from (2.3). However, the very idea of replacing the aggregate with one or more permeable flat or other figures on which there occurs a pressure discontinuity and concentrated power take-off, turned out to be very productive.

Let us consider a typical example in which a velocity U constant at all points of the figure and perpendicular to the surface of the plate is set on a permeable flat figure modeling an energy unit. The unsteady equations of hydrodynamics (Euler) are solved by the method of finite differences. The solution for a figure in the form of a strip-plate (two-dimensional non-stationary problem) turns out to be different with the introduction of the flow symmetry condition and without such a condition (Fig. 2.4). This difference manifests itself in the flow pattern and pressure distribution only behind the plate. It is essential for determining loads.

It is curious that although the difference in loads on the permeable plate in different formulations is quite large (especially at low U), the maximum power factor differs slightly, remaining within

$$(C_P)_{max} = 0.55 \div 0.59 \tag{2.14}$$

In contrast to the Betz-Zhukovsky theory, the drag factor of a flat impermeable plate calculated in this way (without an additional condition on the symmetry of the flow behind the plate) turned out to be exactly equal to the experimental value of $C_D = 2$ for an infinite plate length. With a symmetric flow around the plate, the pressure drop behind it is less (the formation of a backward jet effect) and the calculated result corresponds to experiments with a rectangular plate in which the long side is 15 times larger than the short one ($C_D = 1.38$).

The pressure drop across a permeable plate that simulates a power unit can be found by generalizing, for example, the Kirchhoff problem (Kirchhoff, 1869, 1876)[4] on the jet flow around a plate in the case of filtration through the plate (GGS[5] model). The pressure factor on the impermeable plate in the Kirchhoff and GGS models, naturally, turns

[4]M. I. Gurevich, *Theory of Jets of an Ideal Fluid*, Fizmatgiz, Moscow, 1961, 496 pp.

[5]Gorban' A.N., Gorlov A.M., Silantyev V.M. Limits of the Turbine Efficiency for Free Fluid Flow, *Journal of Energy Resources Technology*, ASME, 2001, v. 123, pp. 311-317.

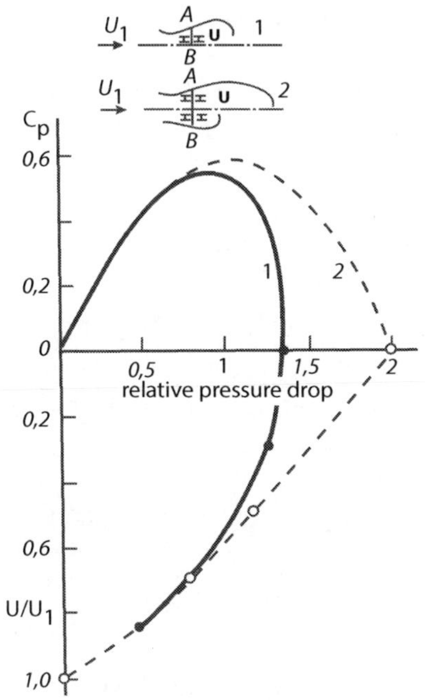

Fig. 2.4 The efficiency of the power unit C_p modeled by a flat permeable plate, depending on the relative flow velocity through the plate of U/U_1 under the hypothesis of symmetric flow (1), and without any additional hypotheses (2). The calculation specifies the relative flow rate through the plate of U/U_1 resulting in a relative pressure drop of $2\Delta p/\rho U_1^2$ and the power (efficiency) factor of $C_p = 2\Delta pU/\rho U_1^3$.

out to be almost the same as 0.88 and much less than the real experimental value. This is explained by the fact that both of these models do not take into account the decrease in pressure behind the plate in real conditions.

The maximum power factor of 0.301 in the GGS model turned out to be so small because the actual decrease in pressure behind the plate was not taken into account and the condition of filtration through the plate formulated when setting the problem, was not actually used in the solution.

The resulting picture, like the Kirchhoff solution, should be close to reality in the case of cavitation flow around the turbine when the pressure behind the plate does not depend on the flow velocity. For ordinary, non-cavitation conditions, the result would be closer to the experimental values if the pressure drop behind the obstacle was taken into account. Such an opportunity for an impermeable plate is provided, for example, by the scheme of D.A. Efros (1946), in which the jet breaking off from a plate returns back to the plate. Ryabushinsky's models with two plates (one after the other) are widely known. In the case of a permeable plate, the model of M.A. Lavrentyev (1958)[6] in which two rotating rings are located behind the plate.

Other models are also possible, which give different estimates and correct results at known limit points ($U = 0$ and $U = U_1$). If, for example, instead of equations (2.3) and (2.4), we simply use the parametric representation of the drag factor of the plate of

$$C_D = C_D^0(1-(U/U_1)^m) \tag{2.15}$$

valid at the indicated limit points at $C_D^0 = 2$, then the expression for the efficiency factor will take a form of

$$C_P = 2(1 - (U/U_1)^m)(U/U_1), \tag{2.16}$$

and the maximum efficiency turns out to be

$$(C_p)_{max} = C_D^0(m+1)^{1+1/m} \text{ at } (U/U_1) = (1+m)^{-1/m} \tag{2.17}$$

In particular, for

$$C_D^0 = 2 \text{ and } m = 0.5, 1.0 \text{ or } 2.0 \tag{2.18}$$

and the maximum efficiency turns out to be

$$(C_P)_{max} = 0.30, 0.50, \text{ or } 0.77 \text{ at } U/U_1 = 0.444, 0.5 \text{ or } 0.577 \tag{2.19}$$

respectively.

If we accept another parametric representation, for example,

$$C_D = 2(1 - U/U_1)^m \tag{2.20}$$

then the maximum efficiency $(C_P)_{max} = 2m^m(1+m)^{-(m+1)}$ to be obtained at $U/U_1 = 1/(1+m)$ which gives for the same values of the parameter of m according to (2.18):

[6]M. A. Lavrentyev, B. V. Shabat. *Problems of Hydrodynamics and their Mathematical Models*, "Nauka", Moscow, 1977.

$$(C_P)_{max} = 0.577,\ 0.500 \text{ and } 0.296 \text{ at } U/U_1 = 0.667,\ 0.5 \text{ and } 0.333 \qquad (2.21)$$

respectively.

It is clear that the widely used Betz-Zhukovsky limit does not provide a reliable estimate. Primitive representations (2.15) or (2.20) can be closer to reality if, instead of the factor of 2, we introduce empirical values of the drag factors of impermeable plates depending on the aspect ratio:

Aspect ratio =	20:1	10:1	4:1	1:1		
$(C_D)_{max}$ =			1.45	1.29	1.19	1.10

Taking these relations into account, instead of estimates (2.19) or (2.21) for a turbine with a rectangular cross-section with an aspect ratio of 4:1, we obtain the maximum efficiency of the order of 45.8÷34.3% (for linear interpolation for m = 1 C_{Pmax} = 0.298) which is close to empirically achieved results. If we use the same reasoning to estimate the maximum efficiency of collinear aggregates replaced in a zero-dimensional consideration by a permeable circular disk, then with a resistance of the impermeable disk equal to 1.25, the maximum efficiency of a collinear turbine (or any axisymmetric system) can be expected within 48÷36%. According to the above-mentioned GGS model, the maximum achievable efficiency of free-flow turbines is 30.1% at a flow velocity through the turbine of 0.613 times the free stream velocity. The problem was posed by the authors for the equation of continuity of the flow of an incompressible fluid flowing around a flat permeable plate located perpendicular to the flow velocity at a distance from the plate. It was assumed that the condition of liquid filtration through the plate, formula (2.5) of the original, should be fulfilled on the plate surface. In a rectangular coordinate system associated with the plate, this condition takes the form:

$$[p] = rV_x \qquad (2.22)$$

Here, as in the original, [p] is a pressure drop across the plate which at the end of the article is assumed to be simply equal to the pressure on the front side of the plate, the density of the liquid is taken to be 1, V_x is the longitudinal component of the flow velocity on the plate (filtration rate through the plate), and r is a filtration drag factor of the plate.

When constructing the kinematic pattern, condition (2.22) was not used. The flow pattern was constructed by generalizing the well-known

Kirchhoff scheme, in which jets of constant velocity equal to the flow velocity at a distance on the approach to the plate, descend from the end points of the plate, and the plate permeability is set by an angle φ in the plane of the velocity potential, i.e., the flow rate through the plate is determined by the kinematic relation of

$$V_x = V\sin\varphi \tag{2.23}$$

unrelated to condition (2.22). This parameter (angle) φ determines the solution of the problem and, in particular, the relative flow rate of the liquid through the plate, i.e., the average velocity of "filtration" through the plate and the entire flow pattern in front of the plate.

The pressure on the plate can be determined by the Bernoulli formula (formula (2.16) of the original) and this record is used to calculate the efficiency of the turbines which can be represented as a function of the speed of "filtration" through the plate (Fig. 2.5). On the other hand, the pressure would have to satisfy condition (2.22). Thus, the parameters φ and r on the plate surface should satisfy the equation

$$\tfrac{1}{2}\,(1 - V^2) = r\,V_x \tag{2.24}$$

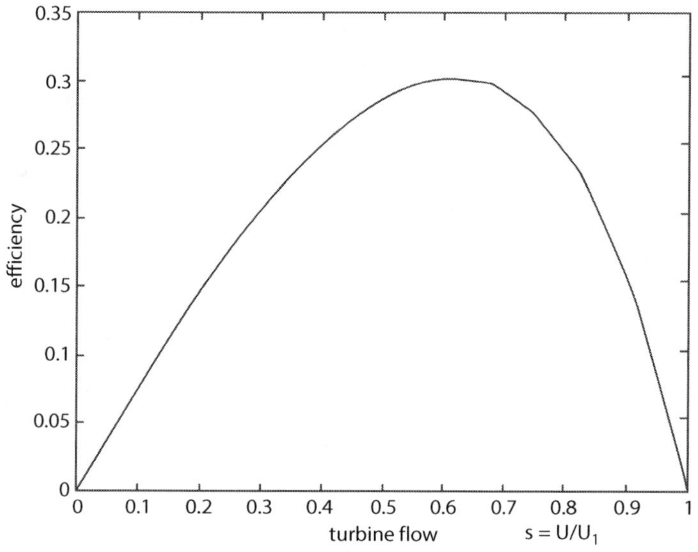

Fig. 2.5 Turbine efficiency vs. "filtration" speed. GGS model.

or

$$V^2 + 2\,r\,V\,\sin\varphi - 1 = 0 \tag{2.25}$$

Hence, it turns out that the modulus of speed on the plate, the drag factor r, and the parameter of φ, should be related by the formula of

$$V = \sqrt{(1 + r^2 \sin^2\varphi)} - r\,\sin\varphi \tag{2.26}$$

in which the values of r and φ do not change along the plate, and the modulus of velocity of V obviously changes from the minimum value on the axis of the plate to the maximum value at the edge of the plate. Thus, equation (2.26) and, therefore, condition (2.22) at a constant r and φ (along the plate) shall not be satisfied in the constructed solution. The average pressure on the plate measured in fractions of the velocity head at a distance from the plate and determined from the original data as the ratio limit of the efficiency to the rate of $s = U/U_1$ of the flow through the plate at $s \to 0$ was 0.77 which is noticeably lower than the Kirchhoff result (0.88), although it would seem that these results should coincide. The difference is explained by the influence of the weight function with which the pressure on the plate is averaged when calculating the efficiency. This function is a flow rate through the plate.

Nevertheless, the resulting estimate of the efficiency of turbines obtained by the GGS model makes sense; only, it does not refer to a physically clear condition (2.22), but to a certain kinematic scheme (2.23) which under some conditions (with variable r) can correspond to reality.

An attempt can be made to refine the GGS model by introducing an additional term into the formula for calculating the pressure drop across the plate as a function of the flow rate through the plate. In its simplest form, we take into account the pressure drop behind the plate in the form of an additional drop which is naturally assumed to be a linear function of s:

$$\Delta C_D = \Delta p = A(1 - s) \tag{2.27}$$

where the factor A is the difference between the actual pressure drop across the impermeable plate and the pressure determined by the Kirchhoff scheme or the GGS model. Both the models actually characterize the pressure on the front side of the plate.

Fig. 2.6 shows the pressure factor C_D calculated from the data on efficiency and velocity through the plate taken from the original for the

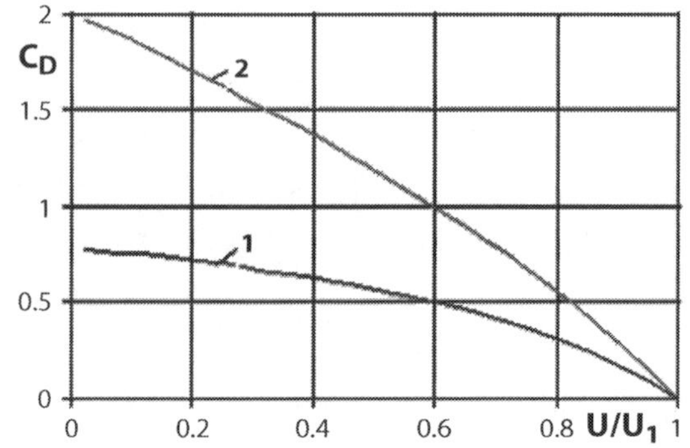

Fig. 2.6 Pressure factor $C_D = C_p/s$ according to the original (1) and modified (2) GGS models.

GGS model (line 1) and corrected by relation (2.27), taking into account the empirical value of the resistance of the impermeable plates (line 2).

The power factor increment corresponding to formula (2.27) is:

$$\Delta(C_p) = A(1 - s)s \qquad (2.28)$$

The maximum increment occurs at s = 0.5:

$$\text{Max}\Delta(C_p) = 0.25\ A \qquad (2.29)$$

Where the factor A depends on the shape of the plate:

The rectangular aspect ratio:	∞	20:1	10:1	4:1	1:1	Circle
C_D =	2	1.45	1.29	1.19	1.10	1.25
A =	1.23	0.68	0.52	0.42	0.33	0.48
$\text{Max}\Delta(C_p)$ =	0.307	0.17	0.13	0.105	0.082	0.12

The GGS model refined in this way for real orthogonal turbines with a rectangular cross-section with an aspect ratio of 4:1 shall give the maximum possible efficiency of about 40% at a relative flow velocity through the turbine of about 0.55.

If we assume that the GGS solution exactly matches the Kirchhoff model, i.e. the pressure on the plate at s = 0 is equal to 0.88, the values of the factor A in the above table should be reduced by 0.11, and the comparison of our numerical model and the modified GGS model for an infinite strip will look like this:

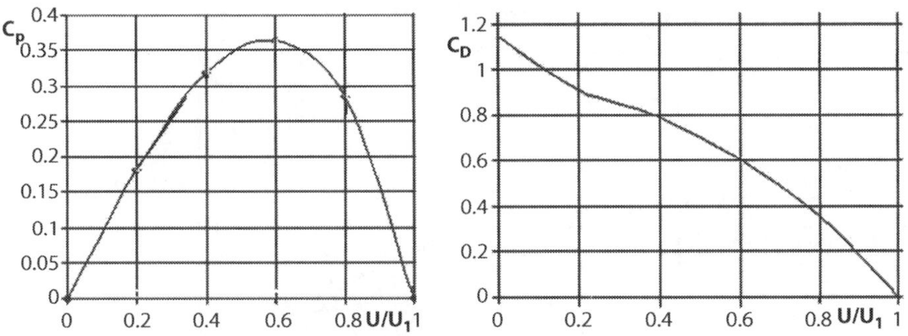

Fig. 2.7 Efficiency and resistance of an orthogonal turbine with a turbine length of 2.5 times the diameter. Modified GGS model.

U/U_1	1	0.8	0.6	0.4	0.2	0.0		
C_D (Fig. 40)			0	0.562	0.945	1.332	1.665	2
C_P (Fig. 40)			0	0.450	0.567	0.533	0.333	0
C_D (GGS modified)		0	0.527	0.947	1.302	1.595	2	
C_P (GGS modified)		0	0.422	0.568	0.521	0.319	0	

There is little difference between the results.

Using the modified GGS model for an orthogonal turbine with a length of 2.5 times the diameter (A = 0.265), we obtain the following values for the efficiency and resistance of the turbine (Fig. 2.7):

U/U_1	1	0.8	0.6	0.4	0.2	0.0.		
C_P (GGS modified)			0	0.285	0.363	0.317	0.182	0
C_D			0	0.356	0.605	0.792	0.910	1.145

The maximum value of efficiency (0.363) is close to that observed in experiments.

Of particular interest are calculations for a permeable circle that simulates an "ideal" collinear aggregate. The pressure factors determined by the Betz-Zhukovsky model and the model by G. Kh. Sabinin[7]) differ to the greatest extent at a low disk permeability (Fig. 2.8).

A numerical model of the flow around a disk with a uniform distribution of velocity through the disk gave intermediate results with exact agreement with the experiment at U = 0.

[7]G.H. Sabinin, The Theory of an Ideal Wind Turbine, Proceedings of the TsAGI (Central Institute of Aerohydrodynamics), Issue 32, No. 200, Moscow, 1927.

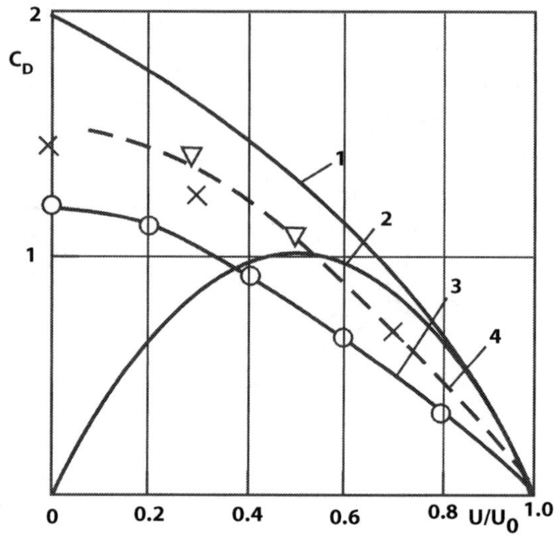

Fig. 2.8 Pressure factor on a permeable circular disc. 1 – G. Kh. Sabinin's model, 2 – Betz-Zhukovsky's model, 3 – points – numerical modeling for unlimited space, line – approximation (16) at m = 1.6 and $C_D^0 = 1.2$,4 – the same as 3, but for wind tunnel conditions with a 16% solidity.

According to this model meeting only one assumption—a uniform distribution of velocities in the flow through the "ideal unit", the pressure factor is approximated by expression (2.15) with the parameters of $C_D^0 = 1.2$ and m = 1.6 which determines the efficiency of the collinear unit in an unbounded flow with a value of $C_p = 1.2(1 - (U/U_1)^{1.6}U/U_1$ with the maximum value of

$$(C_p)_{max} = 0.41 \text{ at } U/U_1 = 0.55. \tag{2.30}$$

It is important to note the great role of external boundary conditions—the calculation for the conditions of a wind tunnel in which the wind wheel model would occupy only 16%, given the drag factor (and, accordingly, the efficiency of the turbine) being 20% more than that in the conditions of unlimited flow.

Both the elementary theory and more complex calculations do not yet reflect the specifics of the power unit which can be disclosed when considering the hydro-aerodynamics of a rotor streamlined at a speed of U. Using one or another mathematical model of the rotor, we can plot the dependence of the pressure drop across the rotor, on the speed U. Combining this solution with a solution simulating the unit with a permeable figure and a discontinuous pressure field, we find the relative flow rate in the unit zone of U/U_1 and all its characteristics. This approach was the basis for many works aimed at finding the optimal designs of power units.

At present, there are mathematical models making it possible to carry out a comprehensive hydro-aerodynamic calculation of power units taking into account flows both in the unit zone and outside it without using a permeable plate model.

When considering turbines in an open boundless flow, the parameters of the problem are the relative geometric characteristics of the turbine as a whole, and of the blades separately, and the relative speed of the blades is V/U. When considering the external problem, the characteristic flow rate U is replaced by the set speed outside the influence of the turbine, U_1 or U_0. The local Reynolds number compiled by the chord of the blade b and by its linear velocity of motion V, may substantially influence this as well.

$$Re_V = Vb/\nu \tag{2.31}$$

ν is a kinematic viscosity factor of the medium.

The turbine power P depending on the incoming flow velocity U_0 and the linear blades velocity $V = \pi Dn/60$ can be represented in the two forms:

$$P = C_P \rho U_0^3 S/2 \tag{2.32}$$

$$P = C_N \rho V^3 S\sigma/2, \text{ where } \sigma = ib/D \tag{2.33}$$

Here S = DH is the area of the axial section of the figure swept by the blades, (Sσ) is the area of the median (chordal) surface of the working blades, $\sigma = ib/D$ is the shading, i is the number of blades, b is the length of the chord of the blade, D is the diameter of the path of the blades, and ρ is the density of the medium (water or air). The factors C_P and C_N depending on the shape of the turbine, the ratio of the speeds V/U_0, the quality of the blades, and the Reynolds numbers are related by the identities of

$$C_N = C_P(U_0/V)^3/\sigma \tag{2.34}$$

$$C_P = C_N(V/U_0)^3\sigma \tag{2.35}$$

In the American literature, expression (2.33) relating the power of the unit to the speed of movement and the surface area of the blades is written differently, keeping in consideration the area of the axial section of the swept surface:

$$P = K_p \rho V^3 S/2 \tag{2.36}$$

The factor K_p is related to the factor C_p by the identity of

$$C_P = K_P (V/U_0)^3 \tag{2.37}$$

Figs. 2.9 and 2.10 show an example of the energy characteristic of a modern collinear unit with a fixed position of the blades presented depending on the "speed" of the wind wheel $C_P(V/U_0)$ or the relative wind speed $K_P(U_0/V)$. Points A, B, C, D mark the same modes in different coordinates.

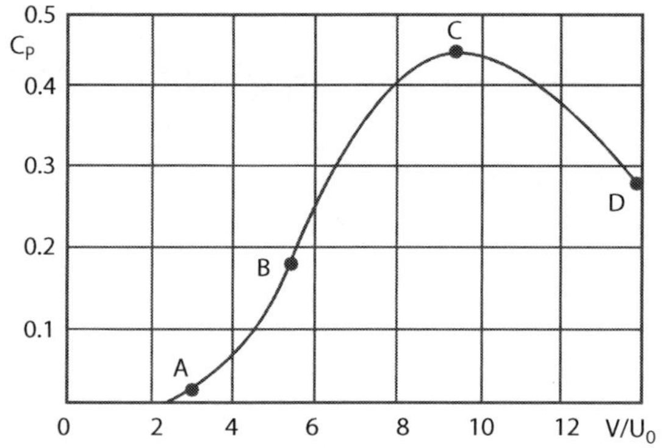

Fig. 2.9 An example of the energy characteristics of a high-speed wind turbine with a horizontal axis and fixed blades.

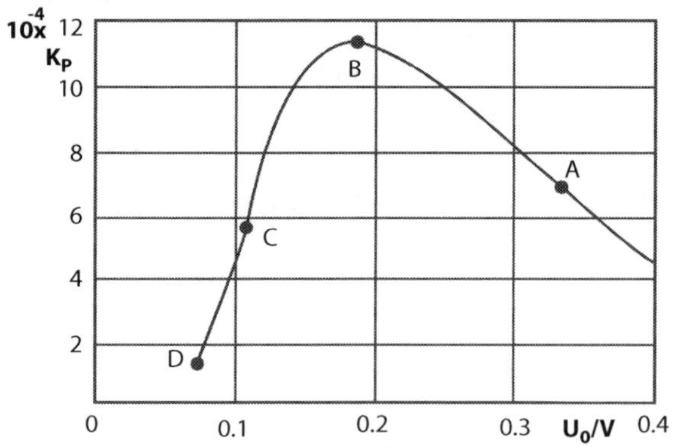

Fig. 2.10 Same as in Figure 2.9, but in the coordinates of wind speed and a power factor K_p.

The characteristic speed of the blades in all cases is taken as the speed of the ends of the blades:

$$V = \omega R \tag{2.38}$$

Power P and torque M on the turbine shaft are related to one another:

$$M = P/\omega \tag{2.39}$$

The relative value of the torque $C_M = M/(\rho U_0^2 SR/2)$ is determined by the factor of C_P and the speed of the turbine – $\lambda = V/U_0$ (in Russian literature λ is sometimes called the "number of modules", in English literature — the "tip speed relation"):

$$C_P = C_M \lambda \tag{2.40}$$

The choice of the notation form (2.36) from the point of view of the theory of similarity is not correct, since the principle of homogeneity of the group of initial dimensional parameters is violated—the area of the blades is not taken, as in (2.33), but the value S = DxH is considered with the area dimension, which value is not, however, directly related to the blades' geometry. This was reflected in the results—for all modes, the K_P parameter turned out to be strongly dependent on the shading σ and was not applied in the analysis of wind turbines (note that in the extensive book[8], in Chapter 5 written by R. E. Wilson, quantitative data on the factor of K_P are given in the graphs with erroneous scales).

The use of representation (2.36), however, turns out to be convenient when analyzing collinear aggregates in which the shading varies along the turbine radius. It is clear that the power taken from the flow on a permeable figure modeling the unit and the turbine power differ by the amount of additional turbulence energy generated by the flow in the rotor.

Representation (2.32) establishes a direct connection between the source reserve energy $\rho U_0^3/2$ carried through a unit of cross-sectional area of the flow and converted by the unit into mechanical power P of the rotating rotor. By taking energy from the flow, the unit slows it down. Therefore, the flow rate in the area of the power unit **U** is always less than the flow rate $\mathbf{U_0}$ at a distance from it, or in its absence. The maximum power that can be taken from the flow corresponds to a proper value of the ratio $\mathbf{U/U_0}$. The elementary theory of Betz-Zhukovsky, as noted above,

[8] *Wind Turbine Technology. Fundamental Concepts of Wind Turbine Engineering.* Editor David A. Spera, ASME Press, 1994, 638 p.

gives the maximum possible value of $C_p = 16/27 = \mathbf{0.5926}$ at $\mathbf{U/U_0 = 2/3 = 0.67}$ — formula (2.13). Calculation of flows through a permeable "plate" simulating an energy unit (the theory of "ideal unit") gives different results depending on what conditions are set on the "plate". If we accept a uniform distribution of flow rates on the plate, then two different solutions are obtained for a rectangular plate—for a symmetric trace, and without this constraint (Fig. 2.4). In both cases, the maximum efficiency of the unit is within $C_p = \mathbf{0.55 \div 0.58}$ at $\mathbf{U/U_0 = 0.55 \div 0.60}$. If we do not require a uniform distribution of flow rates over the surface of the permeable plate simulating the power unit, but set the condition for the proportionality of the local flow rate through the plate to the flow rate modulus near the plate (filter plate) and take into account the decrease in pressure behind the plate (a modified GGS model), the maximum efficiency turns out to be practically the same.

The actual efficiency of the units depends significantly on their design and may differ considerably from the above estimates, both towards lower values, which is explained by the non-optimal design or mode and energy losses, and towards high values, which may be associated with the attraction of the energy of a part of the flow not passing through the section of the unit.

Representation (2.33) may seem physically less intuitive. However, it makes sense, since, in fact, the process of removing energy from the wind occurs when the wind flows around the blades of the unit. The forces arising in this case are proportional to the square of the relative velocity of the flow around the blade and its area, and the local power is proportional to the speed of the blade. Indeed, let us consider the flow around the blade of a wind turbine in a plane perpendicular to the median surface of the blade and parallel to the flow velocity in the zone of the wind turbine (Fig. 2.11).

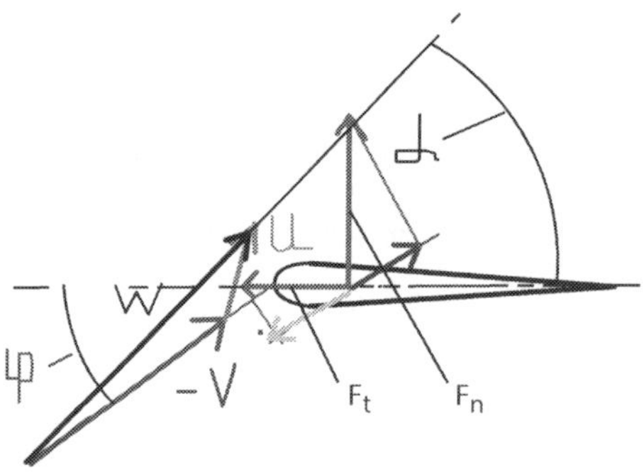

Fig. 2.11 Flow diagram of the power unit blade.

The relative speed of the flow around the blade **W** is defined as the vector difference between the local wind speed U and the linear speed **V** of the blade tip. The force acting per unit length of the blade can be represented as the vector sum of the pulling force $\mathbf{F_t}$ acting along the chord of the blade and the normal force $\mathbf{F_n}$ acting perpendicular to the chord of the blade.

$$\mathbf{F_t} = C_t \rho W^2 b/2 \tag{2.41}$$

$$\mathbf{F_n} = C_n \rho W^2 b/2 \tag{2.42}$$

Each of these forces is proportional to the velocity head of the relative flow velocity and the aerodynamic factors C_t, C_n, which depend on the angle of attack α, the shape of the blade profile, and its relative elongation ("aspectratio" $A = L_b^2 / S_b$ where L_b is the blade length and S_b is the blade surface area. For a rectangular "plan" $A = L_b/b$), blade surface roughness, Reynolds number, and the level of flow turbulence.

Factors C_t and C_n can be expressed in terms of the lift factor C_L and drag factor C_D — in the Russian literature, CY and CX:

$$C_t = C_L \sin\alpha - C_D \cos\alpha \tag{2.43}$$

$$C_n = C_L \cos\alpha + C_D \sin\alpha \tag{2.44}$$

Figs. 2.12, 2.13 show the influence of the elongation and thickness of the blade profile. At large angles of attack, but less than the "critical" value corresponding to the separation of the flow, dynamic effects of delaying the

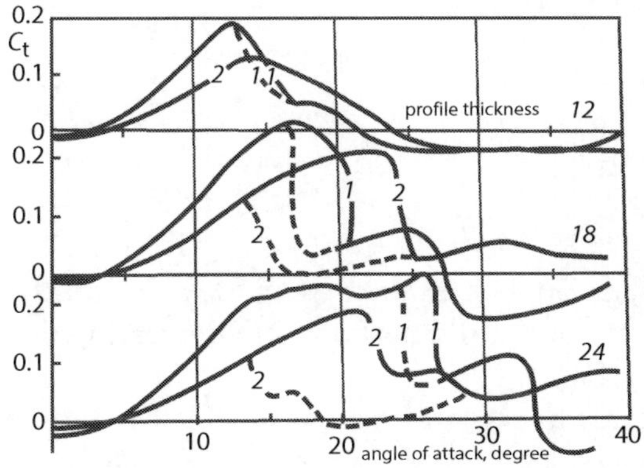

Fig. 2.12 Factor of pulling force for NACA 00XX profiles with infinite blade length (1) and A = 5 (line 2). Dotted line — reverse — decrease in the angle of attack. Re = 0.59 10^6.

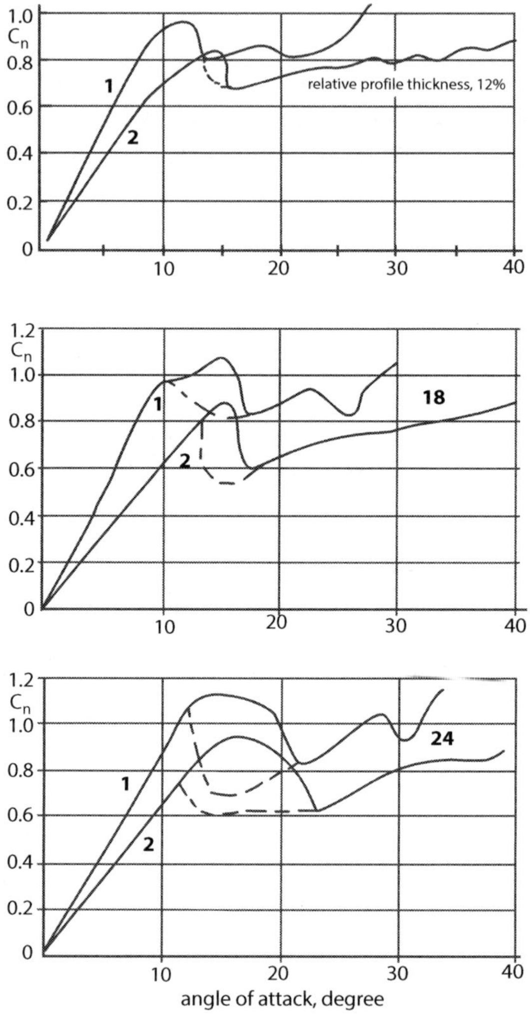

Fig. 2.13 Factor of normal force for NACA 00XX profiles with infinite blade length (1) and A = 5 (line 2). Dotted line — reverse — decrease in the angle of attack. Re = 0.22 10^6.

moment of separation with an increase in the angle of attack and delaying the restoration of the continuous flow with decreasing the angle of attack, are observed.

This is especially noticeable on relatively thick and short blades.

The influence of the Reynolds number and roughness on the lift and resistance of the profiles is large, especially at angles of attack close to the critical ones (Figs. 2.14, 2.15).

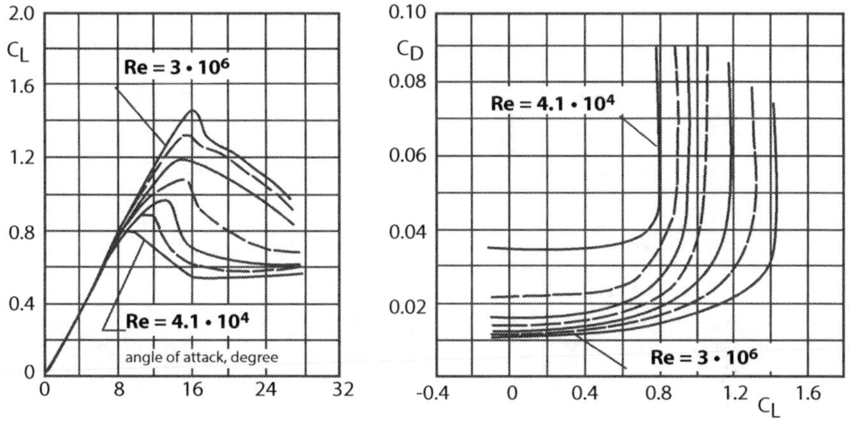

Fig. 2.14 NACA 0018.

Special profile developed for low-speed aircraft of non-commercial aviation at the Langley Research Center (USA) has high and stable aerodynamic characteristics.

The profile was named General Aviation-Witcomb-1, abbreviated as GAW-1 (Fig. 2.16). This profile, with a relative thickness of 17%, gives a record high value for the factor of pulling force: $C_t = 0.356$ at $\alpha = 16^0$and Re = 1.9 10^6. The full characteristics of the NACA and GAW-1 profiles are included in the Appendices.

By multiplying forces (2.41, 2.42) scalarly by the speed of the aerodynamic center of the blade, **V** and summing up the result, we obtain the power taken by the wind turbine from the wind per unit length of one blade. In a conventional collinear power unit, the direction and value of the flow velocity in the unit zone **u** almost does not depend on the position of the blade during the rotation of the rotor, but it can vary for different distances from the center of rotation and for different points in time (an unsteady wind).

In orthogonal machines, on the contrary, the magnitude and direction of the flow velocity relative to the blade are different for each point of the blade track, but are almost the same for different sections of the unit with straight blades. The number of blades, the relative chord length b/D, the mutual influence of the blades, and design features are manifested in the relative value and distribution of the flow velocity **u** within the units.

The energy characteristics of high-speed orthogonal and collinear units are qualitatively different (Fig. 2.17). Although the maximum energy efficiency in both cases is approximately the same (for the new type of

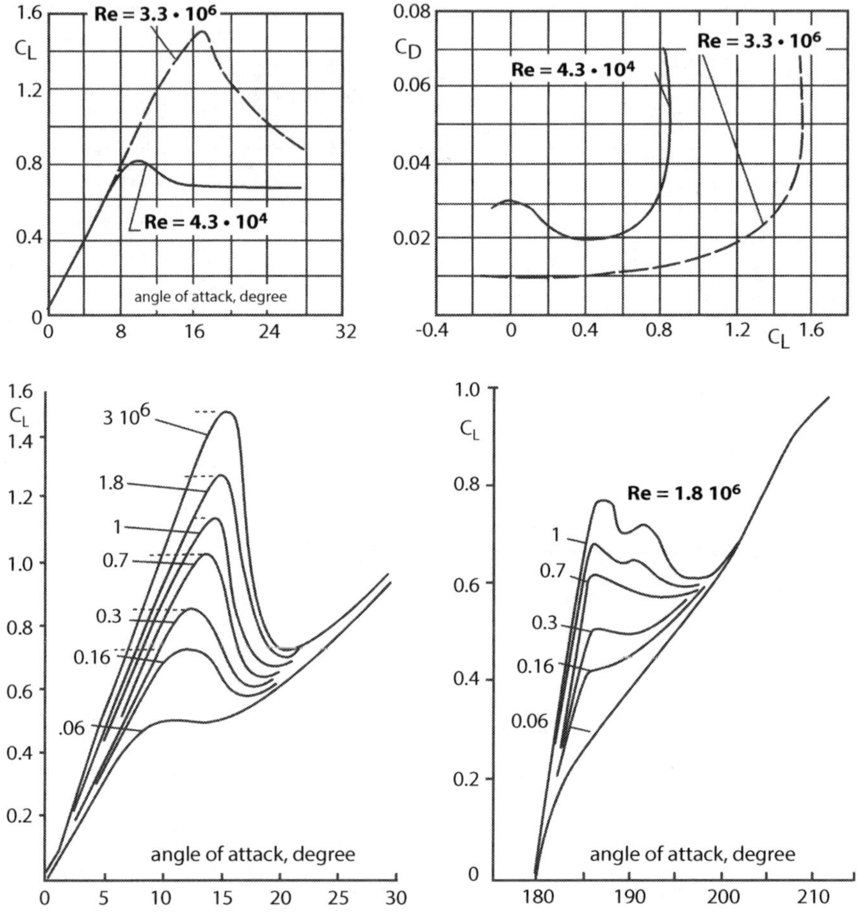

Fig. 2.15 NACA 0015.

orthogonal machines with straight blades and optimal shading $\sigma = 0.3$, as can be seen in Fig. 2.17, it can even be noticeably higher), the type of characteristics is different.

The collinear unit has relatively high blade speeds. The failure of the braking system and the loss of the electrical load of such a wind turbine even at low wind speeds practically may cause the acceleration and destruction of the machine. On the contrary, in an orthogonal machine, the speed of the blades is only 3 to 4 times higher than the speed of the flow, and the accelerating speed is not much higher than the optimal one. The collinear wind turbine should make a lot of noise – the ends of the blades move too fast, and the hydraulic unit will be at risk of cavitation on the

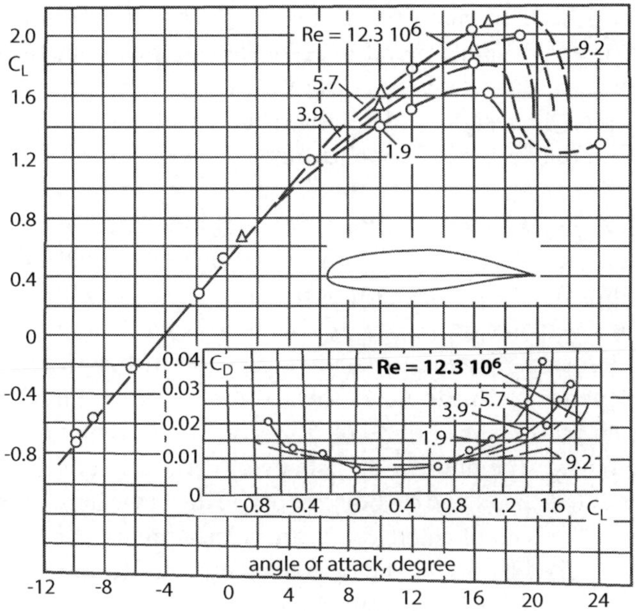

Fig. 2.16 GAW-1.

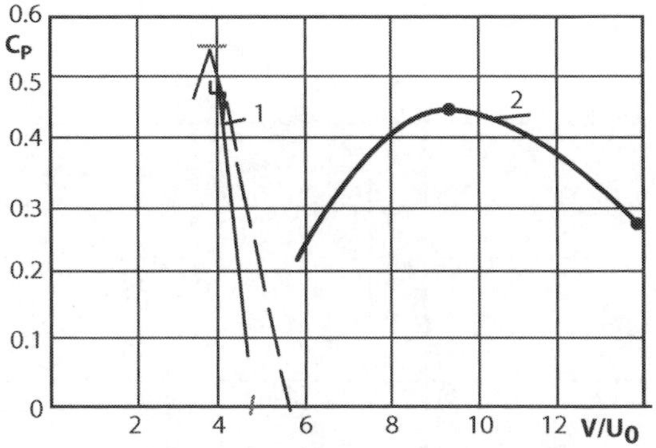

Fig. 2.17 Characteristics of the best orthogonal and collinear aggregates with a fixed blade turn. 1 – orthogonal unit, 2 – collinear unit.

blades. On the other hand, the shallow shape of the maximum efficiency for collinear units makes the requirement for regulating the speed of the blades less rigid.

An orthogonal unit of the Darrieus type ("jump rope – troposken") with low shading can have a characteristic with an extended zone of high efficiency and high accelerating speeds (Fig. 2.18). These units, as noted, require preliminary spin-up and experience significant unbalanced aerodynamic loads causing dangerous vibrations.

In principle, the design of hydraulic turbines for free flows has much in common with the designs of wind turbines. However, there are also significant differences associated with the higher density of the aquatic environment, relatively low flow rates and, accordingly, low inertial forces, and the possible influence of phase transitions (aeration and cavitation) in the flow. The water flow in many cases cannot be considered as unlimited. In rivers and even in sea straits, it may be advisable to use certain stream-guiding structural elements. The energy characteristic of the turbine $C_P(V/U_0)$ significantly depends on the design of the turbine and the conditions in which it will be used. In the same boundary conditions, the influence of the shape of the turbine, i.e., the number and shape of the section of the blades and traverses, and the speed and turbulence of the flow (see items 3, 4), is great. According to the data of calculations and experiments, the power factors C_N or K_p with fixed blades in the operating range of turbines are approximately described by the linear functions:

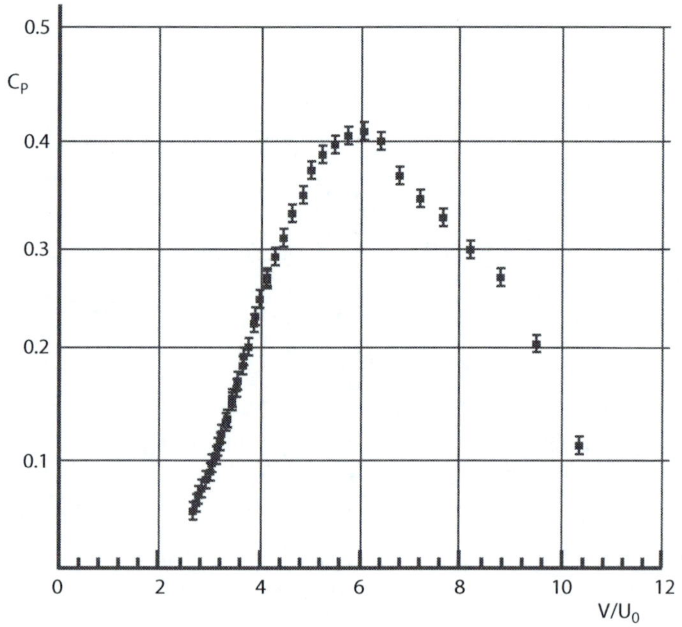

Fig. 2.18 Efficiency of the Sandia two-blade orthogonal turbine. Shading σ from 0.05 to 0.12. The rotation speed is34 rpm. D = 34.2 m, H = 41.9 m. Clean blades.

$$C_N = B\,(U_0/V - B_0),\ \text{if}\ B_0 < U_0/V < B_{lim} \qquad (2.45)$$

$$K_p = B\,(U_0/V - B_0),\ \text{if}\ B_0 < U_0/V < B_{lim} \qquad (2.46)$$

In fact, records (2.45, 2.46) mean that at a constant blade speed V = const, the turbine power is directly proportional to the first (and not the third!) flow velocity. This was found during tests of collinear and orthogonal turbines of various types (Figs. 2.19 to 2.21).

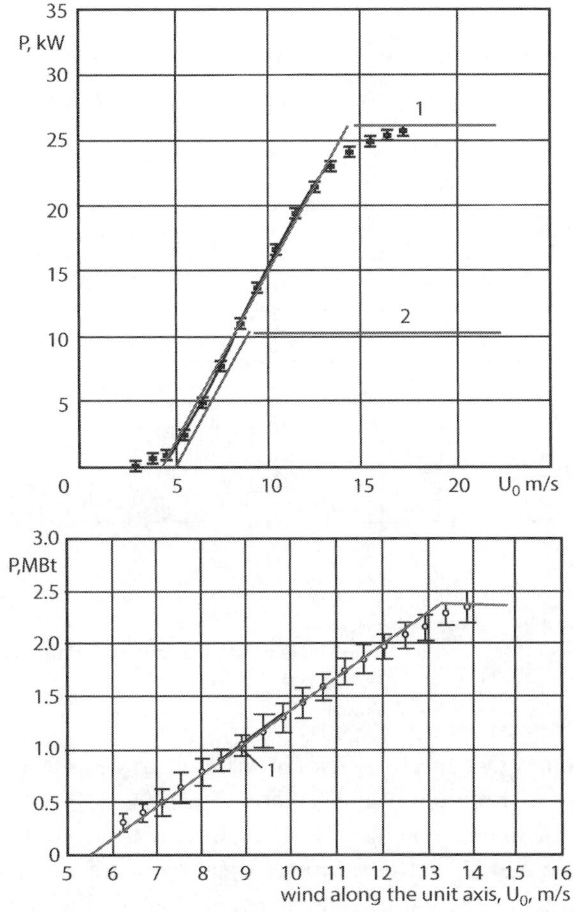

Fig. 2.19 Field tests of collinear units with a diameter of 10 m (top) and 91.4 m (bottom). (1) clean blades, (2) blades covered with dust, insects, plant seeds (after prolonged use).

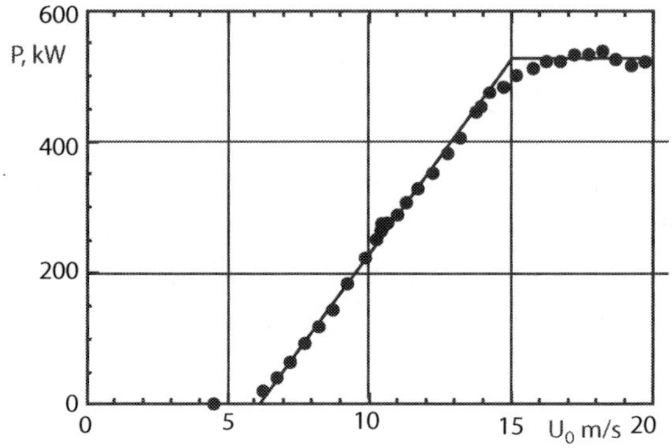

Fig. 2.20 Power of a Darrieus type two-blade orthogonal wind turbine depending on the wind speed (Sandia, D = 34.2 m, H = 41.9 m, σ_{min} = 0.053 with the rotation speed of 34 rpm. Full-scale measurements with cleaned blades, January-April 1989).

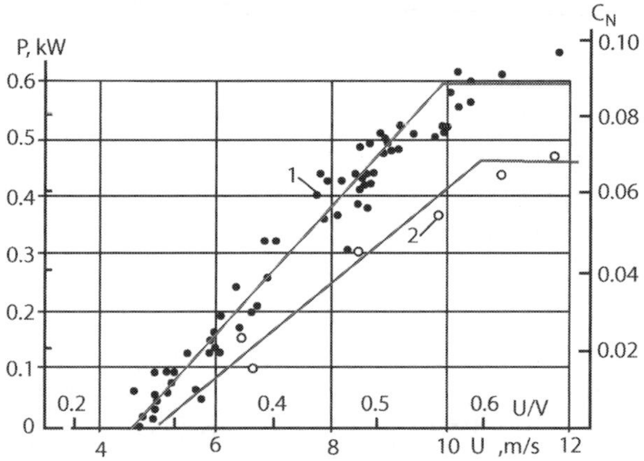

Fig. 2.21 Tests of an orthogonal wind turbine with straight blades performed at the Institute of Atmospheric Physics of the Academy of Sciences of the USSR (1) and in the TsAGI wind tunnel (2). U is the wind speed at a distance from the unit, V is the speed of the blades.

Thus, the energy characteristic of a turbine can be approximately specified by only the three parameters. Parameter $1/B_0$ is the maximum relative speed of the blades reached by the turbine without braking, or, on the contrary, parameter B_0 is the minimum relative wind speed at which the turbine starts to work. Parameters B and B_0 determine the efficiency of the turbine:

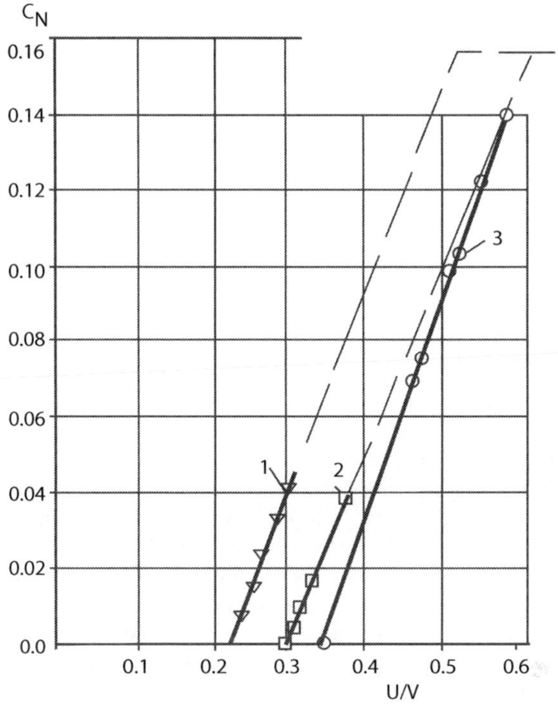

Fig. 2.22 Power factor C_N of orthogonal turbines vs flow rate. Test results: 1 – two-tier optimized turbine with straight blades according to Fig. 0.22, in a wind tunnel, $\sigma = 0.3$; 2 – turbine with curved blades according to Fig. 0.25, in a hydraulic flume at a flow rate of 1 m/s, $\sigma = 0.45$; 3 – optimized "helicoid" turbine according to Fig. 0.26, in the basin at a drawing speed of 0.6 to 2.4 m/s, $\sigma = 0.45$.

$$C_P = B\,(U_0/V - B_0)(V/U_0)^3\sigma \text{ – for orthogonal machines} \quad (2.47)$$

$$\text{Or, } C_P = K\,(U_0/V - K_0)(V/U_0)^3 \text{ – for collinear machines} \quad (2.48)$$

The maximum efficiency is, respectively:

$$(C_P)_{max} = 4B\sigma / 27B_0^2 \quad \text{at } V_{opt} = 2\,U_0/3B_0 \quad (2.49)$$

or

$$(C_P)_{max} = 4\kappa / 27\kappa_0^2 \text{ at } V_{opt} = 2\,U_0/3K_0 \quad (2.50)$$

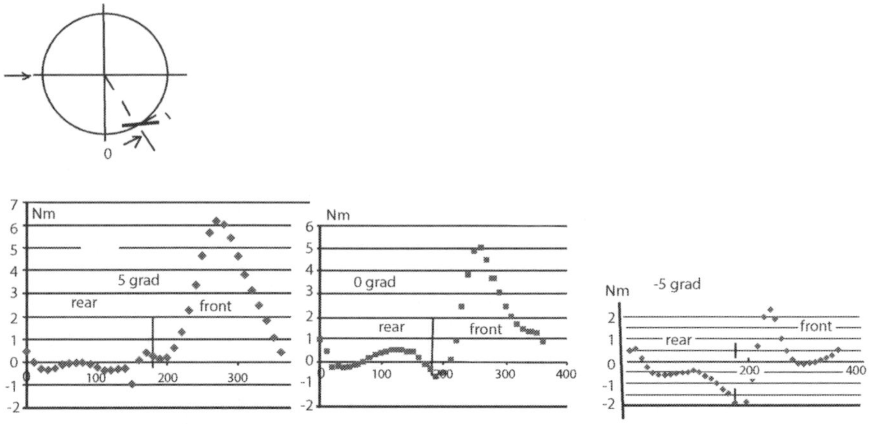

Fig. 2.23 Torques on the turbine axis (Nm) as a function of the angle of the blade position on the track at different angles of rotation of the blade chord relative to the aerodynamic center. On the left there is the tip of the blade outward of the track at 5^0, on the right, the tip of the blade inside the track at 5^0, and in the center—that without turning (the chord of the blade is situated tangentially to the track of the aerodynamic center. The lade chord is 180 mm, track diameter is 640mm, profile is NASA 0018, blade length 900mm, rotation speed 600 rpm, and the wind speed is 10 m/s. V/U = 2. The efficiency of the turbine at installation angles of plus 5, zero, and minus 5 degrees, was C_p = 0.24; 0.192; –0.03.

The peculiarities of the characteristics of orthogonal turbines are largely related to the fact that under the influence of the unit, the flow velocity in the front section of the blade path is lower than the speed on the approach (far from the unit), but is significantly higher than in the rear section of the path. In fact, the rear section of the track is almost inoperable (Fig. 1.23, item 3). This defect of orthogonal machines is eliminated in the scheme of multi-blade units of a relatively large diameter (item 5).

The idea of these units protected by many patents of the author, is to use vertical turbulent transfer of wind energy restoring the wind field in front of the rear blade formation.[9]

Orthogonal aggregate diagrams are applicable to atmospheric currents and water flows. The transformation of the energy of unsteady currents arising in straits as a result of ebb and flow has its own specifics. For the full use of the energy of these currents, there is no need to completely block the basin mouth and create a pressure front. The maximum possible power and

[9]V.M. Lyatkher, Orthogonal Units to Use Energy of Flows with the Highest Efficiency and Power, "Alternative Energy and Ecology" *International Scientific Journal*, Nos. 16 to 18, 2017, pp. 1-11.

the maximum possible power generation are obtained with a well-defined hydraulic resistance introduced by the turbines placed at the entrance to the TPP pool (in its narrowest part). There is some optimal resistance corresponding to the maximum power and possibly another resistance corresponding to the maximum output[10].

Let z_0 (t) and z(t) denote the water level in the sea and in the TPP basin, respectively, Ω is the area of the water mirror in the basin, Ω_p is the area of the free cross-section of the strait (TPP waterway), A is the tide height with the period T, ξ is the hydraulic drag factor of the water circuit of the TPP. Let us consider the case when the pool is not flowing, i.e., it communicates with the sea through one strait only.

In this case, the flow rate through the strait, Q(t), shall be determined by the course of the water level in the pool:

$$Q(t) = \Omega dz(t)/dt \tag{2.51}$$

The head losses in the strait are proportional to the velocity head of the flow in its effective cross-section:

$$| z_0(t) - z(t) | = \xi\, U^2/2g, \quad U = Q(t)/\,\Omega_p \tag{2.52}$$

The energy balance equation in the zero-dimensional approximation for a periodic tide takes the form of

$$\alpha\, (dz/dt)^2 = \ | \text{Cos}\ 2\pi t - \ z(t) \,| \tag{2.53}$$

where

$$\alpha = [\Omega/\Omega_p]^2 \xi A/4gT^2 \tag{2.54}$$

The water level in the basin z(t) and in the sea is measured in fractions of the initial amplitude equal to half the tide height A, and time is measured in fractions of the period T. Thus, all variables in (2.53) are dimensionless. The tide height at sea is considered constant.

Equation (2.53) can be written explicitly with respect to the time derivative of the water level in the basin:

---

[10] V. M. Lyatkher, "Optimal Use of Tidal Energy", *Reports of the Russian Academy of Sciences (DAN RF)*, 2006, v. 411, No. 4, pp. 476-479, "Energy Potential of Tidal Power Plants", *Proceedings of the Russian Academy of Sciences (Izvestiya AN RF), Energetika*, 2006, No. 4, pp. 39-46.

$$z' = \text{sign}\,(\text{Cos}\ 2\pi t \quad z)\ \sqrt{}\ |\text{Cos}\ 2\pi t \quad z|/\ \alpha \tag{2.55}$$

Function

$$\text{sign}x = +1, \text{ if } x > 0$$

$$\text{sign}x = -1, \text{ if } x < 0$$

Equation (2.55) was solved numerically with the initial condition obtained from the analytical approximation for $\alpha \to 0$.

Initial condition at t = 0 is

$$z = \{\sqrt{}\ (1 + 4\pi^3\alpha) - 1\}/2\pi^3\alpha \tag{2.56}$$

The calculations were carried out until the solution z(t) reached the periodic mode. This usually took place starting no later than the fifth oscillation. The steady-state oscillations in the pool can be approximately represented as

$$z(t) = z_m \text{Cos}\ (2\pi t - \varphi) \tag{2.57}$$

In this case, the expression for the TPP power at fixed dimensions of the free cross-section is determined by the expression of

$$P(kW) = 9.8\ \eta z'\ [\Omega]\ \mid \text{Cos}\ 2\pi t - \ z(t) \mid A^2/4T =$$

$$= 8[\Omega]\ A^2/4T\ 2\pi z_m\ |\text{Sin}\ (2\pi t - \varphi)\ (\text{Cos}\ 2\pi t - \ z_m \text{Cos}\ (2\pi t - \varphi))\ | \tag{2.58}$$

Here η is the total efficiency of hydraulic units taken as usual such that:

$$9.8\ \eta = 8 \tag{2.59}$$

Expression (2.58) after transformations takes the following form:

$$P = 2\pi\ z_m \mid - \text{Sin}\ \varphi + B\ \text{Cos}\ (4\pi t - \beta)|\ \ \Omega A^2/T \tag{2.60}$$

where

$$B = \sqrt{}\{1 + Z_m^2 - 2z_m \text{Cos}\phi) \equiv \sqrt{}\{1 + z_m^2 - 2z(k\,2\pi)) \tag{2.61}$$

$$\text{Cos}\beta = (z_m \text{Sin}2\phi - \text{Sin}\phi) / \sqrt{\{1 + z_m^2 - 2z_m \text{Cos}\varphi)}$$

$$\text{Sin}\beta = [\text{Cos}\varphi - z_m \text{Cos}2\phi] / \sqrt{\{1 + z_m^2 - 2z_m \text{Cos}\varphi)} \qquad (2.62)$$

The maximum power value within one tidal cycle occurs at Cos (4πt – β) = –1 and can be represented in the following form of

$$P_{max} = 2\pi z_m \mid \text{Sin}\varphi + B \mid \ \Omega A^2/T \equiv C\Omega A^2/T \text{ where } C = 2\pi z_m \mid \text{Sin}\varphi + B \mid \qquad (2.63)$$

The relative power, phase, and amplitude of the wave in the bay as a function of the parameter α are shown in Fig. 2.24.

The maximum value of the TPP power at a fixed area of the basin, height, and tide period corresponding to the maximum of the factor **C**, is at α = 0.03÷0.04:

$$C_{max} = 7.27 \qquad (2.64)$$

The maximum power value that can be obtained from tides in a fixed area pool is of

$$P_{max} = 7.27\ \Omega A^2/T \qquad (2.65)$$

If the tidal period is T = 12.4h = 4.46 10^4 s, and the basin area is measured in km^2, then formula (2.65) takes the form of

$$P_{max}\,(kW) = 163\ \Omega A^2 \qquad (2.66)$$

Here Ais the maximum tide height (m). The graph of the change in the TPP power under these conditions follows the dependence of

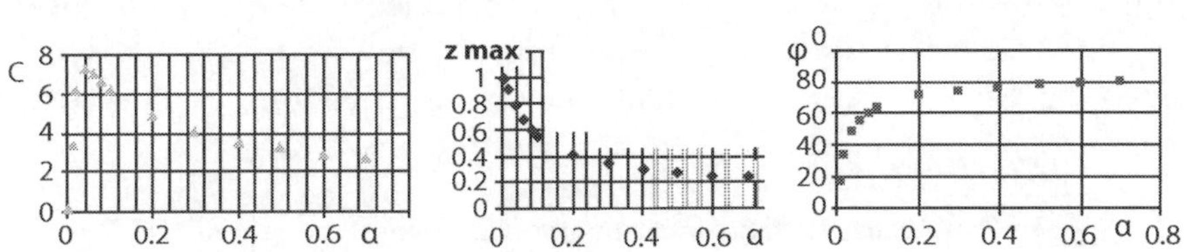

Fig. 2.24 Tidal parameters and power of the tidal power plant as a function of criterion α.

$P = 2\pi z_m \,|-\text{Sin}\varphi + \text{BSin}4\pi t|\, \Omega A^2/T$, where $z_m = 0.775$, $\text{Sin}\varphi = 0.744$, $B = 0.751$ or, approximately

$$P \approx P_{max}|-1 + \text{Sin}4\pi t|/2$$

The average power per tidal cycle is:

$$P_{mean} = 3.64\Omega A_{av}^2 / T \text{ or } P_{mean}(kW) = 81.5\Omega A_{av}^2 \qquad (2.67),$$

where $\mathbf{A_{av}}$ is the average height of the tides.

With the maximum energy withdrawal from the tidal flow, the height of the tides in the basin decreases by only 22.5%, and the phase shift is 48^0 (Fig. 2.24)[11].

Both of these circumstances are very important—a relatively small change in the tide height and the absence of a closed pressure front ensures the preservation of the ecological situation in the basin, and the presence of a phase shift makes it possible, by placing one more or several TPPs on the inner basins, to obtain in total some basic power component.

The value of the drag factor ξ in (2.54) depends on the turbines design and their layout in the strait. If the cross section of the strait is completely occupied, for example, by optimized orthogonal turbines, the drag factor shall be approximately $\xi = 1.3$.

In cases where floating or rigidly fixed hydraulic units installed in one row across the strait do not provide the optimal value of the parameter α and do not make it possible to obtain maximum energy efficiency from the developed basin, then, in such situations, the turbines should be installed in the strait in several rows, with each located from the other at a distance sufficient to restore the normal velocity field.

This solution is also useful for maintaining the ecological purity of the facility. The relatively slow-moving blades of orthogonal turbines which are rarely installed in machines are not dangerous for fishes, but the presence of quiet flow zones between the rows of turbines can be useful to allow for the resting of fishes and does not require additional material costs. The author's turbines, as well as any other orthogonal or collinear machines, are highly efficient only at a certain ratio of flow rate vs blade speed. The tidal current speed is constantly changing. Therefore, it is necessary

[11]Estimated by Chris Garrett and Patrick Cummins,"Generating Power from Tidal Currents" *J. Wtrwy., Port, Coast.*, and *Oc. Engrg.*, Vol. 130, N 3, pp. 114-118 (May/June 2004), when takingoff maximum power from tidal currents, the tidal height decreases by 26%.

either to change the speed of rotation of the turbine or to choose this speed constant in providing maximum output. Let us estimate the optimal speed of rotation of one rotor of an orthogonal turbine[12].

If the flow velocity U on the approach to the turbine changes according to the law of

$$U = U_m \text{ Cos } \beta, \text{ where } \beta = 2\pi t/T \tag{2.68}$$

then the power of the turbine rotor at a constant speed of rotation according to representations (2.32), (2.33), and (2.47) will be determined by the relations of

$$P = B[(U_m/V) \text{ Cos}\beta - B_0]\rho V^3/2 \, A_t\sigma, \text{ if } 0<\beta<\beta_1,$$

$$\text{where } \beta_1 = \text{arcCos } (U_1/U_m) \tag{2.69}$$

$$P = 0, \text{ if } \beta_1<\beta<\pi-\beta_1 \tag{2.70}$$

The average power P_{av} is determined by integrating P over the interval of

$$0<\beta<\pi/2$$

$$P_{av} = B[(U_m/V)\text{Sin}\beta_1 - B_0\beta_1]\rho V^3 \, A_t\sigma/\pi \tag{2.71}$$

Denoting the ratio of the speed U_1 vs the maximum flow rate of U_m by the symbol of

$$y = U_1/U_m = B_0V/U_m \tag{2.72}$$

we rewrite expression (2.71) in the following form:

$$P_{av} = \rho U_m^3 B B_0^2 A_t \sigma \varphi(y)/\pi \tag{2.73}$$

$$\varphi(y) = y^2\{\text{Sin}(\text{arcCos}y) - y \text{ arcCos}y\} \tag{2.74}$$

---

[12] V.M. Lyatkher, Tidal Power Plants Without a Pressure Front, *Hydraulic Engineering Journal,* 2006, No. 4, pp. 26-37.

Taking into account relation (2.49), expression (2.73) can be rewritten differently:

$$P_{av} = (27/4)C_p\max\rho U_m^3 A_t\varphi(y)/\pi \tag{2.75}$$

The maximum value of function (2.74) is 0.088 and takes place at

$$y = y_{optim} = 0.57 \tag{2.76}$$

Therefore, the optimal speed of the blades is

$$V_{optim} = 0.57\ U_m/B_0 \tag{2.77}$$

For one of our turbines studied experimentally, $B_0 = 0.205$ and

$$V_{optim} = 2.78\ U_m \tag{2.78}$$

For a helicoid three-blade turbine with the shading of $\sigma = 0.41$ and a blade profile NACA 0020 in the best configuration, the parameters have the following values (our processing of the A.M.Gorlov's experiments): $B = 0.60$, $B_0 = 0.34$, $C_{p\ max} = 0.35$.

In this case, the optimal speed of the single-speed turbine blades is:

$$V_{optim} = 1.7\ U_m \tag{2.79}$$

At the same time, in any case, the average power will be maximum and equal to 0.378 of the maximum turbine power possible at a given maximum flow rate. If the turbine rotation speed was not constant but was changed in proportion to the flow rate while maintaining the maximum turbine efficiency, then its average power would be 0.42 of the maximum value. Thus, assuming a constant rotation speed, we lose about 10% of the possible energy production. Using a two-speed generator, this loss can be reduced by about 2 times.

When using two-speed generators, the optimal rotation speeds can be estimated according to the condition of the maximum average power which in this case will already be a function of two variables—the speeds of the blades V and V_1 at high and low flow speeds, respectively. The moment of switching the speeds of rotation of the turbine is determined by the equality of powers in these two modes of

$$(U/V - B_0)V^3/2 = (U/V_1 - B_0)V_1^3/2 \tag{2.80}$$

From this equation, we obtain the flow rate value at the moment of switching the turbine rotation speeds:

$$U = U_1 = B_0(V^2 + VV_1 + V_1^2)/(V + V_1) \tag{2.81}$$

The flow rate at the moment of complete cessation of power output by the turbine is:

$$U = U_2 = B_0 V_1 \tag{2.82}$$

The average power is determined by an expression similar to (2.73), (2.75):

$$P_{av} = \rho U_m^3 B A_t \sigma \psi / \pi \tag{2.83}$$

$$\psi = z_1^2 \left[\left(\text{SinarcCos}y_2 - \text{SinarcCos}y_1 \right) - bz_1 \left(\text{arcCos}y_2 - \text{arcCos}y_1 \right) \right]$$
$$+ z^2 \left(\text{SinarcCos}y_1 - bz\text{arcCos}y_1 \right) \tag{2.37}$$

$$z = V / U_m,\ z_1 = V_1 / U_{m,} y_1 = U_1 / U_m$$
$$= B_0 \left(z_2 + zz_1 + z_1^2 \right) / (z + z_1),\ y_2 = U_2 / U_m = B_0 z_1$$

The absolute maximum of the function $\psi(z, z_1)$ corresponding to the optimal speeds of the blades in fast and slow modes, can be found by solving the system of equations of

$$\partial\psi/\partial z = 0,\ \partial\psi/\partial z_1 = 0. \tag{2.84}$$

Calculations according to (2.83), (2.84) show that the maximum optimal turbine rotation speed is quite reliably determined from formula (2.77). The second, lower optimal speed, has relatively little effect on the result and can be assigned according to the design conditions of the generators. So, if the first speed of the blades is taken according to (2.77), and the second speed of the blades is equal to the maximum flow speed, or is 20% higher (z_1 = 1 or 1.2), then the average turbine power will be 94.6% or 96.5%, respectively, from the maximum possible one achieved with a continuous change in the rotation speed of the turbine in proportion to the flow rate on the way to the turbine.

Let us now consider another limiting case, namely, a stationary flow in a channel, at the boundaries of which a level (or pressure) difference is specified. The maximum power that can be taken off from the flow in the channel does not depend on the type of hydraulic units used and is determined from the following considerations.

The power N_0 lost by the flow per unit of water consumption in domestic conditions, is determined by the following expression:

$$N_0 = \rho g \Delta H = (\lambda L / R + 1)\rho U_0^2 / 2 \qquad (2.85)$$

Here λ is the reduced factor of hydraulic resistance of the canal, L is the length of the canal, R is the hydraulic radius of the wetted cross-section of the canal, ΔH is the level difference between the reservoirs connected by the canal, ρ is the water density, g is the acceleration of gravity, and U_0 is the flow velocity in the channel.

After the construction of the power plant in one of the canal sections, the water flow through the canal will change and the flow rate in the canal outside the power plant's alignment will take a U value which is less than U_0. Assuming the difference between the ends of the channel ΔH, caused by external causes, remains unchanged, we rewrite equality (2.85) with the introduction of a concentrated pressure drop Δp (loss of flow power) at the power plant.

3

Collinear Units and Their Modifications

"The storm has passed. Few of us survived.
Many are missing from friendship's roll call.
I've returned to the orphaned region
I haven't seen in eight years.

Who should I call to? With whom should I share
The sad joy that I'm still alive?
Here, even the windmill — that one-winged bird
Made of logs — stands with eyes shut tight."

S.A. Yesenin, *Soviet Russia, 1924*
Translated into English by Anton Yakovlev

Collinear units, the axis of rotation of which is parallel (collinear) to the flow velocity (Fig. 3.1), can be high-speed units with narrow blades of an asymmetric airfoil (1) and low-speed units with a large number of relatively wide blades (2), sometimes supplied with louvers (to reduce the load on blade in a storm). High-speed units are usually more preferable for generating electricity, slow-speed units are better aggregated with various mechanical devices (for example, wind or water mills).

Slow-speed units of the old design develop high torque which was previously used to drive pumps pumping out water (Holland) or millstones grinding grain (Russia).

Wind power generators have appeared relatively recently (Fig. 1.1, 1.2).

Modern low-speed units (Fig. 3.2) are convenient for driving pumps in combination, in particular, with water purification systems based on membrane technologies requiring relatively high pressures. The efficiency of these machines is quite high and, in addition, they have a low "runaway" speed (Fig. 3.3).

Russian "gantry type" and "tent type" mills of the last century, multi-blade unhurried "Americans" and the modern ones—all three-blade giant wind turbines with a horizontal or slightly inclined axis of rotation work

Victor M. Lyatkher and Ziaur Rahman. Wave, Wind and Current Power Generation, (95–158) © 2022 Scrivener Publishing LLC

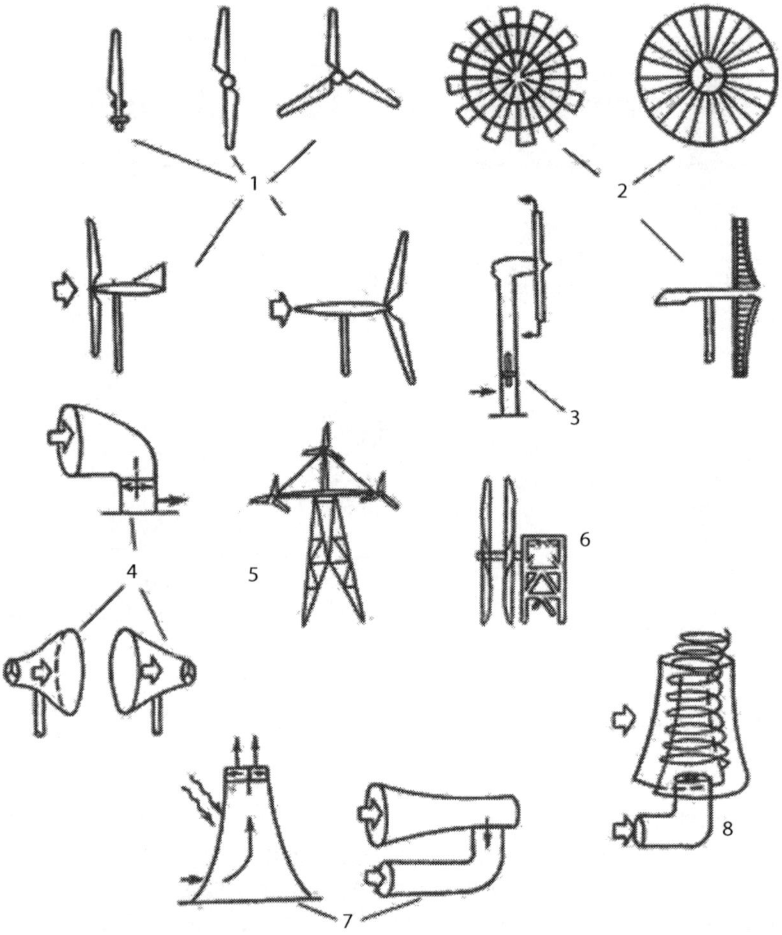

Fig. 3.1 Design options for collinear power units.

according to the same energy principle. According to this principle, the wind turbine within the circle swept by the blades creates wind resistance by causing a concentrated pressure drop Δp on this imaginary (swept) circle, and with a decrease in the longitudinal air velocity component to a value of U which is less than the wind speed U_0 and is far from the wind turbine. Thus, the air flow within the wind turbine loses the power P determined by formula (8) in item 1.

In the "ideal" case, all this power is transferred to the wind turbine. In real conditions, part of it is converted into the energy of a screw movement and turbulent pulsation movements arising behind the wind turbine.

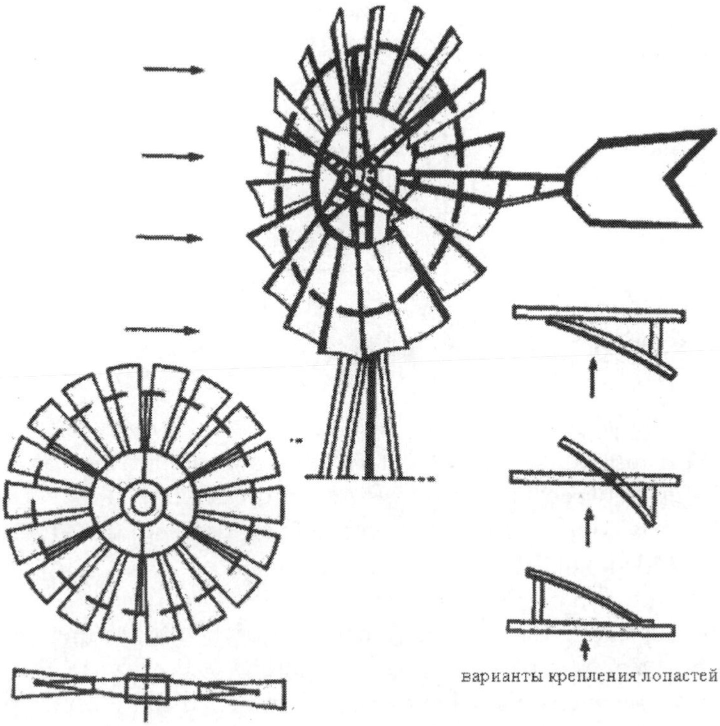

Fig. 3.2 Modern low-speed wind turbine ("American").

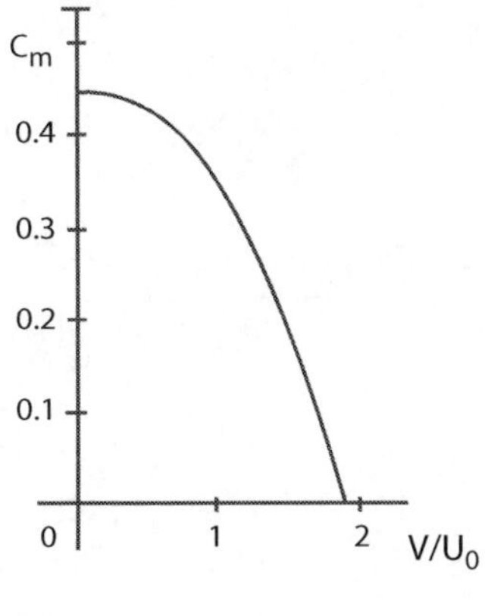

Fig. 3.3 The torque factor of the unit is shown in Fig. 3.2.

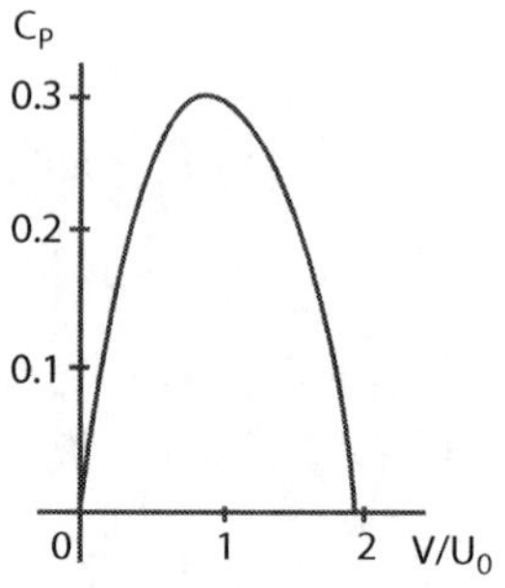

Fig. 3.4 Unit efficiency according to Fig. 3.2.

The correlations given in item 1 make it possible to estimate the maximum value of the unit efficiency, but still does not make it possible to reveal the relationship between the design of the wind turbine and its energy efficiency. This is done on the basis of considering the conditions for the flow around the blades of the wind turbine with a flow of speed *U* directed along the axis of rotation of the unit (Fig. 3.5).

While neglecting the swirl and curvature of the flow in the area of the wind wheel, i.e., assuming $U_\varphi = 0$, $U_r = 0$ and $U_x = U$, we write the expression for the aerodynamic load acting perpendicular to the chord per unit area of the blade, in the following form:

$$q_n = \frac{\rho U^2}{2} C_n(\alpha)\left(\frac{r^2\omega^2}{U^2} + 1\right) \tag{3.1}$$

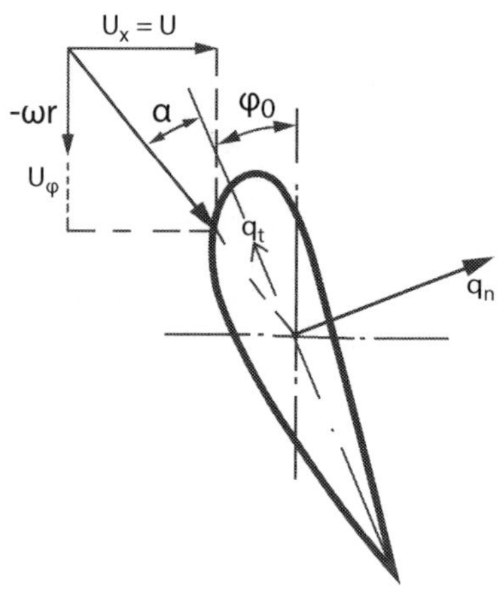

Fig. 3.5

The same along the chord of the blade (in the direction of its movement):

$$q_t = \frac{\rho U^2}{2} C_t(\alpha)\left(\frac{r^2\omega^2}{U^2}+1\right) \tag{3.2}$$

Here r is the distance from the axis of rotation to the considered section of the blade, ω is the rotational frequency of the rotor of the wind turbine (ωr is the linear speed of the blade section under consideration),

$$\alpha = arctg\frac{U}{\omega r} - \phi_0, \tag{3.3}$$

α is the angle of attack at which the air flow flows around the blade of the wind turbine; φ_0 is the angle of rotation of the blade relative to its speed. $\varphi_0 > 0$, if the forward (in motion) edge of the blade is turned towards the oncoming wind flow. C_n and C_t are the aerodynamic factors depending mainly on the shape of the airfoil of the blades and their elongation ("aspectratio" $A = L_b^2 / S_b$, where L_b is the blade length, and S_b is the blade surface area. For a rectangular "plan," $A = L_b/b$).

The projection of the load on the blade acting in the direction of the axis of rotation (in the direction of the wind) is determined by the angle of $\varphi_0(z)$:

$$q_x = q_n \cos\varphi_0 - q_t \sin\varphi_0 \tag{3.4}$$

Integrating the q_x value over the surface of the blades and referring the resulting force to the area of the circle swept by the blades, we find the expression for the pressure factor C_D through the geometric and aerodynamic characteristics of the wind turbine blades:

$$C_D = \left(U/U\frac{0}{\pi}\right)^2 i\left\{\int_{\bar{r}_0}^{1}\bar{b}(\bar{r})C_n(\alpha)\left(z^2\bar{r}^2+1\right)\cos\varphi_0 d\bar{r} - \int_{\bar{r}_0}^{1}\bar{b}(\bar{r})C_t(\alpha)\left(z^2\bar{r}^2+1\right)\sin\varphi_0 d\bar{r}\right\} \tag{3.5}$$

Here $\bar{r} = \frac{r}{R}$, R is the radius of the swept circle, $\bar{b} = \frac{b}{R}$ is the relative blade width, $z \equiv \frac{\omega R}{U}$ is the true speed of the wind turbine; i is the number of blades, and U/U_0 is the relative flow velocity in the unit zone.

Having equated (5) to (11 in item 1), or to the value of the pressure drop according to Figs. 3.3, 3.7 or a similar expression (27 in item 1) corresponding to the modified GGS model, we shall find the value of the relative flow velocity through the wind turbine U/U_0 and the maximum possible power utilization factor C_p corresponding to the adopted rotor design. Calculations show that usually the second term in the brackets on the right-hand side of (5) is much less than the first term, and thus it can be neglected.

Since the maximum value of the power utilization factor is observed at certain values of U/U_0 and C_x, for example, according to Betz-Zhukovsky at $U/U_0 = 2/3$ and $C_D = 8/9$, then, when designing such units, designers, naturally, tend to ensure the fulfillment of the condition of

$$\frac{8}{9} \approx \left(\frac{2}{3}\right)^2 \frac{i}{\pi} \int \bar{b} C_n \left(z^2 \bar{r}^2 + 1\right) \cos\varphi_0 d\bar{r}$$

or

$$\int_{\bar{r}_0}^{1} \bar{b} C_n \left(z^2 \bar{r}^2 + 1\right) \cos\varphi_0 d\bar{r} = \frac{2\pi}{i}. \tag{3.6}$$

The aerodynamic factor C_n which determines the load on the blade normal to the chord, is relatively weakly dependent on the shape of the blade airfoil and differs little from the values corresponding to a flat plate; therefore, the axial load on the rotor and the characteristics of energy removal from the wind flow for collinear machines differ little at different blade airfoils (Fig. 3.6).

Correlation (3.6) shows that for low-speed machines (small z), such as windmills in Russia, a large relative area of the working blades is needed. With the optimal turn of the blades $\varphi_0 \approx 40 \div 50°$, the optimal blade width is determined by the condition of

$$\frac{ib}{R} \approx \frac{2\pi}{1{,}15} \approx 5{,}5. \tag{3.7}$$

If $i = 4$, then $b \approx 1.4R$. Such wide blades were structurally very heavy, and the whole structure appeared to be very vulnerable to strong winds.

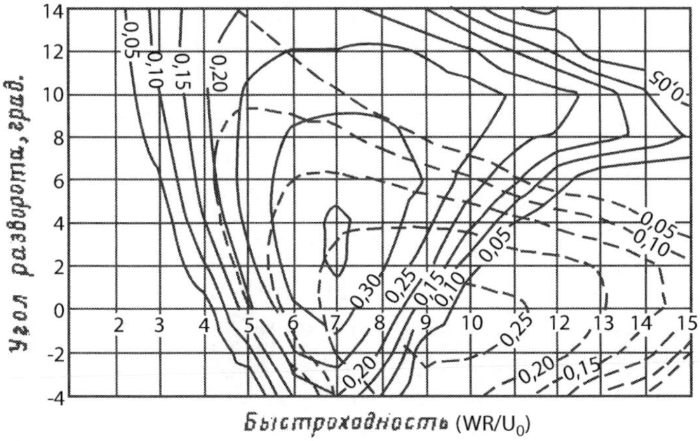

Fig. 3.6 Efficiency contours of power generating units, C_p, with Espero blades (solid lines) and NACA 0012 (dashed lines).

The builders of windmills a hundred years ago, fearing destruction, apparently did not seek to maximize the use of wind power, despite being aware even then of the advisability of widening the tips of the blades. On modern high-speed machines, this constructive solution, the expediency of which is clearly visible from (5), has not yet been implemented.

The value of the integrals on the right-hand side of (5) strongly depends on the speed z and relatively weakly, on the angle of rotation of the blades. Table 3.1 shows the values of the integral of

$$J = \frac{1}{\pi} \int_{0,2} \bar{b} C_n(\alpha)\left(z^2 \bar{r}^{-2} + 1\right) \cos\phi_0 d\bar{r}, \tag{3.8}$$

Table 3.1 The values of the integral J characterizing the relative unit drag.

z **φ0, deg**	**2**	**4**	**6**	**8**	**10**	**12**
–4	0.05	0.12	0.25	0.46	0.78	1.14
0	0.05	0.12	0.29	0.50	0.75	1.02
+4	0.05	0.13	0.28	0.44	0.62	0.82
8	0.04	0.13	0.24	0.34	0.45	0.55

in which connection, the aerodynamic factor $C_n(\alpha)$ is adopted for the Espero airfoil, and the relative length of the airfoil chord $\bar{b}$ accepted equal to $\bar{b} = 0.083$ at $\bar{r} = \bar{r}_0 = 0{,}2$ and $\bar{b} = 0.05$ at $\bar{r} = 1$.

From formula (6) and Table 1 it follows that, according to the Betz-Zhukovsky theory, the adopted relatively narrow blades should be installed in a number of 2 ÷ 3 pieces with a high speed $z_0 = \dfrac{\omega R}{U_0} = (10 \div 20)\bar{U} = 7 \div 8$ in the zone of the maximum of C_p. This simple recommendation is carried out in modern high-speed wind turbines with a traditional layout.

The unit blades' turn control is carried out so that when the wind speed increases above the limiting (calculated) value of U_p, the rotor power remains constant, corresponding to the generator power. The design of the wind speed is determined by a technical and economic analysis: an increase in U_p entails an increase in the energy generation (up to a certain limit), but it also entails an increase in the cost of mechanical and electrical equipment, as well as in the duration of low quality energy delivery (when $U_0 < U_p$, the power varies greatly following changes in the wind speed). A constant value of the power of the wind turbine at $U_0 > U_p$ can be achieved by special profiling of the blades. In all cases, based upon correlation (3.5), the maximum load averaged in time and over the swept surface, acting on the operating unit, should be observed at the design wind speed of $U_0 = U_p$.

If we replace the relative blade width of $\bar{b}$ by the average value under integrals (3.5), then it is possible to record explicitly the ratio between the blade width vs. the optimal speed of the wind turbine corresponding to the maximum use of the flow power:

$$i\bar{b} = \frac{2\pi}{\int_{\bar{r}_0}^{1} C_n \left(z^2 \bar{r}^{-2} + 1 \right) \cos\varphi_0 d\bar{r}}. \tag{3.9}$$

The calculation results according to (3.9) can be approximately obtained from Table 3.2:

$$i\bar{b} \approx \frac{0{,}12}{J}.$$

An increase in speed makes it possible to reduce the blades' width and, it would seem, to reduce the material consumption of the unit rotor (the total load on the rotor with a fixed span of the blades in the maximum C_p

Table 3.2

z_{opt}	4/3	8/3	4	16/3	20/3	8
$i\bar{b}$	2.4	1	0.41	0.24	0.16	0.12

mode, i.e., at a constant relative speed of U/U_0, as can be seen from (3.5), does not change). However, this is true only up to a certain limit, since with an increase in speed, the load on the blade also increases (according to (3.4)), the bending moment increases, and, therefore, the thickness of the airfoil and (or) the thickness δ of the bearing structure of the skin should increase. If, on the basis of technical and economic optimization, we fix the relative thickness of the blade airfoil $\frac{c}{b} = const$, then the drag moment of the blade changes in proportion to b^2:

$$W = k_w \delta \cdot b^2.$$

This means that an increase in speed (with a constant blade length) will entail a decrease in the rotor weight only in the range when the strength of the blades is determined by centrifugal forces and it weakly depends on aerodynamic loads. This condition restrains the increase in the speed of the wind turbine blades which is desirable for reducing the weight of the rotor, gearbox, and electrical equipment.

Let us analyze the problem of hydro- or aerodynamic calculation of the unit and the results of its solution with the sequential elimination of the introduced assumptions, the expansion of the degree of universality and the corresponding complication of the method*.

The equation of momentum conservation recorded for a stream tube formed by streamlines passing through a circle swept by the tips of the working blades of the unit has the following form (Fig. 3.7):

$$pU_0Q + p_0\Omega_0 - (pU_2Q + p_2\Omega_2) + [\beta_1(p_{--} - p_0) +$$

$$+ p_0](\Omega - \Omega_0) + [\beta_2(p_+ - p_2) + p_2](\Omega_2 - \Omega) - (p_{--} - p_+)\,\Omega; \qquad (3.10)$$

where **U** is the flow rate of the medium through the surface of the disk simulating the unit; $\mathbf{U_0}$, $\mathbf{U_2}$, — wind speed, respectively, in front of and behind

*V.M. Lyatkher, Yu.B. Shpolyansky, Loads on Hydro and Wind Power Units in a Free Flow and their Energy Characteristics, *Hydraulic Engineering Journal*, 1986, No. 12, pp. 16-22.

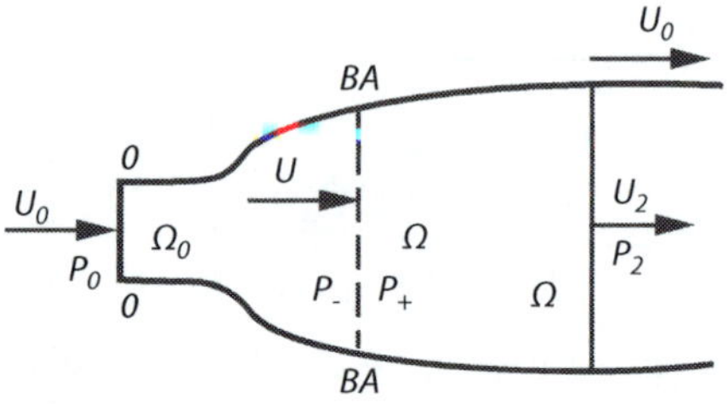

Fig. 3.7 Flow diagram of the unit.

the unit; **p**$_-$, **p**$_+$ is the pressure, respectively, before, and behind the unit; $\mathbf{\Omega_0}$, $\mathbf{\Omega}$, $\mathbf{\Omega_2}$ is the area of the stream tube in sections **0**, **BA**, and **2**, respectively, in Fig. 3.7.

In equation (3.10), as usual, the shear stress acting on the surface of the selected volume is not taken into account, but, in contrast to traditional schemes, the factors β_1 and β_2 take into account the change in pressure along this surface.

Equation (3.10) is supplemented by the equation of conservation of the angular momentum which imposes restrictions on the tangential components of the flow velocity in front of the unit $U_{\varphi--}$ and $U_{\varphi+}$:

$$p \int (U_{\varphi+} - U_{\varphi--}) Ur d\Omega = M, \tag{3.11}$$

where **M** = N/ω is the torque acting on the unit; **ω** is the circular rotation frequency; **N** is the power received by the unit.

Bernoulli's equations are written separately for the flow sections in front of and behind the unit. Neglecting the influence of the swirl in the flow away from the unit and selecting the values of U_0 and $pU_0^2/2$ as scales of velocity and pressure, we obtain the following system of equations:

$$2u - 2u_2u + \beta_1 p - (1 - u) + \beta_2 p + (u/u_2 - 1) = p_{--} - p_+ \tag{3.12}$$

$$u_{\varphi-}^2 + u^2 + p_- = 1; \tag{3.13}$$

$$u_{\varphi+}^2 + u^2 + p_+ - u_2^2 = 0; \tag{3.14}$$

$$2/3(u_{\varphi+} - u_{\varphi-})z = -[u_{\varphi-}^2 + p_- - u_{\varphi+}^2 - p_+], \tag{3.15}$$

where $\mathbf{z} = \omega R/U_0$; **R** is the radius of the swept disk.

The correlation between the flow rotation speeds in front of and behind the unit shall be:

$$\beta_3 = U_{\varphi--}/U_{\varphi+}; \tag{3.16}$$

as well as the values of the factors $\beta_{1,2}$ cannot be found from the reduced zero-dimensional equations — they are determined by solving the complete hydrodynamic problem.

The possible range of values for these parameters, determined from physical limitations:

$$0 \leq \beta_{1,2} < 1;$$

$$-1 \leq \beta_3 < 1$$

For example, for $\beta_1 = \beta_2 = 0$ and $U_{\varphi--} = U_{\varphi+} = 0$ we obtain the result of A. Betz and N.E. Zhukovsky:

$$C_p = p_{--} - p_+ = 4(1 - u)u \tag{3.17}$$

The introduction into consideration of the "vortex solenoid" system with hypothetical properties is actually equivalent to some special hypotheses regarding β_1, β_2, u_{q-}, and u_{q+}, made it possible to obtain a different correlation for C_p†:

$$C_p = 4(1 - u)/2 - u \tag{3.18}$$

Correlation (3.17) gives underestimated results when u→0 (impermeable disk). On the contrary, correlation (3.18) overestimates the pressure drop across the low-permeable disk. In the range u> 0.7, the difference between calculations by (3.17) and (3.18) is small. This can be interpreted as a weak influence of the flow swirl in the specified range of relative velocities.

Given the values of β_1, β_2, β_3 and z, it is possible to construct a series of dependences of $C_p = C_p(u)$ according to (3.12) to (3.15) corresponding to different conditions (Fig. 3.8).

In conditions when the flow swirl can be neglected, a simple model seems to be the most natural, in which a real wind turbine is replaced by a

†Sabinin, G.Kh. Theory and Aerodynamic Calculation of Wind Turbines. *Proceedings of TsAGI,* 1931, issue 104.

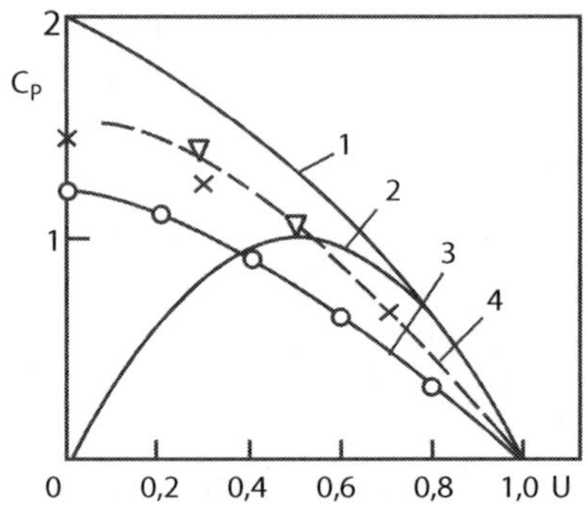

Fig. 3.8 Dependence of the relative pressure drop across the unit $C_P = 2\left(p_- - p_+\right)/\rho U_0^2$ vs the relative velocity of ($u = U/U_0$) of the medium movement through the unit: 1. the G. Kh. Sabinin model (18); 2. Betz-Zhukovsky model (17) — uniform distribution of velocity over the section Ω and constant pressure along the boundary of the stream tube; 3. calculations made by the author, Yu.B. Shpolyansky and D.N. Mileteyev (solution 19) — uniform distribution of velocities over the section Ω, a free flow; 4. calculations for conditions in a wind tunnel.

disk with sources (right) and sinks (left) evenly distributed over its surface. Numerically solving the corresponding problem for the Euler equations[‡], we can find the dependence of $C_p = C_p(u)$ and, in particular, determine the values of the factors β_1, β_2 corresponding to the adopted model. This solution (Fig. 3.8) agrees with the experimental data on the flow around an impenetrable disk (u reflow pp. 102-127 - done 5/6 0), but it noticeably differs from the results according to (3.17) to (3.18) in the entire range of u. The result obtained means that the assumption about the invariability of pressure over the tube surface of the type ($\beta_1 = \beta_2 = 0$), i.e., correlation (3.17), contradicts the assumption about the uniformity of longitudinal velocities distribution in the unit zone.

The results of the numerical experiment shown in Fig. 3.8 can be summarized by the formula:

$$C_P = C_P^0(1 - u^m), \tag{3.19}$$

where $C_P^0 \sim 1.2; m \sim 1.6$.

[‡]Lyatkher, V.M., Prudovskiy, A.M. Hydraulic Modeling. M., Energoatomizdat, 1984, p. 392.
Militeyev, A.N. Numerical Modeling of Pulsation Currents and Heat and Mass Transfer in Shallow Unstratified Water Bodies. *Proceedings of the Gidroproekt*, 1983, issue 91, p. 41-52.
Militeyev, D.N., Shpolyansky, Yu.B. Study of Separation Zones by Numerical Methods. *Proceedings of the Gidroproekt*, 1983, issue 91, pp. 52-59.

The power taken off by the unit from the flow, with a small effect of swirling, is determined by the expression of:

$$P_0 = \Delta p U \Omega = C_P (\rho U_0^3 / 2) \Omega, \tag{3.20}$$

where $C_P = C_D u$; $\Delta p = p_{--} - p_+$.

Usually this power is identified with the power received by the unit, which is represented in the form of:

$$P = C_P (\rho U_0^3 / 2) \Omega, \tag{3.21}$$

where C_P is the utilization factor of the flow power (in domestic works on wind power, the designation ξ was used).

If we take

$$P = P_0,$$

then

$$C_P = u C_D(u) \tag{3.22}$$

In real units, there is no reason to assume (3.22), since part of the flow energy is converted into large-scale turbulence and it is not used by the unit. Equality (3.22) can be taken as the definition of an "ideal" unit in which all the energy lost by the flow (3.21) is transferred to the unit (3.22).

The correlations of $C_P = uC_D(u)$ still do not allow finding specific values of C_P and C_D in the function of the design parameters of the units;however, we can use them to estimate the maximum values $(C_P)_{max}$ and the corresponding speeds of $(u)_M$ (Table 3.3).

The maximum values of the power factor, according to the described model are noticeably lower than that according to the models of

Table 3.3 Maximum values of power factors.

	Model		
Factors	**Betz-Zhukovsky**	**Sabinin**	**(22)**
$(C_P)_{max}$	0.59	0.69	0.41
$(u)_M$	0.67	0.59	0.53

Betz-Zhukovsky and especially, of G.Kh. Sabinin. Extensive experimental data give the values of $(C_p)_{max}$ close to the calculations according to (3.22): in experiments performed in the free flow, no values were recorded larger than $C_p = 0.4$ (usually $(C_p)_{max} = 0.31$ to 0.38*), in wind tunnels, the maximum C_p values reached 0.47, which can partly be explained by the effect of the tunnel walls. Apparently, the results of Betz-Zhukovsky and Sabinin do not give real asymptotic estimates for the parameters of the units.

To completely solve the problem and determine the physical value (u) within the framework of the zero-dimensional model, elementary correlations of the wing theory are used, which determine the components of the load on the blade, acting perpendicularly (q_n) and parallel (q_1) to the chord of the blade airfoil (Fig. 3.5):

$$q_n = ½[C_y(\alpha) \cos \alpha + C_x(\alpha) \sin \alpha](r^2\omega^2/U^2 + 1); \quad (3.23)$$

$$q_1 = ½[C_y(\alpha) \sin \alpha — C_x(\alpha) \cos \alpha](r^2\omega^2/U^2 + 1); \quad (3.24)$$

$$q_{n,1} = 2q_{n,1}/pU^2; \quad (3.25)$$

$$\alpha = \text{arctg}\,(U/\omega r) - \varphi_0,$$

where α is the angle of attack on the blade element located at a distance r from the axis of rotation; φ_0 is the angle of rotation of the airfoil chord in relation to the blade speed.

With known loads q_n and q_1, the average (per revolution) pressure force acting on the unit is calculated, and the pressure factor is determined:

$$C_D = i[F_n \cos \varphi_0 - F_1 \sin \varphi_0]\, u^2; \quad (3.26)$$

$$F_n = 1/\Omega \int bq_n dr; \quad (3.26a)$$

$$F_1 = 1/\Omega \int bq_1 dr$$

where the integrals on the right side are calculated over the entire blade length; i is the number of blades.

Equation (3.26) along with equation (3.19) (earlier it was taken with equation (3.17) or (3.18)) close the system and allow finding the velocities (u) in the unit zone.

The power developed by the unit is determined by the moment of the pulling forces acting on the blades and the unit rotation frequency shall be:

$$C_P = iz_u[M_n \sin \varphi_0 + M_l \cos \varphi_0]u^3; \tag{3.27}$$

$$M_n = 1/R\Omega \int bq_n rdr; \tag{3.27a}$$

$$M_l = 1/R\Omega \int bq_l rdr;$$

$$z_u = \omega R/U$$

The values of F_n and M_n characterize the dimensionless value of the shearing force and moment at the base of the blade, acting from the blade plane; F_l and M_l are the same in the blade plane (Table 3.4 for $r_0/R = 0.2$). To obtain the dimensional values of the forces (moments), the values of the forces F (or the moments M) must be multiplied by $\rho(u^2U_0^2\Omega/2)$ or $(\rho u^2U_0^2\Omega R/2)$.

The stated approach differs from the traditional one only in the way of determining the dependence of $C_D = C_D(u)$. The accuracy of the calculation method was verified using experiments carried out at Vetroen Research and Production Association under the guidance of V.E. Fedotov in a wind tunnel with a 3 m diameter test section. We tested a 1.2 m diameter AVEU-6 wind turbine model with two blades having the Espero airfoil ($r_0 = 0.2R$). The filling of the test section of the tunnel with the model did not exceed 16%.

The test conditions differed from the natural ones by the presence of impermeable tunnel walls, which was taken into account in the calculation. It was on the tunnel walls ($r = D/2$) that the impermeability condition was set of

$$U_{r\,(r=D/2)} = 0 \tag{3.28}$$

The calculations were carried out for units with relative geometric dimensions of the AVEU-6 model serially produced by the Vetroen Research and Production Association, with a variable chord length of the blade airfoil, that is, with a variable blade width from $b/R = 0.083$ at $r = r_0$ to $b/R = 0.05$ at $r = R$. For the Espero airfoil, the relative airfoil thickness varies from $c/b = 0.22$ at the base of the blade to $c/b = 0.13$ at the tip. The coordinates of the NACA airfoil with the same relative thickness of the airfoil along the radius and the values of the aerodynamic factors are taken from known publications[§].

[§]Vachon W.A., Schiff D. Test performance of large wind turbines. *Int/Wind Energy Symp., 5th Annual. Energy – Sources Technol. Conf. and Exhib., N. Orleans. March 1982,* NY., 1982, p. 137-157.

Hutter U. Moderne Windturbinen, 1979. Maekawa H. Application of the vortex theory to high speed horizontal – axis wind turbines. *Bul. of JSME,* v. 27, N 229, 1984, p. 1460-1466.

Table 3.4

Maximum values of forces F and moments M acting on the blades	Speediness $z_u = \omega R/u$											
	2		4		6		8		10		12	
$(100F_n)$	5	5	13	13	29	29	50	50	78	75	114	102
	6	6	15	15	33	31	55	50	82	72	113	95
	5	5	11	11	24	23	42	37	64	49	88	60
$(100\lvert M_n\rvert)$	3	3	10	10	21	21	37	36	57	53	83	72
	4	4	11	11	24	22	40	36	59	51	81	67
	3	3	8	8	17	17	31	26	47	34	63	41
$(100\lvert F_l\rvert)$	9	9	13	8	25	25	52	52	89	80	140	99
	7	6	25	25	54	47	88	67	125	82	161	93
	2	2	7	6	20	20	42	38	73	47	107	53
$(1000\lvert M_l\rvert)$	7	7	13	8	22	22	44	44	70	56	108	70
	6	5	20	19	40	34	63	46	88	55	113	61
	2	2	7	6	18	18	36	30	61	34	83	35

Note: 1. The data in the cells of Table 2 refers to the Espero airfoils, NACA 4415, and NACA0012.

2. The first column of numbers in each cell refers to the case when the angle of rotation of the blades (φ_0) is changed in the range from "–4" to "+10"^{0}; the second column — when (φ_0) is changed in the range from 0 to 6^0.

In the root part ($r<r_0$), the blade is adopted as a cylindrical one, 0.011R in diameter. It is shown that, although the blade chord length is being decreased with distance from the axis of rotation, the resultant of the loads on the blade is located at a distance of 2/3 to 3/4R from the axis of rotation. The reducing the range of angle variation, especially for the NACA 4415 airfoil (Table 3.4), and turn of the blades (φ_0), noticeably reduces the load.

As a result of the calculation, the flow patterns were determined for different (u), and the dependence of C_D (u) was found which lay noticeably higher than the corresponding curve for an unbounded medium. The difference between the calculated and experimental data obtained under the appropriate conditions is small (Fig. 3.9). The same result is obtained by comparing the calculations for the NACA 4415 airfoil and Maekawa, H measurements.

Based on empirical data on the lifting force and airfoil drag factors for the Espero airfoil, the values of the integrals determining the function of $C_D(u)$ were calculated, the function $u = u(\omega R/U_0)$ was found, and the characteristics of the wind turbine were determined from (3.26), (3.27).

The results of calculations for the experimental conditions in a wind tunnel differ from the results obtained for a free wind turbine—the aerodynamic loads and power factor, even with a small relative shading of the tunnel (16%), are almost 20% greater than in an open flow. This is confirmed by experimental data where at 12% shading of the pipe, the tests gave the C_P values 15 to 20% greater than those in nature.

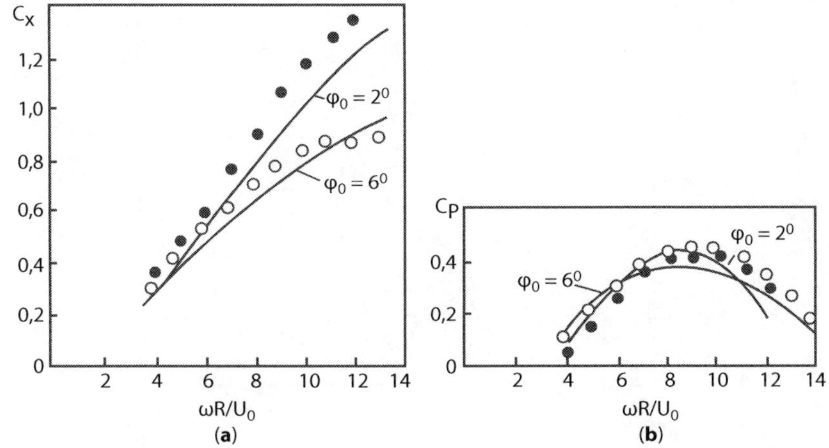

Fig. 3.9 Comparison of the results of calculations (lines) and experiments (points) for the conditions of a wind tunnel: a is the pressure factor on the unit; b is the power factor.

Thus, for practical calculations of wind turbines and hydraulic units with a traditional layout, the developed model with a uniform distribution of longitudinal velocities over the swept surface is satisfactory. Calculations show that the pressure drop across the unit (3.26) and, therefore, the relative speed and power factor C_P according to (3.20) do not depend on the pulling force ($F_n\cos\varphi_0 > F_l\sin\varphi_0$), that is, they almost do not depend on aerodynamic properties of the blade airfoil. On the contrary, the values of $M_l\cos\varphi_0$ and $M_n\sin\varphi_0$ are comparable, and at high speed ($z_u > 10$) and the optimal angle of rotation of the blades, the contribution of the pulling force is most significant and the power factor of the unit, C_P, should depend on the aerodynamic qualities of the blade. For example, for a two-blade unit with the Espero blade airfoils, NACA 0012, and NACA 4415 at $r_0 = 0.40R$ or $r_0 = 0.2R$, are practically identical values $(C_P)_{max} = 0.41$ and are obtained at u = 0.54 to 0.58, but significantly different values $(C_P)_{max}$ of 0.27 to 0.35 are obtained at $r_0 = 0.40R$ and $(C_P)_{max} = 0.28$ to 0.36 at $r_0 = 0.2R$.

The Espero airfoil gives a greater value of the maximum value of C_P, but only in a relatively narrow range of variation of the relative rotation frequency of $z = \omega R/U_0$ (Fig. 3.10). The NACA airfoils give lower maximum C_P values, but the change in C_P in the function of z, is more gradual. Therefore, with a "soft" control which does not provide a strictly optimal speed, the NACA 4415 airfoil can give more output than the Espero airfoil. This feature is also preserved in the characteristics of the unit with the NACA 0012 airfoil; however, the decrease in the maximum C_N here is very large. The unit efficiency will, therefore, be lower.

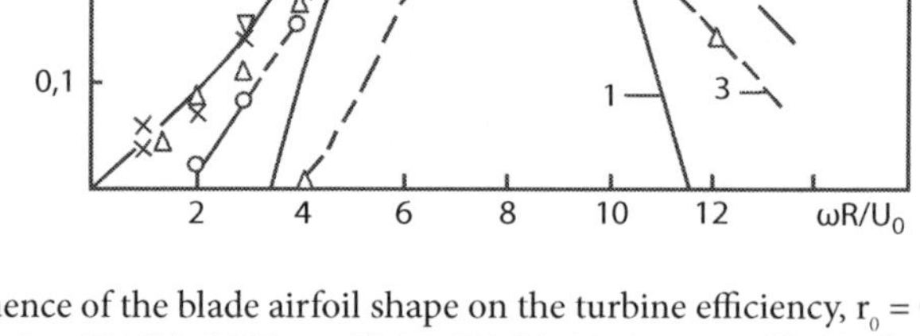

Fig. 3.10 Influence of the blade airfoil shape on the turbine efficiency, $r_0 = 0.40R$: 1– Espero, $\varphi_0 = 2^0$; 2 — NACA 4415, $\varphi_0=2^0$; 3 – NACA 0012, $\varphi_0= -2^0$; 4 is the average characteristic of the "ideal" unit.

The developed technique can be used to compare the assemblies of units that differ little from the traditional one. In particular, the influence of the root part length and the presence of a 0.4R radius hemispherical flow were revealed. It was found that the maximum values of C_p and the left branches of the characteristics change insignificantly (Fig. 3.11). The right branch of the characteristic in the presence of a fairing is noticeably lower. With a sufficiently tight regulation, this does not affect energy generation.

The presence of the fairing allows the layout to be changed qualitatively by placing generator poles along the fairing diameter and, while maintaining the maximum blade length of 64 m, ensure the radius of the swept circle of 107 m and the maximum power of the wind turbine $N_{max} = 7.2(107/64)^2 \sim 20$ MW at the same design of the blades which is designed for 7.2 MW machines.

The influence of the length of the root part turned out to be insignificant (at $0.2 \leq r_0/R \leq 0.40$)) – Fig. 3.11. This can be explained by the fact that the root part of the blades generally has little effect on the energy generation and, in addition, at $r_0/R = 0.40$, at a greater distance from the axis of rotation, there was a relatively large blades' chord length than at $r_0/R = 0.2$. This result indicates the advisability of detailed optimization of the blade shape.

For unconventional schemes, it may be necessary to complicate the problem formulation in order to take into account the uneven distribution of speeds over a disk or a ring simulating the unit. In the developed model, the unit is replaced by an axisymmetric stationary figure, the local pressure drop of which is determined by correlation (3.26), in which, however,

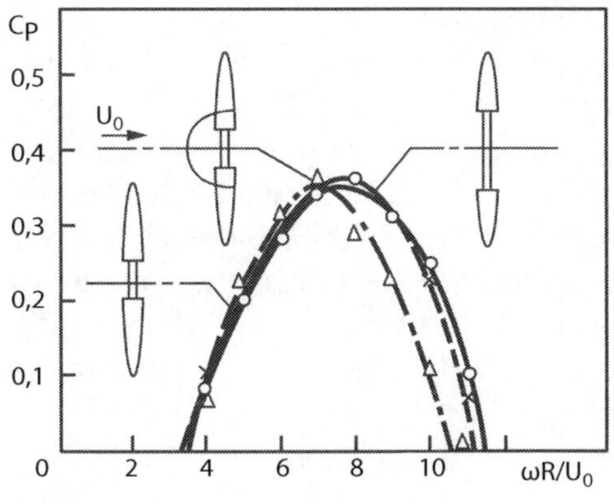

Fig. 3.11 Influence of the design of two-blade units. The Espero airfoil, $\varphi_0 = 2^0$.

the longitudinal component of the velocity U_x is assumed to be unknown, varying along the coordinate r:

$$\Delta p = \lambda(r\,\omega / U_x, \varphi_0)(pU_x^2/2) \tag{3.29}$$

Here the parameter λ is close in meaning to the parameters p and q_n:

$$\lambda \approx (iq_n b/2\pi r)\cos\varphi_0 \tag{3.30}$$

where q_n is defined by (3.23).

In this model, the problem also splits into two: solving the hydrodynamic problem for the Euler equations with condition (3.29) on the surface replacing the unit, determining $u_x(r)=U_x/U_0$ and calculating the torque and power by a formula similar to (27), but with the relative velocity u_x introduced under the integrals:

$$C_P = iz[m_n \sin\varphi_0 + m_1 \cos\varphi_0] \tag{3.31}$$

where $m_n = 1/R\Omega \int bq_n u_x^2 rdr$; $m_1 = 1/R\Omega \int bq_1 u_x^2 rdr$

If the swirl angle φ_0 varies in the function of r, then the corresponding functions, as in (3.27), should be introduced under the integrals.

If, to simplify the analysis, we introduce the value of λ averaged with the weight function of r over the surface swept by the blades, and the equality is thus realized:

$$\lambda \approx iP\cos\varphi_0 \approx iP \tag{3.32}$$

where P is defined by (3.26a).

In the zone of maximum C_p, usually P ≈ 0.5 at the largest z_u P ≈ 1, at small z_u and $C_N \neq 0$ P ≈ 0.2 to 0.3.

Fig. 3.12 compares the distributions of longitudinal and transverse velocities at $0.5 \leq \lambda \leq 2$, which, for a two-blade wind turbine (i = 2) covers the range of possible values. The parameter λ, naturally, significantly affects the relative value of the velocity u_x. In this case, however, the unevenness of the velocity distribution is small. At large λ, the radial velocity components are relatively large, which justifies the use of end devices. In the presence of a fairing, the speeds near it decrease slightly. However, at the blades' tips, the speeds increase which ensures the preservation and even some increase in the utilization factor of the wind power. Fig. 3.13 shows the velocity vectors and relative pressures contours (in fractions

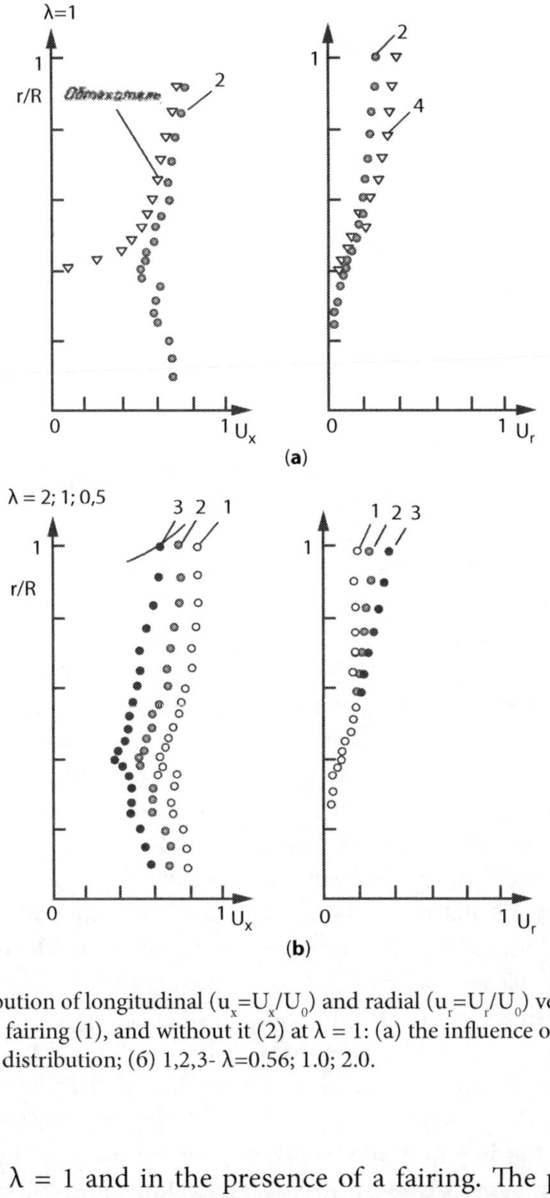

Fig. 3.12 Distribution of longitudinal (u_x=U_x/U_0) and radial (u_r=U_r/U_0) velocities in the unit plane with a fairing (1), and without it (2) at λ = 1: (a) the influence of the parameter λ on the velocity distribution; (б) 1,2,3- λ=0.56; 1.0; 2.0.

of $pU_0^2/2$) at λ = 1 and in the presence of a fairing. The picture below indicates, in particular, the advisability of turning the blades towards the flow by about 1^0 to 2^0. The developed technique can be effectively used to assess aerodynamic loads and energy characteristics of various types of horizontal units.

The foregoing shows that the classical estimates of the maximum power of the units are based on contradictory hypotheses about the constancy of

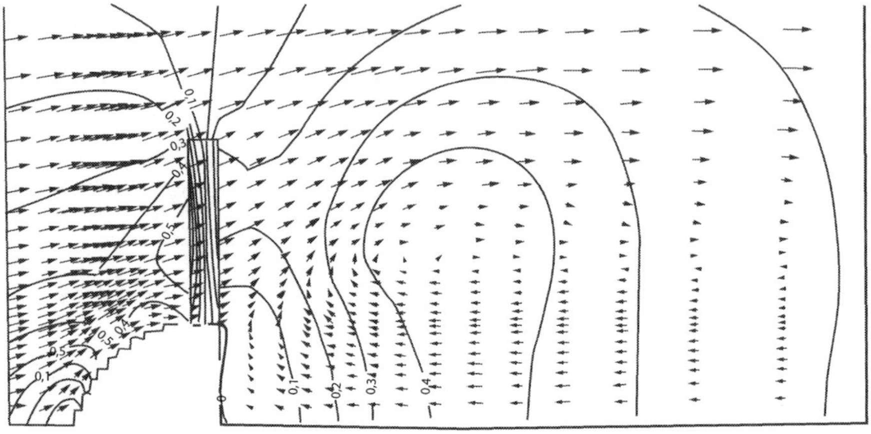

Fig. 3.13 The vectors of flow velocities in the plane passing through the unit rotation axis λ = 1………

pressure on the surface of the stream tube enclosing the unit, and about the uniform distribution of velocities in the unit plane. If we keep only one assumption about the uniform distribution of velocities (the first model), then it turns out to be possible to calculate the hydro- or aerodynamic loads and power factors that are consistent with the experimental data. According to this model, the maximum power factor of units with a horizontal axis in a free flow without concentrators **cannot exceed 0.41** which is 1.44 times less than the popular Betz-Zhukovsky estimate.

The hypothesis of a uniform distribution of velocities in the unit plane does not allow the study of unconventional layouts of horizontal units. To eliminate these limitations, a complicated second model has been developed and implemented on a digital computer, according to which the local pressure drop across the unit is taken proportional to the square of the local speed with the proportionality factor depending mainly on the speed and the relative area of the unit blades.

The new models were used to compare the distributions of velocities and pressures, the energy characteristics of units with, and without, a central fairing. It has been established that the introduction of a fairing of up to 0.8R in diameter, as well as the corresponding shortening of the blades with an airfoil while maintaining the outer diameter (2R) of a circle swept by them, practically does not cause deterioration of the energy characteristics of the unit.

The influence of the aerodynamic characteristics of the blade airfoil has been studied. It has been established that the best performance is achieved by two-blade units with the Espero airfoil or NACA 4415 airfoil. The maximum value of the power factor for these airfoils in the open flow is 0.35 to 0.36.

When testing hydro and wind turbines in wind tunnels or chutes, the loads and power factor of the units may be overestimated compared to the conditions of a free uniform flow. In particular, if the wind tunnel section is blocked by the unit by 16%, the maximum power factor may turn out to be more than 20%.

The energy characteristics of the units change slightly with an increase in the blades' root part length or with the introduction of a central fairing (with a radius of up to 0.4R). This makes it possible to consider the arrangement of units with a central powerhouse and relatively short blades which, in turn, allows the unit power of horizontal wind turbines to be increased (in the future) to up to 20 MW while maintaining the maximum length of the blade console achieved in modern wind turbines.

Calculations show, however, that for machines with a rotor diameter of about 100 *m* and a power of about 3 MW, a single-blade design with a very high speed could be quite competitive. An additional limitation of the blades' rotation speed can be the acoustic radiation of the blades in the natural wind flow and especially in the places where the blades cross the turbulent wake behind the support tower. An example is known when, due to the environmental limitations of acoustic infrasonic radiation, the rotation speed of a large unit and its power were reduced. The crossing of the turbulent wake behind the tower by the blades of the wind turbine also causes additional dynamic loads on the blade. Therefore, in modern large units, designers prefer to place the rotor in front of the support tower (Fig. 3.14), although for small and medium-sized units, the location behind the tower may be preferable, since it frees them from special flow orientation devices.

High-power single-blade machines did not become commonplace. On the contrary, such 0.2, 0.5, 1, and 1.5 kW wind turbines with a hinged blade (Fig. 3.15) are, or were produced by the VINDEK R&D Center (Moscow).

Collinear hydraulic units which have no restrictions on acoustic radiation, are located behind the supporting structure, even at high unit capacities (Fig. 3.16).

For hydraulic units, it is desirable to have a maximum rotor speed. Therefore, rotors with two blades are most often used here. The optimal design of the wind turbine with two blades behind the tower, whi.ch was serially produced in the 1990s by Advanced Wind Turbines (USA) which provided the lowest energy cost at that time (Fig. 3.17), can be used in water turbines. An additional limitation here can be only cavitation on the blades.

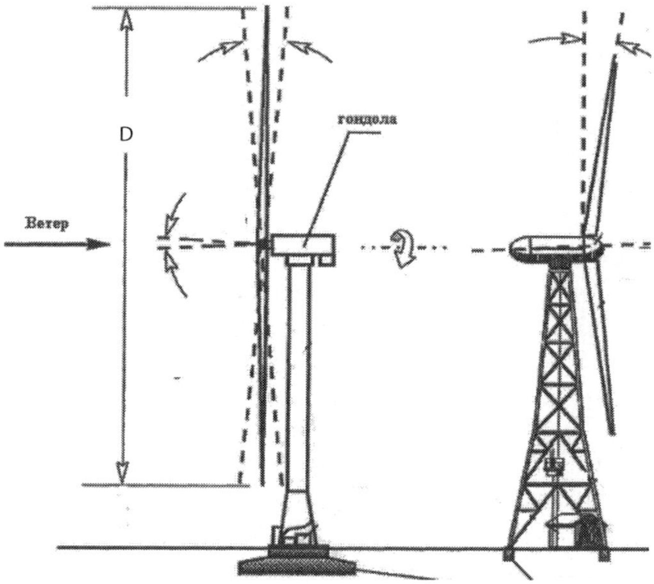

Fig. 3.14 The position of the rotor in front of the tower (upwind) – to the left and behind it (downwind).

Fig. 3.15 Single-blade self-aligning unit.

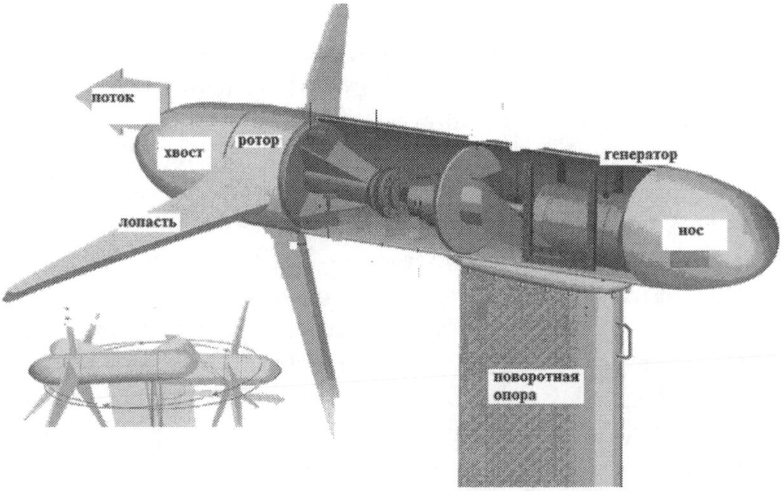

Fig. 3.16 A Vordant hydraulic unit turning downstream by the flow. Power 750 kW (35 rpm) at a current speed of 2.7 m/s. The turbine diameter is up to 10.8 m. The rotation speed changes in proportion to the flow rate.

Fig. 3.17 An AWT turbine behind the support tower, the diameter is 26.2 m, power is 275 kW at the wind of 15.3 m/s, initial wind of 5.4 m/s, maximum wind for operation of 24.6 m/s, hurricane of 59 m/s. A 47.3 m tower.

The optimal sizes of units are determined by the influence of two opposite tendencies. Specific material consumption and capital investments per unit of power (or output) of wind and hydroelectric units tend to increase with an increase in the size and power of the unit. This is determined by the fact that the weight of the rotor power material and of the entire unit increases in proportion to the third power of the linear size, and the power and output in a flow which is uniform along the height, increases in proportion to the square of their linear sizes. Thus, theoretically, the weight of the load-bearing material per unit of power (output) should increase with the size of the unit. In practice, however, it is observed only in small (P< 1 kW), and very large (P> 1,000 kW) wind turbines.

On medium-sized wind turbines, structural (non-power) elements are responsible for the large proportion of the material weight. Therefore, the specific material consumption in the function of the size of the unit does not change monotonically. For example, for 60, 330, and 750 W Howden units (Fig. 3.18) with a 15, 31, and 45 m three-blade rotor, the specific weight of the unit is 112, 135, and 107 kg/W calculated per unit of power capacity, or 0.052, 0.089, and 0.073 *kg*/*kWh* per unit during energy generation for the year at the average wind speed of 6 *m*/*s* at the machine axis.

An increase in the unit capacity of wind turbines shall significantly reduce the operating costs per unit. With the same relative distance between units in a wind farm (for example, 2 *D*), the area of land per unit of power or generation of a plant with large units shall be reduced by using

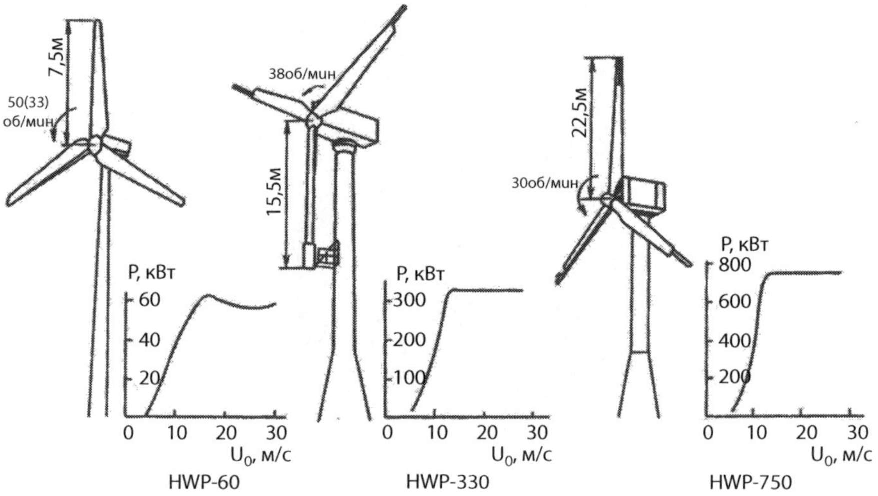

Fig. 3.18 A series of Howden optimized wind turbines (England).

wind at high speeds and high altitudes. All these facts justify the growth of the unit capacities of wind turbines which are now practically limited only by the strength of modern aviation materials, technological capabilities of enterprises, as well as environmental requirements for low noise limiting the speed of the blades. A modern wind turbine that meets the specified requirements of Mod-5B-107 (Fig. 3.19), has a 97.5 m diameter two-bladed rotor with a maximum blade tip speed of 89 *m/s* and a power capacity of 3,200 kW. The company has designed the Mod-5B-105 unit with a 126 *m* diameter rotor and a power capacity of 7,300 kW.

A further increase in console outreach is not predicted yet. An increase in the unit power of a wind turbine can be achieved while maintaining the maximum blade length of 45 ÷ 60 *m*. This is possible due to the introduction of a central turbine hall covered by a profiled fairing (Fig. 3.20).

An even greater technical and economic effect can be obtained by using a central toroidal or hollow cylindrical support structure offset relative to the

Fig. 3.19 Mod 5B.

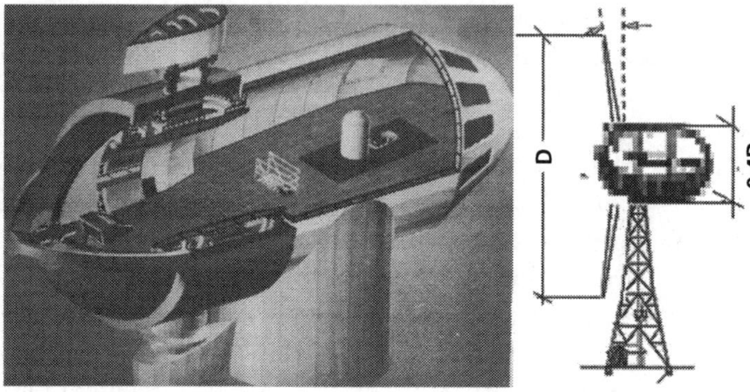

Fig. 3.20 Central turbine hall (right) of a high-power wind turbine with shortened blades.

axis of the vertical support. Two options are possible here. Either the location on the torus or on the cylinder of one row of blades fixed on the rim which is the rotor of the electric generator, or (second option) location on the support of two rows of blades rotating in the opposite direction (Fig. 3.21)[¶].

The use of counter-rotation of the blades facilitates the design of the generator; however, it increases the total weight of the rotating parts. Due to the fact that the blades' rotation plane is displaced in the direction of the wind from the axis of the vertical support, the unit itself is oriented in the direction of the wind by choosing the energetically most efficient position. This unit compares favorably with traditional machines also in that the central air jet causes an ejection effect thus contributing to an increase in power.

Calculations show that the presence of a turbine hall or a 0.4 *D* diameter hole in the center of the unit does not impair the energy characteristics in the zone of maximum power factor C_p and low speeds corresponding to the most important strong winds (Fig. 3.22).

The presence of a central turbine "hall" of one shape, or another (cylinder or torus), but of a relatively large diameter, makes it possible to qualitatively change the layout by placing generator poles along the outer diameter of the "hall" and keeping the maximum blade length equal, say, to 63 *m*. In this case, the diameter of the swept circle will be equal to $D = 0.4D + 2\text{x}63\ m$, i.e. $D = 210\ m$, and the possible installed power

¶Lyatkher, V.M. and Shpolyansky, Yu.B. "Wind power plant", USSR Inventor's Certificate No. 1361371 dated August 22, 1987, Bulletin No. 47, December 23, 1987. Lyatkher, V.M., Aleksandrov, A.S., Shpolyansky, Yu.B., Yuferov, V.I. "Wind power plant", USSR Inventor's Certificate No. 1357611 dated August 8, 1987, Bulletin No. 45, December 7, 1987.

Fig. 3.21 Wind turbines without multipliers, with a central ejection jet and a linear (arc) or multi-pole induction generator. On the right is the simplest scheme with one row of blades fixed on the squirrel-cage rotor of the generator, on the left is a counter-rotor version with two rows of blades moving in opposite directions.

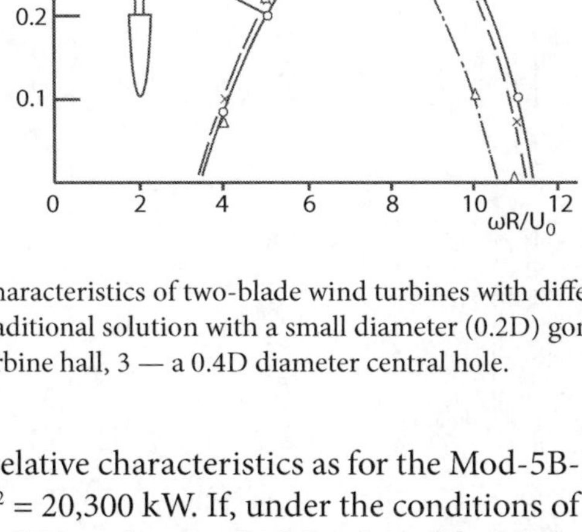

Fig. 3.22 Energy characteristics of two-blade wind turbines with different layout of the central part. 1 — traditional solution with a small diameter (0.2D) gondola, 2 — a 0.4D diameter central turbine hall, 3 — a 0.4D diameter central hole.

(with the same relative characteristics as for the Mod-5B-105 unit) shall be 7,300·(210/126)^2 = 20,300 kW. If, under the conditions of limiting acoustic radiation, we shall keep the speed of the tips of the blades the same as that of Mod-5B-105, i.e., 89 m/s, then the speed of the rotor in the linear (arc) generator in the schemes according to Fig. 34, 35 will be of 35.6 m/s. This speed of intersection of magnetic fields in an induction generator with a squirrel-cage rotor is sufficient to create a highly efficient electric machine with a weight of no more than 2.35 kg/kW. There is no need to make the second row of counter-rotating blades in the wind turbine. Thus, using the layouts shown in Fig. 34, 35 makes it possible to increase the unit power of wind turbines up to about 20 MW while maintaining the dimensions of the blades mastered by industry and tested in practice (Fig. 3.23).

In order to avoid low-frequency noise, the blade system can be placed in front of the support tower, and the orientation of the unit is performed by an additional drive on command from the wind direction sensors (like all large modern machines). To reduce the material consumption of the unit while also maintaining the principle of counter-rotor machines, the second rotor of the system can be located in the central hole of the structure according to Fig. 3.21. On the rim of this inner rotor there is a winding of inductors with an exit through a collector in the center of the system (Fig. 3.24). The short-circuited rotor of the generator is located on the inner rim of the outer wheel.

This arrangement does not increase the capacity of the unit, but may be useful in improving the performance of the generator. These ideas are partially implemented in the design of the hydraulic unit shown in Fig. 3.25. In counter-rotor circuits, it may be promising to use a new generator with two squirrel-cage rotors moving in opposite directions and one stationary

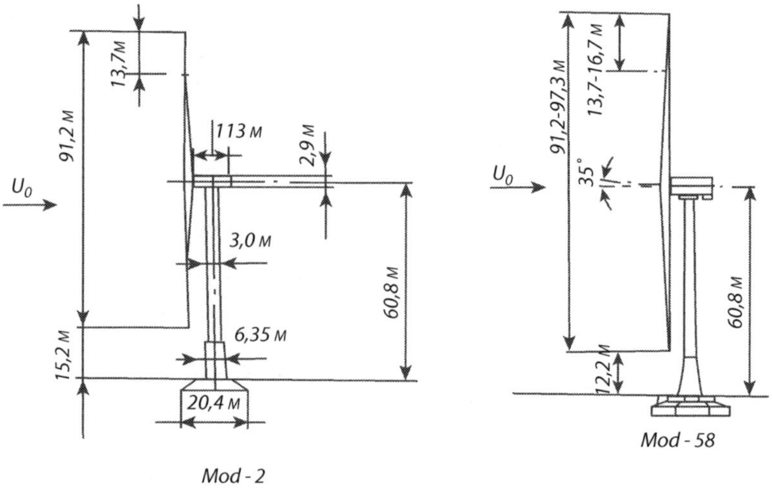

Fig. 3.23 Largest American wind turbines tested.

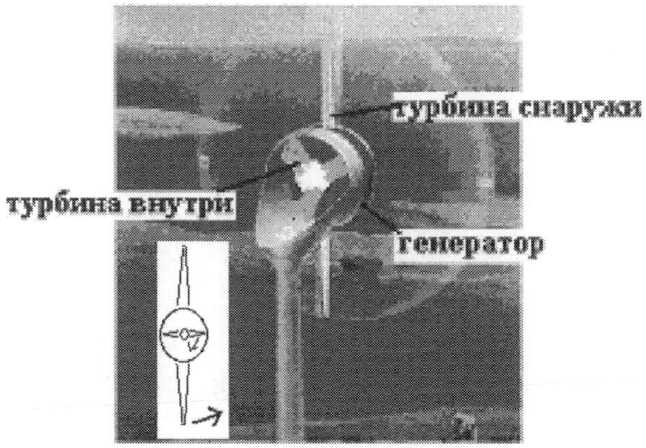

Fig. 3.24 Control wind turbine without a multiplier with a linear (arc) generator between the turbines.

Fig. 3.25 Open-HydroGroup Hydro Turbine. The diameter of 12m, central hole of D = 4.8 to 5.4 m, power of 300 kW.

inductor ring between them, from which electric power is output without collector devices. This solution is especially important for subsea hydraulic units.

It would be convenient to start the practical implementation of the proposed solutions with the modernization of Lincoln Electric large serial machines. These machines have three blades sweeping a circle of about 100 m in diameter. The blades are 54 m long and are fixed on the power unit turning in the wind by a special drive. The power unit is located on a 75 m high tower. The maximum power of the unit is 2.5 MW. Having retained the blades and the support tower, it is possible to change the design of the power unit—to use a linear (arc) generator complete with a magnetic suspension of the rotor on a ring forming a 6 m diameter central ejection jet (Fig. 3.26). Such a scheme allows increasing the power to 7 MW and increasing the reliability and efficiency of the system due to the rejection of the heavy multiplier.

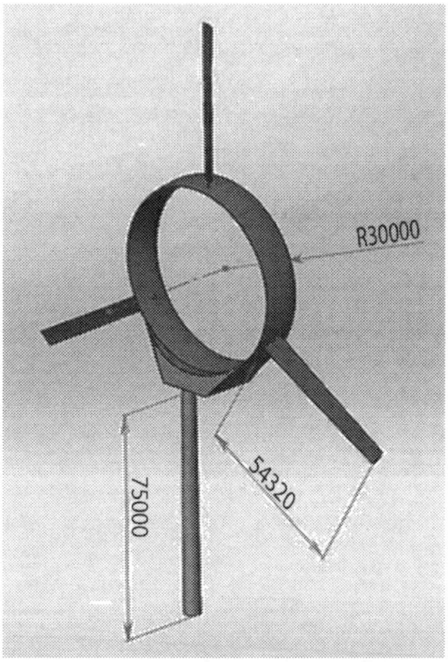

Fig. 3.26 Lincoln Electric wind turbine modernization scheme. The dimensions in mm.

The blade tips of such a unit describe a circle, the diameter of which is 1.67 times greater than the diameter of a standard machine; however, the speed of the blade tips can be left the same, which leaves the unit's acoustic noise unchanged. At an 8.4 rpm rotor speed, the speed of crossing the magnetic fields in a linear (arc) generator will be about 29 m/s which ensures its high energy and economic efficiency. This scheme protected by the USSR copyright certificate (No. 1361371, August 22, 1987), does not reduce the energy efficiency of the rotor (Fig. 3.21), but significantly reduces the weight and cost of the power unit while increasing the power and output of the unit. Wind turbine Enercon E-126 with blades 60 m long reconstructed according to the scheme in Fig. 3.26 could have a power capacity of over 20 MW.

Another approach to increasing the power of large machines is proposed by a system in which the central hall is made in the form of a balloon rotating with blades fixed on it and transmitting torque to the ground through a tethered cable (USSR Author's certificate No. 1250697). To compensate for the torque, this solution can be modified by using two balloons (or submersibles) rotating in the opposite directions against the generator.

However, all units of the type under consideration with an axis of rotation parallel to the flow retain a significant drawback of traditional designs, which consists of the impossibility of taking into account the non-coplanarity of the flow (air or water) in height. The difference in wind direction at different heights that is imperceptible on units with a small blade span, on large units can seriously reduce their energy efficiency and cause destruction of the blades. This unfavorable feature of the units can be partially compensated for by a special design of the blade which provides for the possibility of turning different sections of the blade as a whole or of its flaps at a different angle with respect to the plane of rotation. This solution, however, further complicates the design and reduces its reliability.

These circumstances, as well as the high speeds of the blade tips cause intense acoustic radiation, significant material consumption, the complexity of regulation, and hence the low reliability of the units, especially in strong winds, limit the prospects for using collinear wind turbines on a large scale in large power systems, although such units are produced in the largest quantities at present. Hydraulic units are still produced in small series. The parameters of the machines produced are comparatively modest. This is due to the fact that the operating loads on the hydraulic units are significantly greater. With the same power and the same dimensions, the design speed of the water flow can be $(1000/1.25)^{0.33} = 9.26$ times less than the design wind speed. However, in this case, the hydrodynamic operational loads on the structural elements will be greater by the same amount, albeit with there practically being no "storm" regimes in tidal power plants. The inertial loads on the rotating elements of the turbines will also be significantly lower.

The energy characteristics of power units presented in the form of the efficiency factor C_p dependence on the speed, $z = \omega R/U_0$, are varied. Yet, the actual dependence of the power capacity normalized by the maximum (calculated) value of P_p on the wind speed on the unit axis for various high-speed collinear units with a constant rotation speed do not go beyond the scatter of data obtained when the same machine was tested in different weather conditions (Fig. 3.27):

$$
\begin{aligned}
P/P_p &= 0 \quad \text{at } U<U_H \\
P/P_p &= (U - U_H)/(U_p - U_H) \text{ at } U_H<U<U_p \\
P &= P_p \text{ for } U>U_p
\end{aligned}
\tag{3.33}
$$

The initial speed of U_n at which power begins to be produced, is usually $5 \div 6$ *m/s*. The design wind speed U_p at which the wind turbine reaches its rated (maximum) power, is $12 \div 15$ *m/s*, on average, 13 *m/s*. The maximum power taken from a unit of swept surface on modern machines also varies within relatively narrow limits:

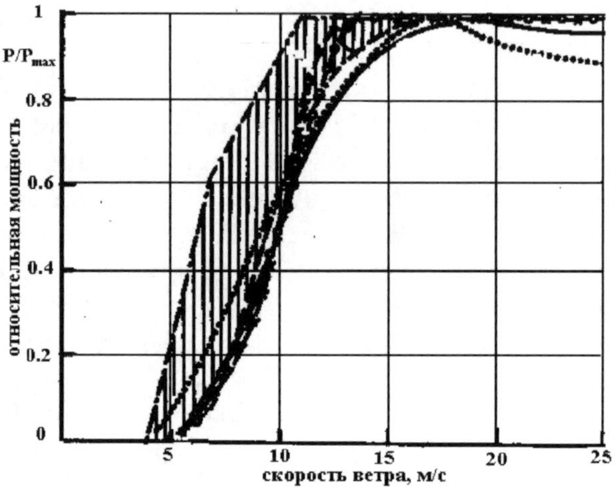

Fig. 3.27 The relative power of prototypes of serial collinear wind turbines is of P/P_{max}. The band of results variation for different wind patterns is shaded.

$$\frac{4P_P}{\pi D2} = 0.35 \div 0.47 \frac{\text{кВТ}}{M^2} \tag{3.34}$$

This value of the power of wind turbines corresponds to a rather low factor of use of wind energy: at $U = U_p C_P \leq 0.33$ actually reflects the influence of the above negative factors. At $U > U_p$, the utilization factor of wind energy is naturally even lower. Therefore the best modern wind turbines convert on average no more than 25 ÷ 30% of the energy of wind flow through the swept surface. When the flow rate is less than the calculated value, the energy efficiency of real units increases, tending to the limit for the "ideal unit" determined by the correlation (3.30) in item 1, Fig. 3.28. In the most advanced machines (Vestas), the initial speed is reduced to 4 m/s, which, however, adds only to the external effect of the machine operating in lightweight winds, practically without increasing energy generation. The introduction of variable speed and variable angle of rotation of the blades on these machines slightly distorts the characteristic. For Vestas units, this characteristic can be set by the three linear sections:

$$P = 0 \text{ at } U < U_n$$

$$P = B\,(U - U_n) \text{ at } U_n < U < U_1$$

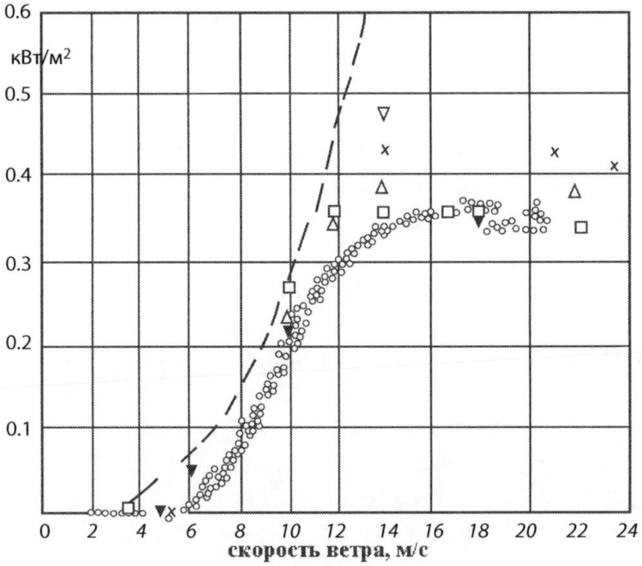

Fig. 3.28 Energy received by different wind turbines per unit area of the swept circle (the points refer to different machines). Line — energy according to condition (30) in item 1 – the maximum possible efficiency of collinear machines in unlimited flow = 41%.

$$P = B_1(U - U_1) + P_1 \text{ at } U_1 < U < U_p \tag{3.35}$$

$$P = P_p \text{ at } U_p < U < U_{max} \quad P = 0 \text{ at } U_{max} < U$$

Vestas machines have the following parameter values:

Unit type	V 90-3	V 90 -1.8&2	V 52 – 850
P_p, kW	3,000	1,800/2,000	850
$P_{1,}$ kW	500	500	500
U_n, m/s	4	3.5	4
U_1, m/s	6	6	6
U_p, m/s	14(15)	11(12)/12(13)	15(16)
U_{max}, m/s	25	25	25

(in brackets are the values of speeds corresponding to the data sheet).

There are known proposals aimed at improving the technical and economic efficiency of collinear wind turbines. One of the simplest ways to reduce the influence of the flow non-coplanarity and turbulence simultaneously with an increase in the unit power capacity is to use concentrators—a short confuser and a diffuser (Scheme 4 in Fig. 1). Experiments and calculations show that the efficiency of the confuser-diffuser circuit can be quite high. The optimal increase in power for a given diameter of a wind or water turbine can be 6 ÷ 8 times. However, the overall assessment of the material consumption of such systems, carried out earlier on the basis of experimental studies at TsAGI and abroad, did not yield positive results. This was due to the use of long diffusers in which the wall friction losses reached 30% of the total losses. At the same time, it was found that in the presence of confusers and diffusers, the flow angularity does not affect the power at angles of up to 30° (the difference is no more than 10%). The power of a conventional wind turbine in free flow at a 30° slope decreases by 1.7 times. In modern designs of concentrators, due to the control of the boundary layer in the diffuser, its length can be relatively small (the funnel angle is up to 80° with the ratio of the areas at the outlet vs. inlet of about 4).

Optimization estimates made by the author gave approximately the same results—it is desirable to have a cross-section ratio in the concentrator of about 2.8. The principle of operation of the concentrator is to create a high-pressure zone in front of the turbine, and a rarefaction zone behind it. By numerical and physical experiments, it was found that the supply of air to the vacuum zone (behind the concentrator) under certain conditions does not decrease, but increases the depth of the vacuum behind the concentrator. This result, which is similar to the ejection action of the central jet, is associated with the development of additional vortex surfaces in the zone behind the concentrator. In terms of energy, the introduction of a concentrator means that the attraction of air masses do not, due to the intensified turbulent exchange of energy, pass directly through the impeller of the wind turbine. The maximum power of the unit with a concentrator according to the scheme in Fig. 3.28 is determined by the same formula (1.4), but with a large factor C_P determined by the formula of

$$C_P = 0.385 \frac{(CD + 0.64)^{3/2}}{\sqrt{0.066 + \xi}} \frac{\Omega}{\Omega_c} \tag{3.36}$$

where C_D is the relative pressure drop across the concentrator. For a single axisymmetric concentrator or a pair of such, $C_D \approx 0.8$. For a system of concentrators forming the conditions of a plane flow (Fig. 3.29) $C_D \approx 1.2$ (see Chapter 4); ξ is the factor of resistance of the draft tube of the wind turbine; Ω, Ω_c is the calculated section and compressed section of the energy path. Taking $\xi = 0.05$; $C_D = 1.2$; $\frac{\Omega}{\Omega_c} = 2{,}8$; ,we get $C_P = 0.75$.

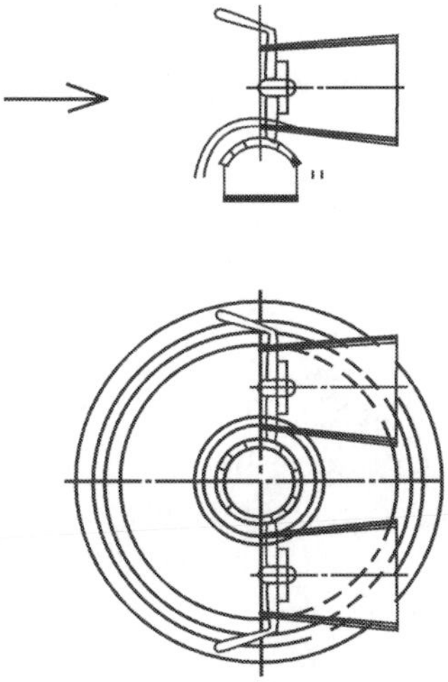

Fig. 3.29 Diagram of concentrators including a confuser in front of the impeller and a diffuser behind it.

The scheme of the French engineer Andreau and the Enfield company (England) has fundamental differences in which the wind flow on a wind turbine built into a vertical shaft is created due to a vacuum at the tips of the hollow wings of the wind turbine located on a tower. This scheme was

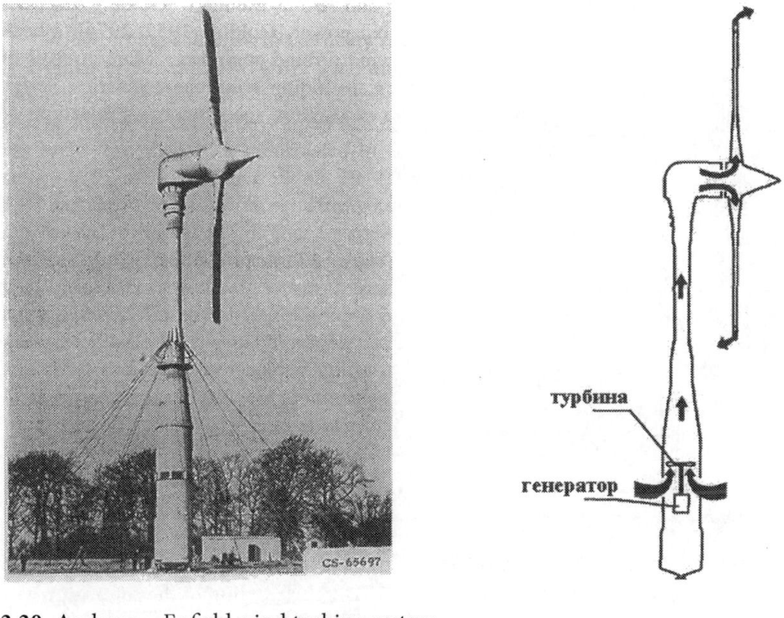

Fig. 3.30 Andreau – Enfield wind turbine system.

implemented in 1955 in kind with a 100 kW generator (the blade span of 24.4 m, U_p = 13.5 m/s) – Fig. 3.30. It showed a very low efficiency which was confirmed by additional theoretical analysis [57]. Similar ideas for the "indirect" use of wind turbine wings continue to appear, the "aerodynamic" power transmission scheme with the placement of high-speed generators and small-span rotor blades at the tips of the main blades being one of these.

Cylinders can be placed at the tips of the blades as rotors, and the whole system works due to the Magnus effect (Fig. 3.31).

A significant drawback of this scheme deserves substantial careful analysis, which is the need to transfer significant power through the collector, large centrifugal loads on the cylinders, and electromechanical systems associated therewith. This scheme, however, has some fundamental aerodynamic advantages.

The idea of concentrators has recently gained particular relevance, thanks to the systematic work of the Lunar Power company to create hydro turbines with short diffusers and confusers (Fig. 3.32).

Tests by the designers of the hydro turbine in an oblique flow showed a systematic increase (!) in power at some angles (Fig. 3.33) and confirmed the overall efficiency of the concentrators.

Collinear units with converging diffusers are also offered in large-scale tidal projects in New Zealand (Fig. 3.34).

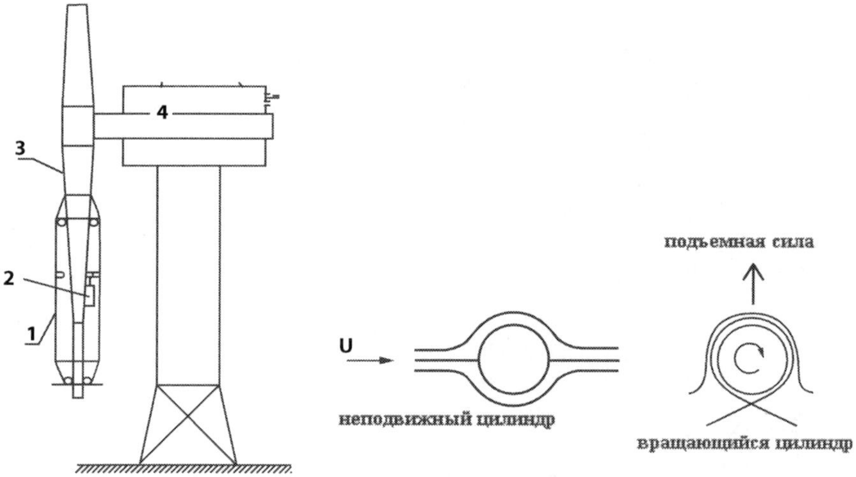

Fig. 3.31 A wind turbine with cylinders (1) rotated by electric motors (2) fixed at the tips of the rotor blades (3) with a traditional nacelle (4). The rotation of the rotor is provided by the lifting force arising from the flow around the rotating cylinder (the Magnus effect).

Fig. 3.32 Hydraulic turbine with a Lunar Power concentrator system. The power capacity of 1.5 MW at a flow rate of 3 m/s, the inlet diameter of 21 m, turbine diameter of 16 m, length of 27 m. The required depth is 35 m. The turbine weight is of 1,200 t. 1. — fairing, 2. — turbine blades, 3. — concentrator, 4. — removable part of the block, 5. — pump, 6. — hydraulic motor with a generator.

Systematic studies of wind and hydroelectric units with diffusers shortened due to the feeding a jet into the boundary layer, and without confusers, are carried out by Vortec Energy World Power. This company has already built and tested a number of wind turbines with a power capacity of up to 1 MW, and it names a 5 MW universal land and sea-based unit with a rotor only 66 m in diameter, as a promising machine (Fig. 3.35).

According to its developers (Emily Rudkin, Derek Phillips), the cost of energy from these machines will be 20% lower than that of the best

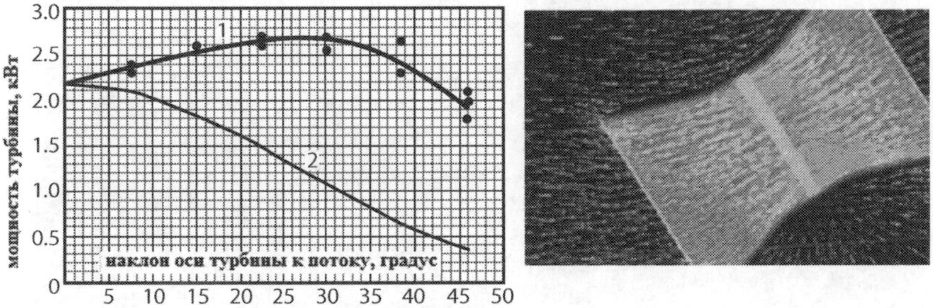

Fig. 3.33 Results of turbine tests in oblique flow in the presence of a concentrator (1), and in its absence (2).

Fig. 3.34 Anthony Bellve turbines (Crest Energy, Ltd) in the Kaipara Energy Project, NZ D 22m, the rotation speed of 5 rpm, unit power of 1 MW, total power of 200 MW.

plants with a traditional layout. Similar low- power machines have been successfully tested in France (Fig. 3.36) and in Germany.

The idea of using the aerodynamic interaction of several units to increase the efficiency of each may also be fruitful. Thus, by placing two wind turbines side by side, one can expect a deeper decrease in pressure behind them and, consequently, greater efficiency of the system. According to the idea of A.G. Ufimtsev and V.P. Vetchinkin, which has been developed in many modern patents and inventor's certificate, several high-speed wind turbines can be located on one support, sometimes even with partial overlapping of the blades.

This arrangement causes operational difficulties — the repair of one of the units complicates the operation of the others. This circumstance becomes

Fig. 3.35 Vortec wind turbines system with diffusers, self-orientated in the wind. The rotor diameter is of 66 m, power of 5 MW (project).

Fig. 3.36 Wind turbine with a diffuser concentrator. The power capacity is 30 kW. France.

less significant with a large number of units. The rotor blades can be arranged on two or more coaxial shafts rotating in one direction, or in the opposite direction. In the latter case, there is an additional gain in fewerel ectrical equipment being needed which, however, turns out to be more complex and less reliable (additional collectors appear). The prospect of preliminary concentration of the wind flow directed to the wind wheel is very tempting. Particular attention is still drawn to the idea of a tower-type unit in which, with the help of controlled louvers, a vortex movement is created with a deep pressure drop in the center of the vortex where the outlet from the suction pipe of a conventional wind turbine rotating in a horizontal plane is located.

In all these schemes, the impellers of wind turbines are faster and the electrical equipment is relatively lightweight; the unit operating costs are also reduced due to the increase in the unit capacity of the units but, in general, the material consumption, specific capital investments, and energy cost usually do not drop. The unit (item 8 in Fig. 3.1) compares favorably with others with the possibility of creating units with large dimensions and relatively small moving weights. However, as in other cases, even with optimal aerodynamics providing a power factor of 0.2 referred to the area

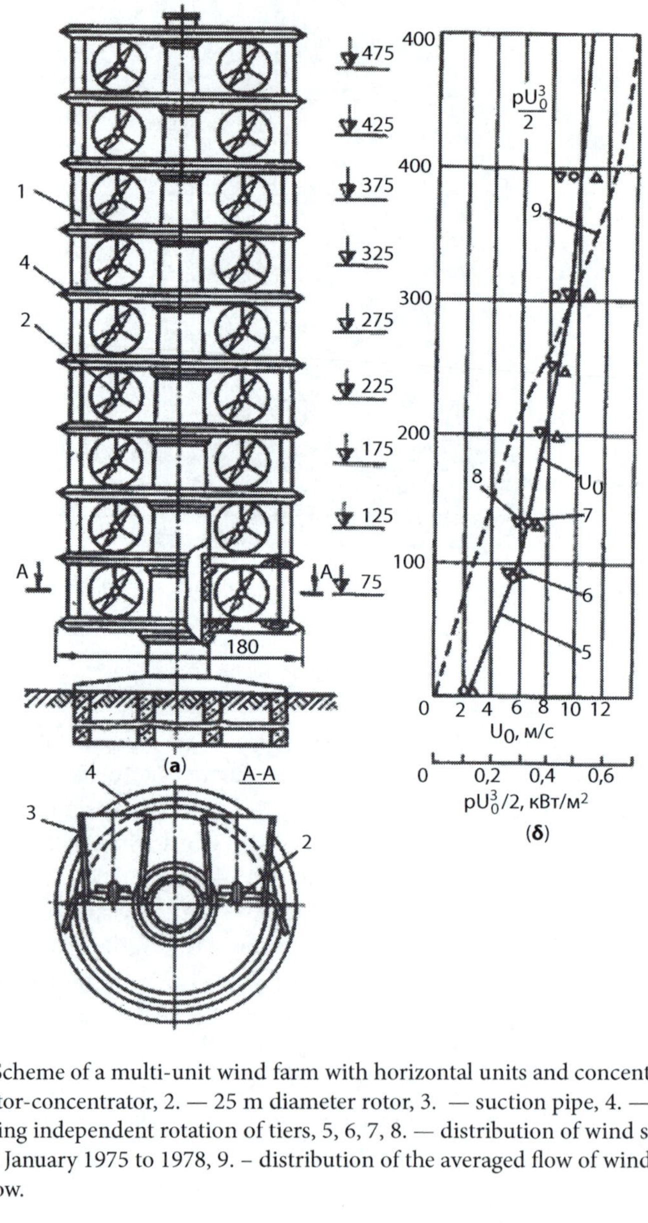

Fig. 3.37 Scheme of a multi-unit wind farm with horizontal units and concentrators. 1. — reflector-concentrator, 2. — 25 m diameter rotor, 3. — suction pipe, 4. — dividing discs allowing independent rotation of tiers, 5, 6, 7, 8. — distribution of wind speed over Moscow in January 1975 to 1978, 9. – distribution of the averaged flow of wind energy over Moscow.

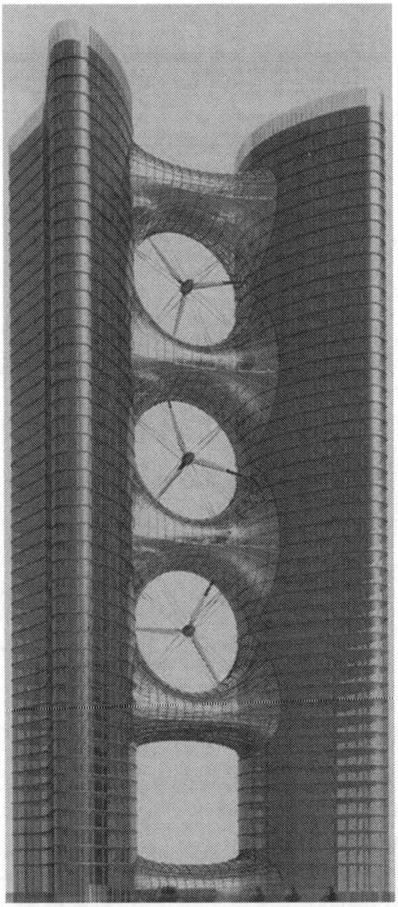

Fig. 3.38 Multi-tiered wind farm between the concentrator houses. Vortec Energy Project. The simplest version with two rails and one unit in between was built and tested in New Zealand.

in the dimensions of the tower, the cost of energy in a 2.5 MW plant turns out to be 10 to 12 times greater than that of the Mod-2 unit.

The ideas of a multi-tiered arrangement of wind turbines in pairs on one bearing tower (Fig. 3.37) or as an element of urban architecture — between houses (Fig. 3.38) have been repeatedly put forward.

The installed capacity of the system of wind turbines with concentrators according to Fig. 3.37 mounted on a 500 m high reinforced concrete tower, for example, for the meteorological conditions of Moscow is about 65 MW and, for the Volgograd region it is twice as much. The diameter of the wind turbines is assumed to be 40 *m*, and the design wind speed should be 13

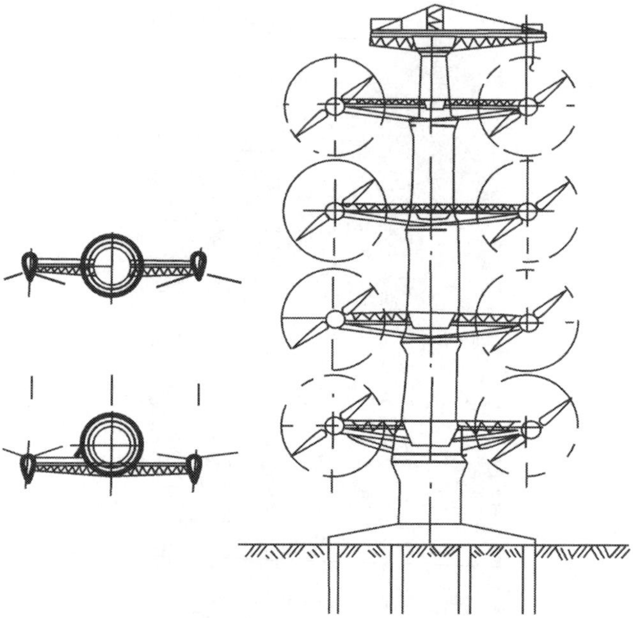

Fig. 3.39 Multi-unit wind farm on a support tower. Automatic wind orientation.

m/s. The discs between the tiers of the units are optional. Rotors can be placed on the tower in pairs, and without concentrators.

The ability of such a system to independently navigate the wind is also important — for this, the support point of the horizontal traverses is located in front of the units, and the units themselves are slightly turned outward. The self-orientation effect can be used in a floating wind turbine additionally protected from storms by turning the wind turbines to a horizontal position**). Having placed pairs of self-aligning wind turbines, for example, 126 m in diameter, with the Mod-5B-105 mechanical and electrical equipment, on a 500 m high reinforced concrete tower, we shall get a compact wind power plant with a capacity of 7.3·8 = 58.4 MW (Fig. 3.39). The energy generation of such a station for the conditions of Moscow will be determined from the following estimates. The average annual energy flow at an altitude of 100, 200, 300, and 400 *m* is 357, 599, 802, and 946 W/m^2. Taking the factor of wind energy use to be equal to 0.25, we find the average annual capacity of wind turbines at the indicated marks: 1.11,

**Milkovitsky, S.I., Lyatkher, V.M., Chumachenko, V.G., Uvarov, L.A., Litavrin, V.A. "Floating Wind Turbine", USSR Inventor's Certificate No. 1221982, dated December 1, 1985.

1.85, 2.49, and 2.94 MW. Thus, the number of hours of use of the installed capacity of each 7.3 MW unit at an altitude of 100, 200, 300, and 400 *m* will be 1,335, 2,230, and 3,526 hours, respectively. The total energy generation of wind turbines is expected to average 147,160 *MWh*/*year*, i.e., about 2,520 hours of use of the total installed capacity. As can be seen, the energy efficiency of the considered wind farm, even in Moscow, is high enough. In Volgograd, output could be 33% greater.

The same idea of self-orientation and increasing the efficiency of neighbouring hydroelectric units is used by MCT in its projects for damless tidal power plants.

Collinear wind turbines in combination with concentrators can be used in a wide variety of wind turbine layouts. We note here only two schemes making it possible to create large units:

— of a tower type;
— of an ejector type.

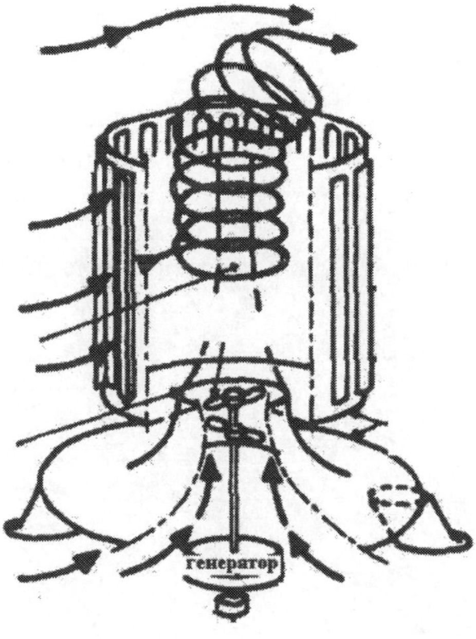

Fig. 3.40 "Tornado" system for converting the energy of currents. In the air flow, the power unit is located at the bottom (on the ground). When used in a stream of water, the power unit can be on top and the generator can be located, for example, on a floating or piled platform above the water surface. The main advantage is the concentration of energy and an increase in the speed of the turbine.

In the "Tornado" tower-type schemes (Fig. 3.40), the air flow is swirled with the help of guide louvers fixed in the walls of the vertical tower.

At the same time, a deep vacuum is created in the center of the tower — above the impeller. The air supply to the vertical-axis impeller goes through air (or water) receivers in the lower (or upper) part of the tower. Numerous studies have determined the energetically optimal ratio between the height H and the diameter D of the tower: $\frac{H}{D} \approx 1,25$. At the same time, the maximum utilization factor of the wind power per cross section of the entire structure is 0.14 ÷ 0.20. The advantage of this scheme is the simplicity of power equipment, independence from the direction of flow, and the possibility of obtaining large unit capacities. Additional vacuum can be created behind the impeller and other means, for example, by flowing around the dome, in the upper zone of which the unit is located (Fig. 3.41).

Ejector-type circuits (Fig. 3.42) make it possible to use the effects of concentrators with a horizontal axis on the ejection line, and to use the effect of the density gradient on the vertical section of the ejected flow route. This can be done when heating the ejected air flow by passing it through the cooling towers of a thermal electric power station or nuclear power plant (in the future, direct cooling of condensers of the thermal electric power station or nuclear power plant is possible without the use of cooling towers).

The power developed by the turbine located on the ejected flow path is determined by (1.4), where the largest cross-sectional area of the suction diffuser or concentrator can be taken as Ω in the optimized design model. The factor C_p in this case shall be about 0.5. Several ejectors increasing the energy efficiency of the unit can be located on one supporting tower. The

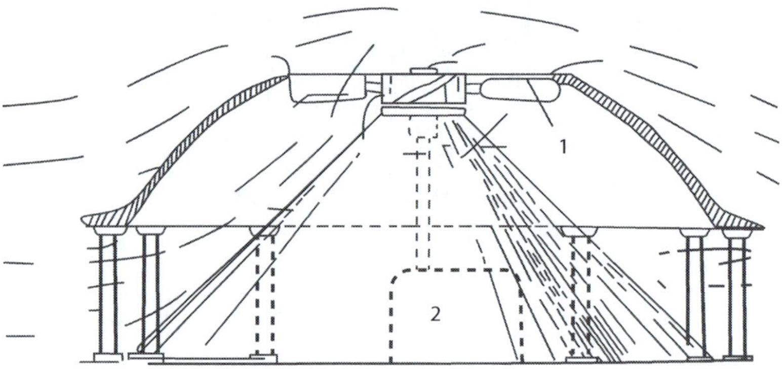

Fig. 3.41 Dome wind farm. Unit 1 is located in the vacuum zone, generator (2) — on the ground.

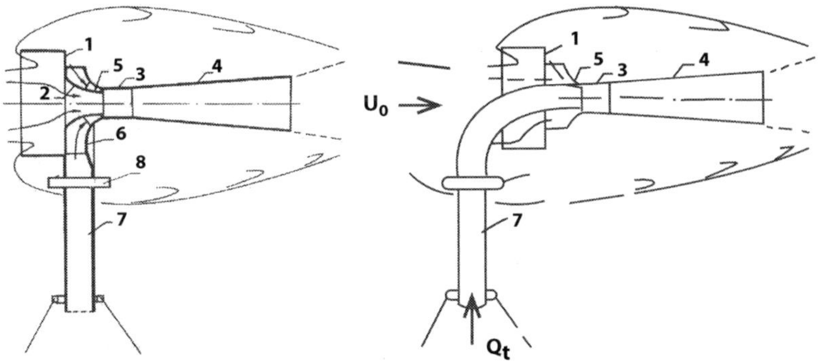

Fig. 3.42 Ejector wind turbine. 1) — accumulator reflector, 2) — concentrator (confuser), 3) — mixing chamber with deep vacuum, 4) — diffuser restoring pressure and orienting the system along the flow, 5, 6) — guiding spiral channel of the turbine path, 7) — turbine air duct which accommodates high-speed turbines with electric generators at its lower marks, 8) — ejector pivot support. On the left there is ejection at the wall, on the right — that in the center of the diffuser.

ejected flow can be supplied at the walls of the diffuser through the volute chamber, as shown in Fig. 3.43 (left), or as a solid jet in the center of the diffuser, which can be even more effective, since the pressure on the axis of the diffuser is usually less than that at its walls.

The same system can be used in a water stream also. In this case, its efficiency can be increased by using not water, but air on the ejected path (Fig. 3.43). The meaning of such a replacement becomes clear if we compare two turbines of the same size operating on fluids of the different densities ρ_1 and ρ_2, but having, for example, the same power of

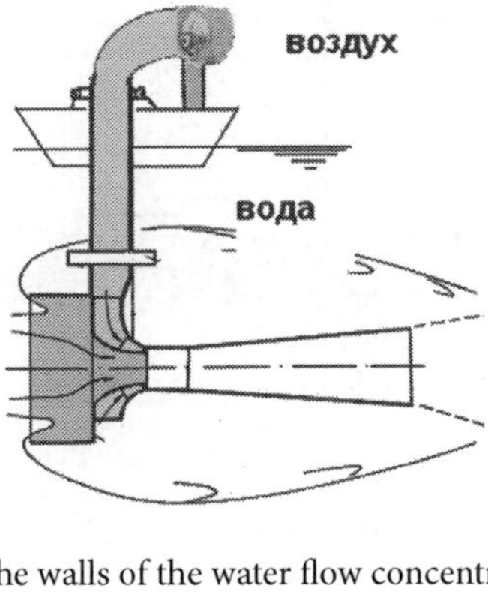

Fig. 3.43 Ejection of air to the walls of the water flow concentrator.

$$P = C_p \rho_1 U_1^3 \Omega/2 = C_p \rho_2 U_2^3 \Omega/2 \tag{3.37}$$

The ratio of the flow rates in this case should be:

$$U_1/U_2 = (\rho_2/\rho_1)^{1/3} \tag{3.38}$$

As we replace water by air, then the flow rate must be increased by about 10 times (more precisely, by 9.3 times). The forces and moments acting on the turbine from the flow side will decrease by the same factor, but the turbine rotation speed will increase by the same number. The latter means a sharp increase in the inertial loads on the turbine, but at the same time a sharp increase in the efficiency of the electric generator rotated by the turbine. Hence, the idea arises to immerse the concentrator in the water flow, and place the turbine and generator on the ejected air flow supplied to the walls of the confuser-concentrator (Fig. 3.43). The ejected (attracted) medium can, of course, also be water, if the velocities in the flow are high and the static pressure (deepening under the water surface) makes it possible not to fear cavitation . However, the air ejection can be more interesting (Fig. 3.44).

The turbine (Fig. 3.44) includes a concentrator (1) equipped on the front face with a grating with well-streamlined rods, a converging inlet channel (2), a working chamber with a diffuser neck and an air duct (3), as well as a diffuser-shaped outlet channel (4), a rotary hinge (5), air turbine (6) with generator (7), and supply confuser (8). All this is located on a floating craft (9). If the gratings become clogged, the center of pressure moves into a plane in front of the axis of rotation and the turbine itself turns and dumps debris. It also acts when the direction of the flow changes, always being

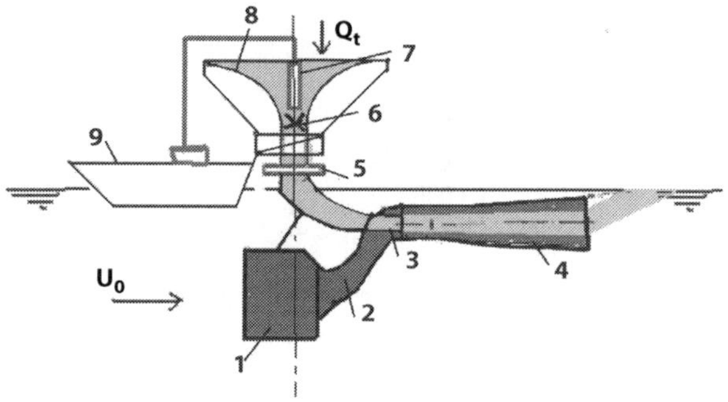

Fig. 3.44 Ejector hydroelectric complex with a high-speed wind turbine on a water stream at low speeds.

guided by the flow. The working chamber (3) can be made in the form of a spiral swirling the water flow and pressing it against the walls of the diffuser, which is useful for increasing the pressure on the turbine (6) and reducing the risk of cavitation in the diffuser. If the velocities in the water flow are high, then another scheme may be better — with an air supply to the walls of the diffuser (Fig. 3.43). Calculations show that the maximum power generated by a system with an ejector with an optimal design can be 2.6 times greater than the power of water flow through a circle with a diameter equal to the diameter of the initial section of the diffuser in Figs. 3.42 or 3.43 and the incoming flow velocity of U_0. In this case, the diameter of the concentrator should be approximately 1.5 times greater than the diameter of the initial section of the diffuser, and the sectional area of the turbine channel should be 0.25 ÷ 0.30 of the area of the diffuser initial section (the diffuser neck) – Fig. 3.45.

The pressure drop across the turbine is measured in fractions of the velocity head of the ejected flow in the diffuser neck. We note that the

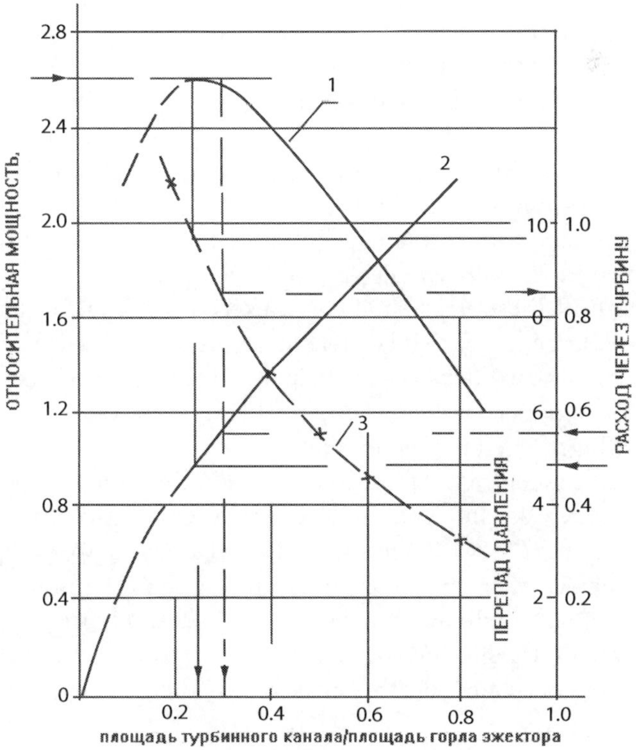

Fig. 3.45 Optimization of the ejector system.

generated power at a constant volumetric flow rate through the turbine, Q_t and the differential pressure Δp does not depend on the fluid density[††]):

$$P = \eta \Delta p Q_t \qquad (3.39)$$

Here $\eta \sim 0.9$ is the efficiency of the "turbine + generator" unit.

The proposed scheme is good for normal, low water flow velocities, when other solutions are very difficult and expensive. Let, for example, the flow velocity be $U_0 = 2$ m/s. The input concentrator has a cross section of 2x1 = 2 m^2. The area of the ejector neck is 0.2 m^2. The air flow through the optimal turbine is 1.2 m^3/s, the area of the air duct is 0.06 m^2, the air speed is 20 m/s, and the rotational frequency of a 300 mm diameter two-blade turbine (speed 8) is 10,200 rpm. By increasing the number of blades, we can reduce the speed to 2, and the rotational frequency, to 2,500 rpm. It is also possible to use an orthogonal turbine with the shading of 0.3 which will give a rotational frequency of 2,500 rpm. The generator power can be 2.6x 0.2x8/2 = 2.1 kW. If the flow rate is 1 m/s, then the power is reduced to 260 W and the rotation frequency is halved and the problem of providing a vacuum in the ejector neck arises — it is necessary to place the diffuser neck as close to the water surface as possible.

Tower-based wind turbines and concentrators have a high material consumption, and they must be designed for maximum storm impacts. A similar result of increasing the efficiency of the wind wheel can be achieved by a more elegant design — special flaps at the tips of the rotor blades. These flaps, perpendicular to the plane of rotation of the blades, form a rotating permeable confuser-diffuser (Fig. 3.46).

The length of these flaps should not exceed 0.1*D*, and their energetic effect can be large. In particular, recent studies have confirmed the high efficiency of outboard flaps on the blades of high-speed units [79]. These flaps are located perpendicular to the blade axis, outlining a kind of confuser-diffuser when the blade rotates.

The idea of concentrators found a kind of development in the inventions of B.V. Voitsekhovsky and M.B. Voitsekhovsky, USSR Inventor's Certificate Nos. 1280183 and 1229418), who suggested placing the working blades on the lateral surface of an imaginary truncated cone or pyramid. The axis of the cone (pyramid) is slightly tilted so that its top and bottom points of the base guide are on the ground (or on the support tower). The authors place a large number of blades on the surface of the cone, suggesting the use

[††]N.Ya. Fabrikant, "Aerodynamics", "Nauka", 1964, p. 127.

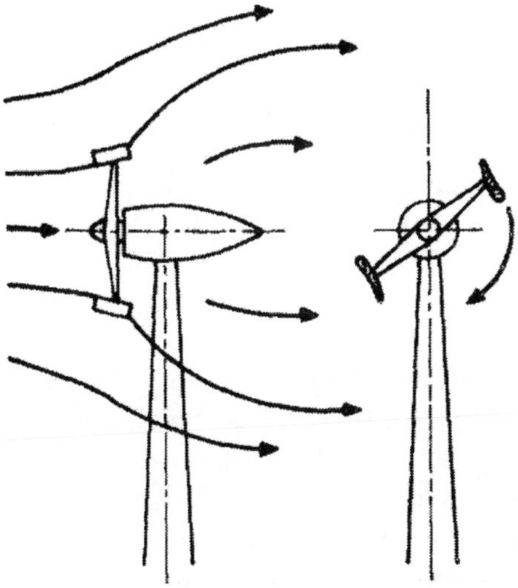

Fig. 3.46

of the unit in the mode of a low-speed machine equipped with pumping equipment. However, obviously, the proposed design scheme can be used for high-speed power-generating machines. The energy efficiency of such machines can be quite high.

Slow-speed wind turbines usually have a high specific material consumption. A quantitative comparison of modern machines is difficult, since machines of different types are designed and built by different firms, according to different standards and with different dimensions. The unique material for comparing is given by wind energy projects of the Central Wind Energy Institute (TsVEI) which developed the multi- blade (18 blades) and four-blade low-speed and high-speed motors with the same diameter of 5 m and with the same height of the tower — of about 15 m, in 1930s. Both units were made at the Petrovsky Kherson Plant and the Medvedev plant situated in Orel City. The optimum rotational frequency at a wind speed of 8 m/s was 40 and 180 rpm, and the power was 1.8 and 2.2 kW.

The weight (kg) of individual elements of the units (wood 60%, steel 40%) in a low-speed unit is significantly greater (Table 3.5).

Improving the designs of high-speed machines has further increased the gap in specific material capacities: the creation of a 5.2 kW low-speed multi-blade unit (TsVEI-D8) 8 m in diameter has reduced the material consumption to 1,200 kg/kW, and a high-speed, three-blade wind turbine of the same diameter tested at Cape Chelyuskin with an output power of

Table 3.5

Unit	Rotor	Head	Tower	Tail	Total	Specific weight, kg/kW
Low-speed	325	454	1,676	112	2,527	1,400
High-speed	135	270	1,815	93	2,315	1,050

5 kW (at $U = 8$ m/s) and its enlarged analogue 12 m in diameter with a generator power of 14.8 kW (at $U = 11$ m/s) had a specific material consumption of 370 kg/kW. The VTL-3 low-speed wind turbine last produced in the USSR (in the 1960s) of the Toguzak Plant Special Design Bureau (SKB) (KazSSR) with a power of 0.8 kW had a specific material consumption of 560 kg/kW. High-speed wind turbines with a capacity of up to 15 kW produced in the same period had a significantly lower material consumption (see Table 3.6).

Material consumption of large wind turbines:
The first large experimental wind farm in the USSR, built in 1931 at the Karanskie Heights near Balaklava, had a rated power of an induction generator of 93 kW (600 rpm), with a span of three blades of 30 m, and a rated speed of 30 rpm. The weight of the wind turbine along with the supporting structure was 49.1 tons (without the support, 34.5 tons), and the volume of the foundations was 60 m^3. The rated power was estimated to be achieved with wind speeds over 10 m/s. According to field tests at an average wind speed of 10.2 m/s, the average output power was 101 kW. Thus, the actual specific material consumption of the Balaklava wind farm was about 490 kg/kW. With the number of hours in power use being about 2,000 (at 1,805 h, the wind speed was greater than 9 m/s, $U_{av} = 5.75$ m/s), and the specific metal costs per unit of energy generation per year for this station of 0.24 kg/kWh were found to be less than those on later units of this type. For example, a 30.1 m diameter Louis Vadotte's French three-bladed wind turbine built in 1958 had a weight of 160 tons with an installed power of 800 kW (200 kg/kW), but had a high design wind speed (16.7 m/s) and, accordingly, had low average annual generation: for Balaklava conditions, where the average duration of winds with a speed of over 16 m/s is 300 hours per year, the annual output of L. Vadotte's unit would be no more than 240 MWh and the specific gravity would be of 0.67 kg/kWh. In 1942 the Balaklava wind farm was destroyed.

Table 3.6

Unit type	VBL-3	VB-3T	Vikhr'	Veterok	Berkut	AVEU-2	VIESKh-D4	AVEU-6*	Sokol
Diameter, m	3	2.8	4	4	4	2	4	6	12
Number of blades	3	2	2	3	2	2	2	2	3
Rated power, kW	1	0.7	1.7	1.7	1.7	0.15	1.6	3.4	12
Specific weight for a unit complete with a 5 to 10 m high tower, kg/kW	430	420	246	276	263	340	150	302	158

* Available now with improved performance. For all units, the specified power is observed at a wind speed of more than 8 to 10 m/s.

A more advanced three-blade unit with a blade span of 50 m and a power of 1,000 kW (at U_p = 14 m/s) was designed at TsVEI shortly before this institute was had been closed (1935).

The total weight of the unit under the project was 160 tons (160 kg/kW) with an annual output of the unit of 600 MWh in areas with an average annual wind speed of 7 to 8 m/s, which corresponds to the specific weight of metal of 0.27 kg/kWh.

Built in the USA in 1941, the Smith-Putnam first two-bladed wind farm of the megawatt class, with a 53 m diameter wind wheel and a 1,250 kW induction generator (U_p = 13.5 m/s) had a weight of 300 tons (240 kg/kW) and, under the Balaklava conditions, it would give an annual output of 750 MWh (0.4 kg/kWh). A 5 MW wind farm (219 kg/kW at U_p = 16.5 m/s) should have become even more economical, the draft design of which was created at the TsVEI Institute in 1932.

The modern most economical large high-speed wind power units (Mod-2, Mod-5B — USA, — NASA, General Electric, Boeing), the development of which created the basis for subsequent mass production, had two steel blades with a span of 91.5 to 97.6 m and the NASA airfoil 230-XX (Fig. 3.34) located in front of the tower. The 13.7 to 16.7 m long blade tips are capable of turning that way providing for aerodynamic control. A planetary gearbox increasing the rotational frequency from 13 to 17.5 rpm to 1,337 to 1,800 rpm, is provided for on the rotor shaft which can swing up to 2.5° in the Mod-5B model. Both models use induction generators, but Mod-5B has variable speed. The generator power in the Mod-2 model of 2.5 MW (3.12 MV·A) is achieved at a wind speed of 12.2 m/s. The weight of the unit, complete with a 61 m high tower, is 309.5 t (124 kg/kW). In the Balaklava conditions, such a unit would give an annual output of about 2,000 MW·h (0.15 kg/kWh).

As for the size of high-speed wind turbines, if we approach the analysis of the efficiency of wind turbines formally, assuming that with an increase in the diameter (D) of the wind turbine, its weight increases in proportion to the third power of the diameter (D^3), and the power increases in proportion to the area of the swept circle $\left(\frac{\pi D^2}{4}\right)$, i.e., second degree of diameter, then one can come to a simple but erroneous conclusion that, with an increase in the size of the unit, its specific material consumption (and cost) per unit of power (or per unit of energy generation) should increase.

In fact, with an increase in dimensions, a rationally designed structure does not retain geometric similarity — the role of inertial forces does affect this relatively little, and fewer off-design (structural) material is consumed. In addition, with the enlargement of the units, the construction and operating per unit labor costs would be sharply reduced. As a result, even with the traditional scheme of a high-speed wind turbine unchanged, the cost of the generated energy considered as a function of its diameter, turns out to have two local minima — with the blade span of 32 to 14 m and 90 to 120 m. The second minimum of the energy cost turns out to be deeper. This result was obtained by N.V. Krasovsky in 1922. At present, the production of large wind turbines has become widespread and practice has shown that the Vestas V90-3, V90 -1.8&2, V52-850 complex wind turbines models proved to be the most popular. The composition of the equipment and the layout of the unit can be considered already traditional (Fig. 3.47) — a rotor with three blades equipped with an individual drive for turning them, and multipliers for increasing the rotational frequency of an induction generator with a thyristor frequency converter which makes it possible to optimize the rotor speed, as well as a slow-speed side brake system that duplicates the capabilities of the aerodynamic braking by turning the blades, an individual drive for turning the nacelle downwind based on the readings of wind sensors located on the roof of the nacelle, auxiliary oil cooling systems for multipliers, and a generator cooling system.

The weight of all this equipment (except for the rotor) changes little on large machines with the change in power:

Unit type	**V90-3**	**V90-1.8&2**	**V52-850**
Rotor diameter, m	90	90	52
Rated power, kW	3,000	1,800/2,000	850
Rotor speed, rpm	8.6 to 18.4	9 to 14.9	14-31.4
Design wind, m/s	15	12/13	16
Rotor weight, t	41	38	10
Gondola weight, t	70	68	22
A 80 m high tower weight, t	160	150	103
Total material consumption, kg/kW	90.3	142/128	158.8
Material consumption for the rotor and nacelle, kg/kW	37	58.9/53	37.6

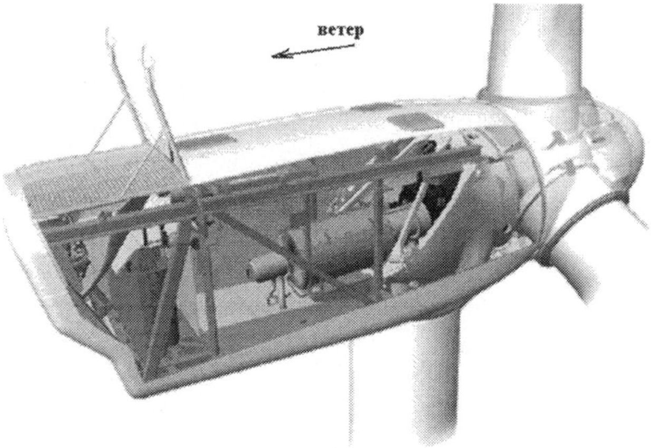

Fig. 3.47 V90-3 modern wind turbine.

Fig. 3.48 NedWind 40.

Fig. 3.49 NedWind 25.

It is useful to compare the weight index of the best machine in this series with what could be obtained using the layout in Fig. 3.36 with a three-bladed wheel (the length of the blades in V90-3 is about 45 m) and a linear (arc) generator actually tested in full-scale conditions, the weight of the inductors of which was 2.35 kg/kW. The diameter D of the proposed machine is determined from the equation D = 0.4 D + 2*45 which gives D = 150 m and the rated power of the unit is $3x(150/90)^2$ = 8.3 MW. The weight of the rotor will remain unchanged (41 tons), and the weight of the generator will not exceed 19.5 tons. If the height and weight of the tower are kept unchanged, then the total consumption of materials will be about 220.5 tons, and per unit of power — only 26.6 kg/kW. Such an indicator can qualitatively change the situation of wind energy. All Vestas machines have a three-blade rotor. NedWindRhenenbyRhenen (Netherlands) has long been producing three-blade and two-blade machines with the same blade design (glass-plastic reinforced polystyrene) and the same design wind speeds of 15 m/s with an operating range of 4 to 21 m/s. The NedWind 40 two-blade machine has a 40.77m diameter rotor, power of 500 kW, and a rated speed of z = 8 (38.7 rpm at a current frequency of 60 Hz). The weight of the rotor is 12 tons, the weight of the nacelle is 22 tons, and the weight of the 38 m high tower is 30.5 tons. The specific consumption of materials for the rotor and the nacelle, is 68 kg/kW.

A three-blade machine of the same company with a rotor diameter of 25.5 m, a power of 250 kW, and a speed of z = 7 (40.66 rpm) has a rotor + nacelle weight = 12 t, i.e. 48 kg/kW.

The tower is 29 m high and weighs 13.2 tons.

Large high-speed wind turbines of the traditional type with a capacity of up to 2.5-5 MW were the main objects of large-scale wind energy until 1982. For example, in the United States, 6 megawatt-class units were built (five 2.5 MW Boeing Mod-2 and one 4 MW Hamilton — Standard WTS -4). In the Hawaiian Islands, a 7.3 MW unit (the blade span of 122 m) was tested.

Mod-2 pilot industrial wind turbines installed in Goldendale (Washington state) in May 1981, by August 1983 had been working for a total of 3,842 hours (in the system, — 3,537 hours) and generated 4,227 MWh with having revealed a number of shortcomings factored in the Mod-5B model.

The wind turbine WTS-4 had : the diameter of the two-bladed rotor of 78.4 m, the power of the generator of 4 MW; fiberglass blades with adjustable turn are located behind a lattice steel tower 80 m high (up to the unit axis), the constant rotor speed is 30 rpm, design wind speed U_p = 16.4 m/s, and maximum operating wind speed — 27 m/s. The total weight of the unit together with the tower is 353.2 t, the specific gravity is 88.3 kg/kW; but, if we take the power at the design speed of 12.2 to 13.9 m/s (as, for example, in the Mod-2 different modifications), the specific gravity will be between 214 to 145 kg/kW. The location of the wind turbine behind the tower was unfortunate, as it caused intense low-frequency noise due to the blades crossing the aerodynamic wake behind the tower. At the request of the public, the plant was stopped, and the rotation speed and power were reduced.

An increase in the unit capacity of the units from 200 to 7,300 kW causes a monotonous decrease in the weight of the superstructure (above the foundation) referred to the unit of power (at a fixed U_p). If we take as a unit the material consumption at a power of 200 kW (Mod-OA) — 204 kg/kW, then the transition to the powers of 2,000 kW (Mod-1), 2,500 kW (Mod-2), and 7,300 kW (Mod-5A) gives the following relative material costs: 0.73, 0.53, and 0.43. Of course, these numbers also reflected advances in technology. On the other hand, the calculations do not take into account important meteorological factors influencing in the opposite way — an increase in wind speed in the function of height and the occurrence of a wind shear (non-coplanarity) in height. The first circumstance justifies the desire to increase the diameter and height of the unit; the second is limiting the span of the blades, and since the main influence on the power take-off is exerted by the end sections of the blades, the efficiency of the unit is especially greatly being reduced.

An expansive program for the creation of large wind turbines was carried out in Sweden, England, Denmark, and Germany. In Sweden, since 1982, two 3 MW and 2 MW units have been operating at wind speeds of 14.2 and 12.5 m/s bearing blades with spans of 78.2 and 75.2 m. The wind turbine in Nasudden, Gotland Kamewa WTS 75 has two blades with a span of 75 m. The rotor is located in front of a reinforced concrete tower 77 m high to the unit axis inclined to the horizon at an angle of 10°. The unit starts to deliver power at a wind speed of 6 m/s, at U_p = 12.5 m/s, a nominal power of 2 MW is achieved and it is kept constant (due to the turn of the blades at a constant rotation speed of 25 rpm) up to a wind speed of 21 m/s. The unit in Maglarp has a similar scheme. The unit in Nasudden was put into trial operation in August 1983.

Wind turbines with a horizontal axis, according to actual construction and operation data at the end of the 20th century, could produce energy with a cost of about 8 cents/kW·h in areas with an average annual wind speed of 5 to 6 m/s and 5 cents/kW·h in areas with an average annual wind speed 8.3 m/s (1987's prices).

According to Boeing, an improved model of the Mod-5B-107 unit with a blade span of 97.6 m and a power of 3.2 MW (at U_p = 12.1 m/s) having a specific gravity of 128.3 kg/kW provided the cost energy of 3 cents/kW·h.

The Monopteros single-blade wind turbine (Germany) had good economic indicators, which, however, turned out to be insufficiently reliable in operation: the formation of cracks in the material under the influence of high loads in the area of the pendulum hub and traverse prompted the supervisory authorities to decide to stop the repair work, complete the research program, and subsequently dismantle the wind turbine.

The speed of the tips of the blades V, power of the units, and the weight of their individual elements referred to the unit swept area, are important energy and economic indicators of wind turbines. The parameters of large wind turbines with a horizontal axis are given in Table 3.7.

A slight increase in material consumption per unit area on the Mod-5B unit as compared to Mod-2 is explained by the increased reliability of the unit and an increase in output due to the introduction of a variable rotation speed and the associated increase in the weight of electrical equipment, with an increase in the maximum permissible speeds of the unit use (up to 27 m/s for Mod-5V compared to 20 m/s for Mod-2).

Table 3.7

Unit type	U_p, m/s	D, m	H^*, m	Power, kW/ (kW/m²)	Weight (total) t/(kg/m²) Of the rotor	Weight (total) t/(kg/m²) Over the tower	Weight (total) t/(kg/m²) Tower	Weight (total) t/(kg/m²) Total
Mod-2	84	91.4	61	2,500/0.38**	90.8/14.1**	170.3/26.5	113.4/17.3	283.6/43.8
				4,000/0.84				353.2/73.5
WTC-4	123	78.2	80	3,200/0.43	143.6/19.2	258.4/34.6	152.2/20.4	410.6/55
				3,000/0.38		100/25.5		
Mod-5B	66 –	97.5	61	750/0.47	11/6.9	51/32	29.4/18.5	80.4/50.5
	89			330/0.44	5.4/7.2	22.4/30	22.3/29	44.7/59
GROWJAN	94.2	100	100	60/12	0.6/3.4	2.7/15	4/23	6.7/38
				0.34				
HWP-750	71	45	35	25/0.32	0.35/4.5	1.43/14	ferroconcrete	
HWP-330	62	31	25				1,500	
HWP-60	78	15	20	2,000/0.45	42/9.5	77/17.5	ferroconcrete	
							1,500	
AeroTech	39 –	10	22.5	1,000/0.62	7/4.4	31/19.5		103/64.5
10 to 25	42						72/45	
WTS 75-2	98	75	80	600/0.41	4.6/3.1	16.6/11.3		56.6/38.5
(Boving — KMW, Sweden)							40/27.2	
Newecs 45 (Stork,				3,000/1.05	34/12	44/15.5		
Netherlands)	106	45	60				ferroconcrete	
WW-0600 (Westinghouse,								
USA)	97	43.3	36.6					
LSI (Wind Energy, England)	107	60	45					

Axle height above the base.** In the denominator — per unit of swept area.

When using three blades, the material consumption of the units slightly increases; however, the speed of the blades and acoustic radiation are noticeably reduced, which turned out to be a decisive factor in the mass production of modern large wind turbines (Figs. 3.95, 3.97). All these machines, however, react poorly to the inhomogeneity of the natural wind field. Their regulation in a real turbulent atmosphere is often ineffective. For example, due to the non-coplanarity (shear) of the wind along the height, the loads on the blades can significantly exceed the calculated ones corresponding to the design regulation regime. This was one of the reasons for the destruction of the first large machine in the United States (March 1945) and damage to the modern machine (Wind Energy Report, June 1981). Unsurprisingly, with traditional wind power conversion schemes in mind, many energy professionals are rather restrained in their optimism about large-scale wind power.

Small and medium-sized wind turbines:

Small and medium-sized traditional wind turbines with a capacity of up to 750 kW have become more widespread in recent years. In Denmark, wind turbines with a capacity of up to 300 kW have become one of the main export items (mainly in the USA). These are distinguished by their simplicity and reliability; many units do not have blade control. The most famous companies, Vestas and Wincon, produce 22 to 110 kW wind turbines. The design wind speed corresponding to the rated power is usually taken to be about 14 m/s. Specific material costs per unit, for example, for Wincon's modern three-blade machines with a capacity of 22, 60, and 110 kW, are respectively 100, 75, and 50 kg/kW.

The simplified small Danish machines with fixed blades made of fiberglass have even better weights. For example, Alternegy wind turbines AeroTech 10-25 with a three-blade rotor diameter of 10 m and a generator power of 25 kW have a rotor weight of 354 kg, Weight of the gearbox is 355 kg and the total weight of the wind turbine (excluding a tower) is 1,080 kg (43.2 kg/kW). The rotation speed is 75 to 80 rpm. Already at a wind speed of 5 m/s, a power of 2.5 kW is output. The installed capacity is reached at a wind speed of 11 m/s. Reducing the average size of wind turbines below the optimal values corresponding to the power of 25 to 50 kW causes a sharp increase in the specific consumption of materials and in the cost of the units. According to the Koster estimates (Germany), the transition from 50 kW units to 5 and 2 kW units causes a decrease in output by 17 and 91 times and an increase in the cost of energy by 2.1 and 6.7 times, respectively. Similar conclusions were formulated on the basis of generalizing data on small- and medium-sized high-speed wind turbines with a traditional layout.

The distribution of weight among individual elements of modern wind turbines of average power of the traditional three-blade scheme manufactured by Howden (Great Britain) (Fig. 3.50), is given in Table 3.8.

In Russia, small units produced by the Vetroen Research and Production Association are obviously not optimal. Yet, they can compete in the world market with units of the same size. The installed power of these units is 4 kW, the blade span is 6 m, and the design wind speed is 10 m/s. The weight of the unit including the 7 m high tower and the embedded parts of the foundation, is 1,100 kg (275 kg/kW).

The development of new small machines and micromachines has begun in the Siberian Branch of the Russian Academy of Sciences: the low-speed wind turbine has broad prospects for aggregation with hydraulic piston pumps or pneumatic pumps; it does not require an electrical converter and is extremely simple and reliable. This machine can be widely used as an individual energy source for water lifting.

Medium, small machines, and especially micro-turbines with a horizontal axis, do not have some of the drawbacks inherent in the large machines, as these are less sensitive to wind "shear,"and adapt more easily to changes in wind direction. Moreover, the spatial heterogeneity of the instantaneous wind speed field is less dangerous for them. The difference in wind direction at different heights, which is imperceptible on units with a small blade span, on large units can seriously reduce their energy efficiency and cause destruction of the blades.

It should also be noted that, in the traditional arrangement, the speeds of movement of the tips of the blades of large machines are high (close to the speed of sound), which causes intense acoustic radiation which, in some cases, is dangerous to human health. In optimized units (Mod-2), the noise level of 50 dB is set at a distance not exceeding 500 m.

Features of small wind turbines are mainly associated with the regulation of the rotor speed, the orientation of the unit in the wind and, most importantly, with protection against acceleration when the load is disconnected, and protection against storm winds. On large machines, these problems are solved by an electronic control system for the rotation of the blades, the speed of the unit rotation, and the mechanical (backup) brake. This whole system is powered by wind speed and direction sensors. On small machines, such a system is usually not installed, but regulation is carried out by mechanical converters responding to the rotor speed (centrifugal regulators) and the maximum force acting on the rotor. Modern tendencies are to simplify the design as much as possible, abandoning speed control and introducing mechanical protection against a hurricane by overturning the rotor, or on the contrary, introducing an electronic-mechanical control

Table 3.8

Machine type	Nominal power, kWh	Power generation at $U_{av} = 6$ m/s, MW·h	Rotor diameter, m	Rotation speed, rpm	Rotor weight, t	Nacelle weight, t	Tower weight, t	Specific weight, kg/kW
HWP-60	60	130	15	50	0.6	2.1	4	112
HWP-330	330	500	31	38	5.4	17.0	22.3	135
HWP750	750	1,100	45	30	11.00	40.0	29.4	107

system similar to control systems for large machines. In one aspect, perhaps, the peculiarity of small machines remains — orientation downwind is carried out exclusively due to aerodynamic forces — either by the location of the unit behind the tower, or by the creation of a powerful "tail-weather vane", the plane of which is often carried out beyond the main rotor. Previously used fantail devices are almost never found in modern designs.

4

Orthogonal Power Units

4.1 High Speed Orthogonal Turbines in the Infinite Flow

Free-flowing turbines that convert the kinetic energy of the flow into electricity are called orthogonal if the axis of rotation of the turbine blade system is perpendicular to the direction of flow. Consider the most energy-efficient high-speed turbines, which have a linear velocity of the blades greater than the velocity of the oncoming flow (the idea of the patent Darrieus [1]). The turbine axis is usually vertical (Fig. 4.1) or horizontal (Fig. 4.2).

The idea of the Darie patent and the first projects of wind turbines was to use curved blades, possibly variable cross-section with supports at the

Fig. 4.1 Orthogonal wind turbines with a vertical axis.

Victor M. Lyatkher and Ziaur Rahman. Wave, Wind and Current Power Generation, (159–194) © 2022 Scrivener Publishing LLC

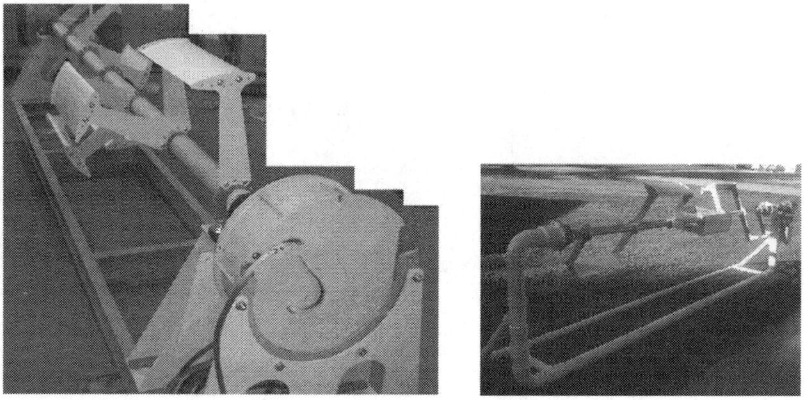

Fig. 4.2 Orthogonal hydro turbines with a horizontal axis.

ends of the blades. The optimized turbine of this type, developed and tested in SANDIA, has shown very high efficiency (Fig. 4.3).

Turbines of this type (Fig. 4.4) actively built and sold.

Dimensions and capacity of the designed turbines have reached impressive values (Fig. 4.5).

However, the production and operation showed adverse features, these being the difficulty of manufacturing it, the insufficient strength of

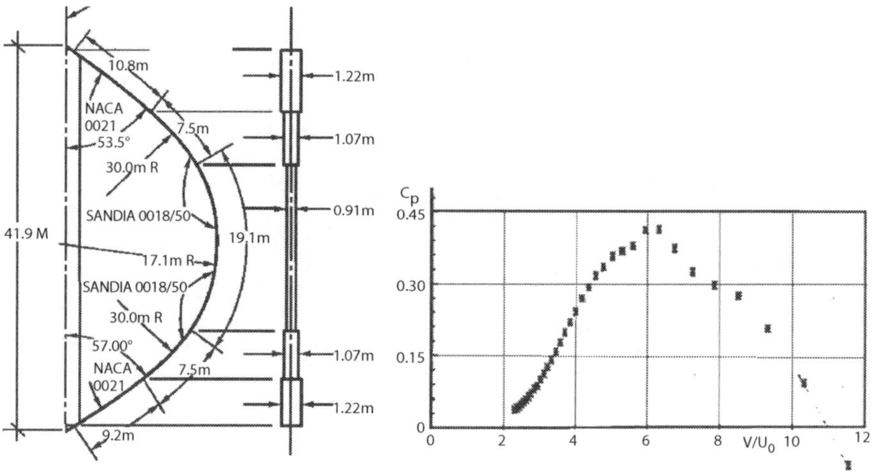

Fig. 4.3 Two-blade Sandia-34 turbine. Efficiency Sandia-34 turbine.

Fig. 4.4 Commercial sample of the high-speed troposkein wind turbine with a changeable ratio of height and diameter of the turbine (Flow Wind Corporation). Height 42-63 m, diameter 17-21 m, power 300-400 kW, wind during the operation from 4.5 to 27 m/s, calculated storm – 58.5 m/s. About 900 sets were installed in California in 1995.

its curved blades, high vibration and inadequate reliability of the rotor. Popularity was acquired with a balanced turbine with multiple straight blades in one or more tiers, as well as turbine blades, curved in a spiral (helicoid).

Turbines of this type work due to the action of the pulling force, developing on the wear of the blade that has an aerodynamic profile when the flow of the blade with a flow angle of attack is less critical. Usually, turbines are designed so that their length is much larger than the diameter of the turbine. This allows for conditions of limitless flow to consider a two-dimensional picture in a plane perpendicular to the axis of the turbine. Numerical models of such turbines [2] allow to optimize their design

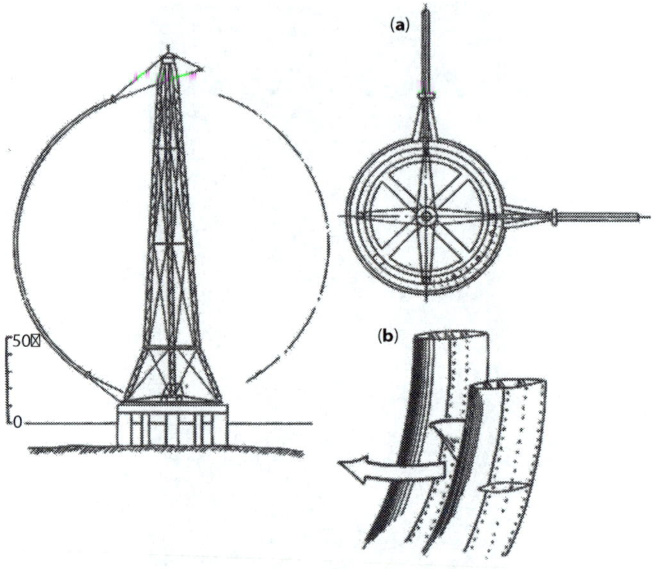

Fig. 4.5 "L-180 Poseidon" according to the Ljungstrom project. Canadian Patent 337.120, Oct.6, 1978, Danish 4146/79. Diameter of the turbine 180 m, power 20 MW, development 45 GWhour/year at average wind 9 - 9.5 m/s. 1980. Shift of working blades at 90° reduces pulsation of loadings and torque.

scheme for the simplest case of application of straight blades. Such optimization, of course, is subject to experimental verification. At the same time, additional difficulties arise due to the fact that the conditions for the flow around the blade wear significantly depending on the viscosity (on the Reynolds criterion) – Fig. 4.6.

Reliable (self-similar) results for real large aggregates can be obtained only on large models at sufficiently high flow rates:

$$Re= Vb/\nu > (7 \div 8)x\ 10^5 \tag{4.1}$$

Here V is the linear velocity of the blade, b is the chord of the blade, and ν is the kinematic viscosity of the medium.

If the tests are carried out on a hydraulic model at a flow rate of, for example, 1m/s and blades speed of 3m/s, the chord models should be at least 25 cm, which requires large dimensions of the experimental stands. In normal practice, self-similarity is not achieved and experimental (model) results can only be used to compare designs. The design methodology, based on empirical data on profile characteristics, can provide a more reliable prediction.

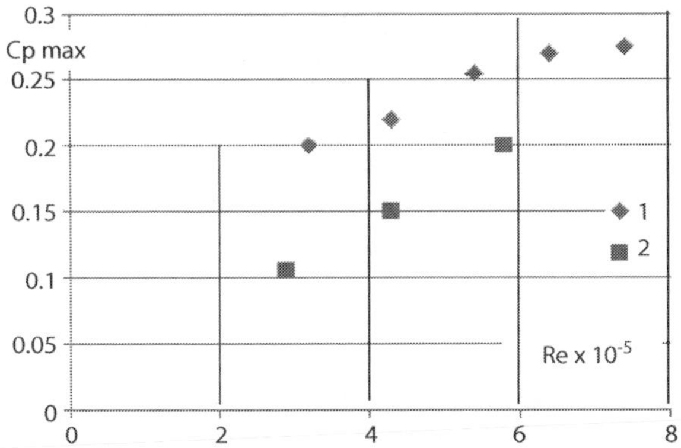

Fig. 4.6 Maximum efficiency of orthogonal power unit depending on the Reynolds criterion. 1-turbine with 3 blades, shading 0.45, profile NACA 0015, 2-single-blade unit with struts and counterweight [2a].

4.2 Efficiency Turbine with Different Parameters

In the considered scheme of orthogonal machines the variable parameters can be:

1) the profile of the blade and its orientation (the rotation of the blade relative to the tangential to the trajectory of the blade);
2) relative chord length of the blade b/D;
3) Number (i) and arrangement of identical blades placed on the same diameter of the track D;
4) the solidity of the turbine σ= ib/D
5) number of groups of blades belonging to different diameters of tracks (one or two tracks);

Consider the impact of these factors.

Fig. 4.7 shows changes in the efficiency of the double-blade turbine in the function of the relative speed of the blades with a profile NASA 0018 without turning the blades and a turn of 3 or 6 degrees (sock outside).

The best results are obtained by turning on 3 or 6 degrees. This is consistent with the experiments in TSAGI [3]. Relatively low values of maximum efficiency of aggregates are associated with low Reynolds numbers Re=Vb/ν=2.6x10^5. The efficiency markedly increases as a result of the solidity

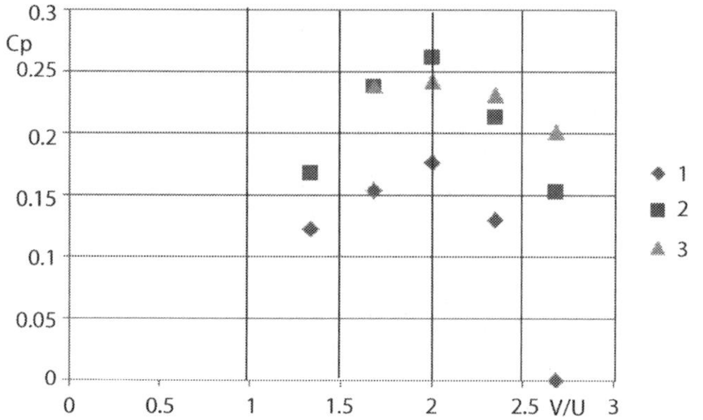

Fig. 4.7 The efficiency of the wind turbine in the function of the relative velocity of the blades. Profile NASA 0018, i=2, b=0.18 m, D=0.64m, L=0.9 m, U=10 m/s, blade turning angles for points 1, 2, 3 are 0, 3, 6 degrees. Solidity σ=0.36/0.64=0.562.

being made greater due to the diameter of the turbine being reduced, for example (Fig. 4.8).

The shape of the profile is relatively weak. Changing the thickness of the NASA standard profile in the range from 15 to 22% practically does not change the characteristics of the unit (Fig. 4.9).

The use of the best modern aviation profile GAW-1 [4] does not much change the characteristics of the unit. More significantly, it affects the

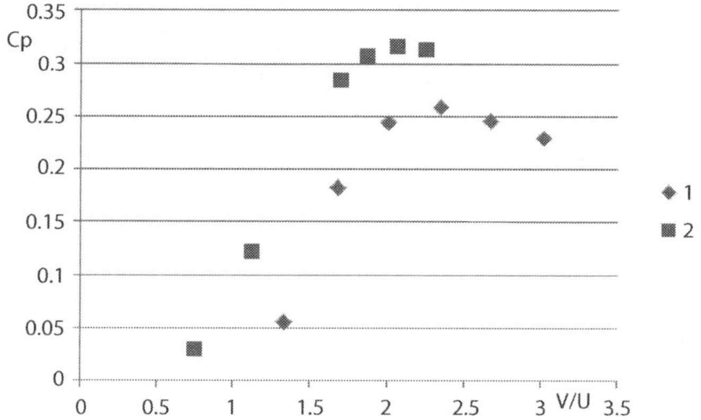

Fig. 4.8 Turbine with the same one blades on the radius 0.32 and 0.16 m (points 1) or on the radius 0.18 and 0.09 m (points 2) – two groups of blades in turbines of different diameters. NACA 0018, b=180mm, 2group, D=0.64m, 5^0, 2 travers b_{tr}=0.1m, t_{tr} =0.02m, U=10m/s, L=0.9m; NACA 0018, D=0.32m, b=0.09m, 5^0, position 180^0, b_{tr}=0.1m, t_{tr}=0.02m.

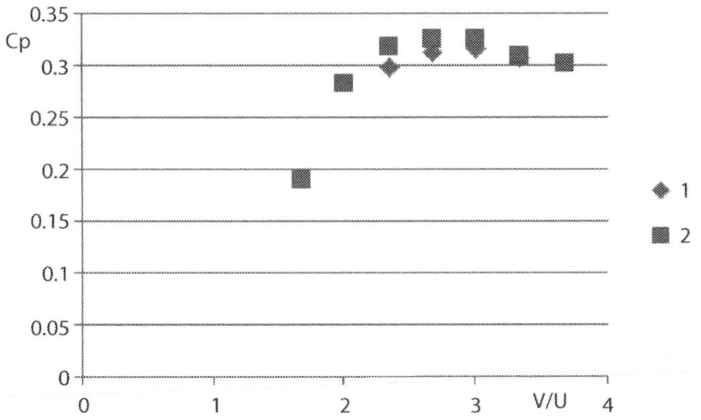

Fig. 4.9 Efficiency two-bladed wind turbine with a profile NASA 0018 (1) and NASA 0021 (2). Turn the blades on 6^0 D=3.2 m, b=0.5m, σ= 0.312, U=10 m/s.

number of blades and solidity of the unit – at a constant chord, blades are most effective when they are single-blade units (Fig. 4.10).

When using two blades in one tier, it is best to place the blades with a shift of 45-60 or 120^0. In this case, the positive torque and maximum efficiency of the turbine for the conditions in Fig. 4.10 reaches the same C_P= 0.26 at V/U=2. It is clear that, in this case, it is necessary to use special devices or several tiers to balance inertial loads.

With the same blades as in Fig. 4.10, two-bladed windmill reaches its maximum efficiency C_P= 0.3 at diameter D =1.8 m and a thin traverse (Fig. 4.11).

The two-tier double-blade turbine was tested in detail on models in the laboratory and in full-scale (nature) conditions (Fig. 4.12).

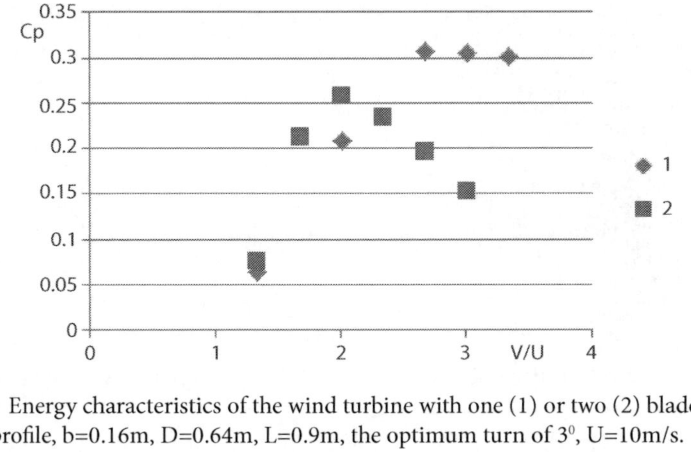

Fig. 4.10 Energy characteristics of the wind turbine with one (1) or two (2) blades. The GAW-1 profile, b=0.16m, D=0.64m, L=0.9m, the optimum turn of 3^0, U=10m/s.

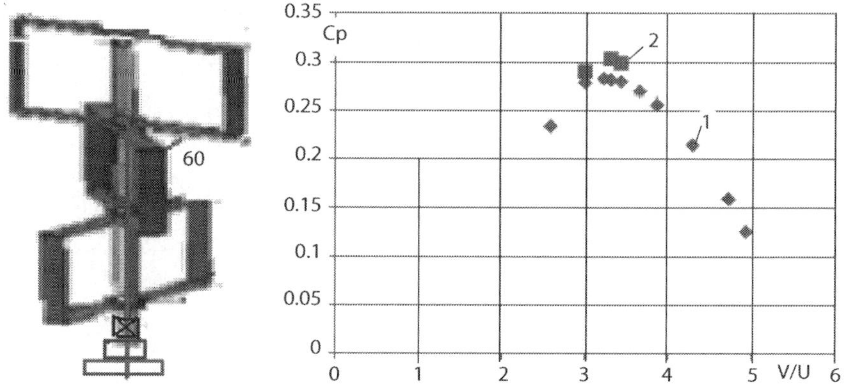

Fig. 4.11 Recommended options for the design of the rotor nominal power of 1 kW at a wind of 11 m/s. The Generator (1) at the base of the rotor (left). Rotor diameter 1.8 m, height of the layer (length of blade) 0.9 m,chord blades are 0.16 m, the profile of the blade GAW-1, a reversal of the chord 20 relative to the aerodynamic center, traverse streamlined cross-section 40x2 mm^2 (point 2) or 80x5 mm^2 (point 1). When the wind is 11 m/s – the optimal rotational speed is 390 rpm (V/U=3.3), and the power is not lower than 1.2 kW. The efficiency is not lower than 30% in the wind speed range from 3 to 12 m/s when the rotational speed is proportional to the wind speed. At a wind more than 12 m/s, the rotation frequency constant is limited to 450 rpm. At a wind of 26 m/s, the unit will stop on the mechanical brake. The rotor shaft can be fixed at the two end supports placed in a light frame. The whole structure in this embodiment is portable and dockable through the use of a frame without foundation.

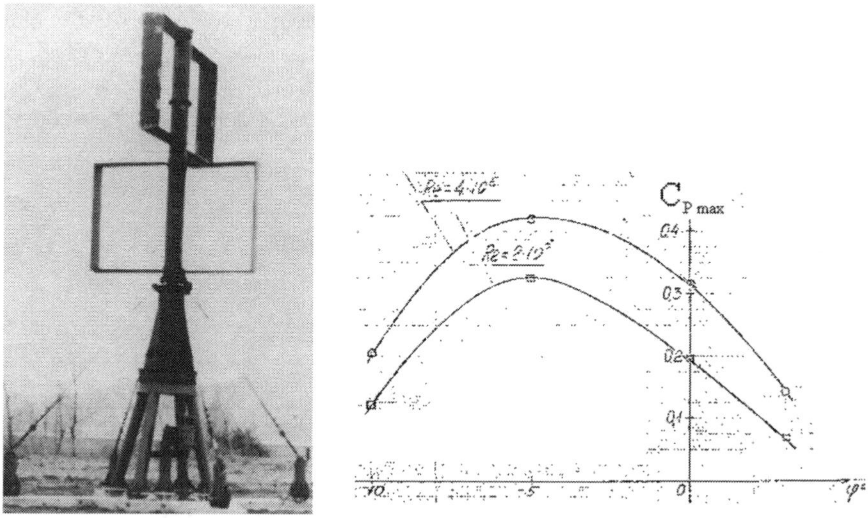

Fig. 4.12 A 16 kW two-tier wind turbine at the site in Dubki (Dagestan)-on the left and its maximum efficiency depending on the angle of rotation of the blades – on the right (according to the model at different Reynolds numbers).

Fig. 4.13 Two-blades orthogonal unit on tests in TSAGI.

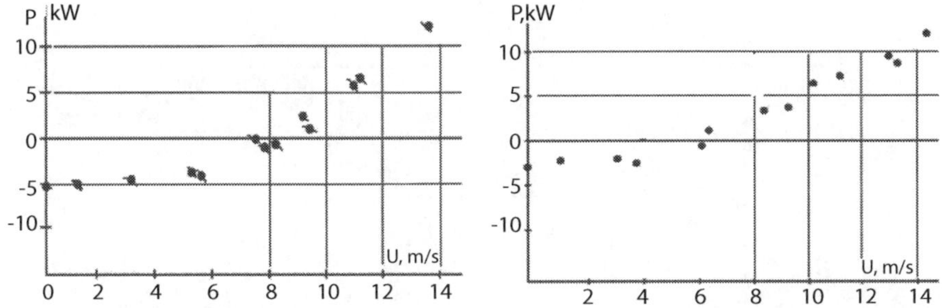

Fig. 4.14 Capacity at the terminals of the generator 4-blade VAWT (left) and capacity at the terminals of the generator 2-blade VAWT (right), depending on the wind speed. A rotor diameter of 9 m, the profile of the blades NASA0021, the average chord of the blade is 0.9m; the length of this part of 3.76 m, the average chord of the end parts of blade of 0.65 m, the total length of 4.64 m. The chord traverse is 0.465 m, and the thickness traverse is 0.088 m. where the speed of the rotor is 38.8 rpm. 4-blade single rotor, one pair of blades is shifted relative to the other for 45^0.

At the constant diameter of the turbine, the increase in the number of blades in the infinite flow strongly inhibits the flow (removes the flow from the turbine) and does not increase the maximum power of the unit, if the chord of the blade is greater than a certain value (Figs. 4.14, 4.15, 4.16).

On the contrary, with a relatively small chord of the blade, the introduction of a second or even a third blade can be useful as this results in a raise in the maximum power (Fig. 4.17).

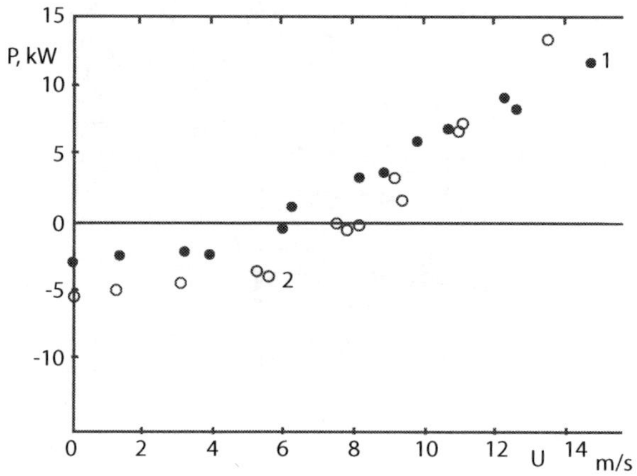

Fig. 4.15 Capacity at the terminals of the generator, depending on the wind speed in the pipe for testing two-blade (1) and four-blade (2) rotors with a diameter of 9 m at a rotational speed of 38.7-38.9 rpm (blade speed 18.2-18.3 m/s).

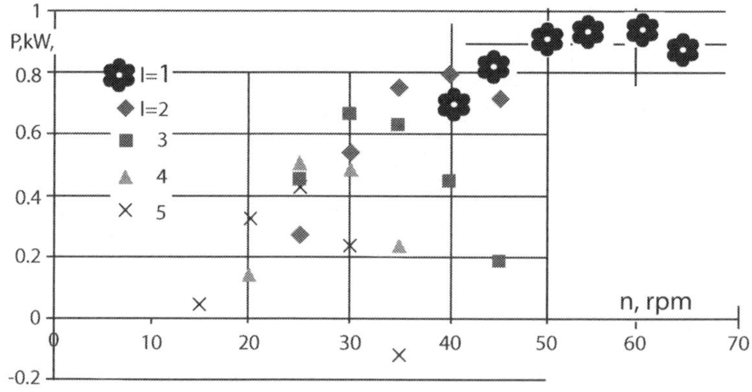

Fig. 4.16 Capacity of straight blade hydro turbines with length L=**5.4m**, Blades GAW-1, b=0.16m, α=0÷2⁰. D=1.28m, Water speed 1 m/s.

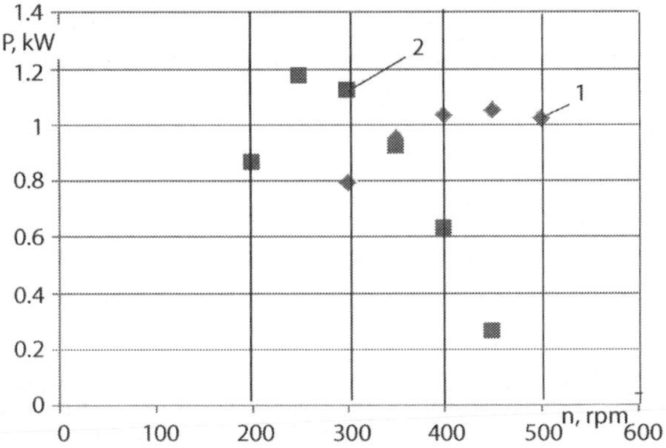

Fig. 4.17 The output of a wind turbine of diameter D=2m, b=0.16m, length L=3m: 1-one blade, 2 - 3 blades in one layer, GAW-1, 3^0, U=10m/s.

The rotor with 3 blades starts better (higher torque at low speeds), but the rotor with one blade even at a relatively small chord (b/D = 0.08) has a higher optimal rotational speed and a flatter characteristic. In a limited stream (in the river) the situation is different. Due to limited depth and width of flow, the capacity of the three-blade turbine will be significantly greater.

A single-blade turbine is the most effective solution if its chord is between 0.5 and 1 turbine radius (Fig. 4.18). Reducing efficiency by increasing the diameter of the turbine, however, does not prevent an increase in the maximum power of the turbine with the same blade.

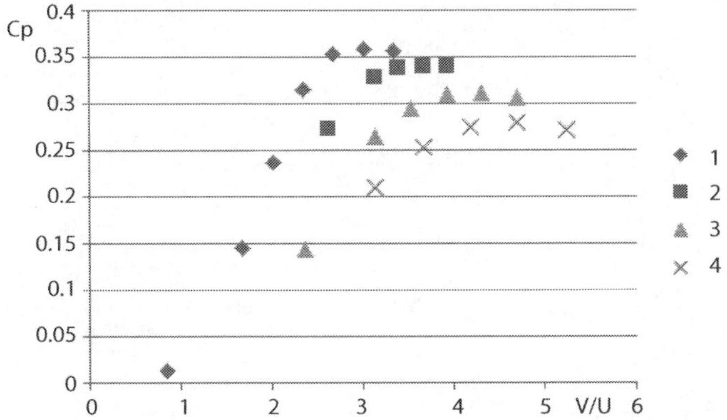

Fig. 4.18 The efficiency of single-blade wind turbines of different diameters D=0.64; 1.0; 1.5; 2.0 m (points 1, 2, 3, 4); chord blades and cross-section traverses unchanged b=btr=0.16 m, . i=1, L=3m; GAW-1; 3^0, U=10m/s.

Single-blade turbines with a different blade profile also reduce efficiency by increasing the diameter above the optimal value (Fig. 4.19), but still provide an increase in the maximum power of the turbine.

On turbines with GAW-1 blade profile without reversal with increasing diameter, the optimal relative velocity of the blade increases without reducing the maximum efficiency (Fig. 4.20).

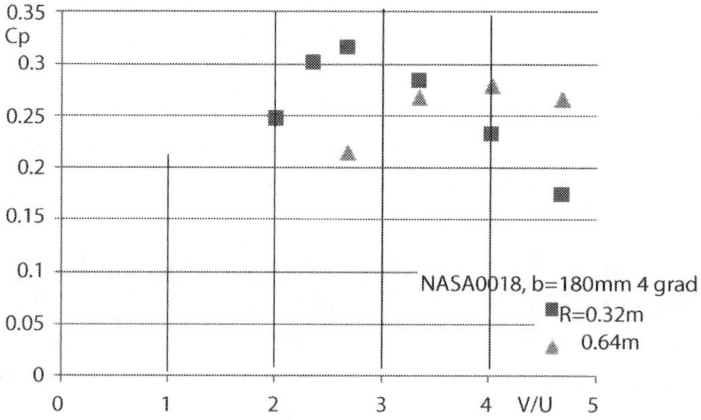

Fig. 4.19 The efficiency of single-blade wind turbine with NASA 0018 profile.

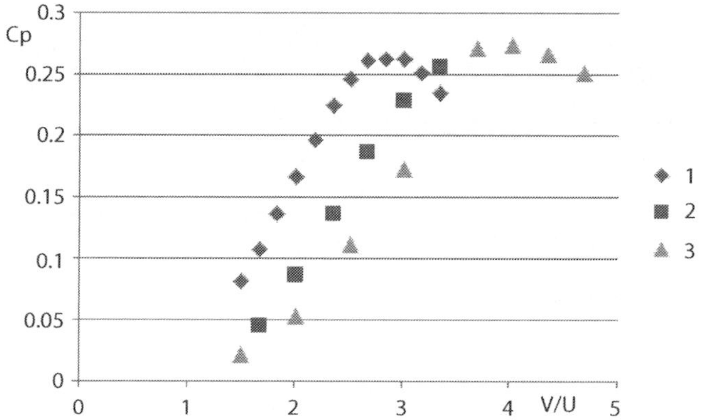

Fig. 4.20 Turbine diameter effect D=0.64 ; 1.28; 1.92m (points 1, 2, 3), GAW-1, 0 grad; b=0.16m; U=10m/s; b_{tr}=0.003 m; L=0.9m; i=1.

As the solidity increases, the efficiency of the turbine increases, reaching a maximum when solidity is around σ= ib/D=0.3. With constant solidity, maximum efficiency is higher with the smaller numeral blades (Fig. 4.21). One blade with a chord 0.48 m (1) at solidity 0.24 gives greater efficiency than three blades with a chord 0.16 m (2). Conclusion: with the same solidity, the smaller numeral blades, the better.

The effect of solidity is noticeable if it is little . With a turbine diameter of 2m, a single-blade turbine with a chord of 0.16m (solidity 0.08) gives a slightly lower maximum capacity than a three-blade turbine (Fig. 4.21), but same as a two-blade turbine with the same blades.

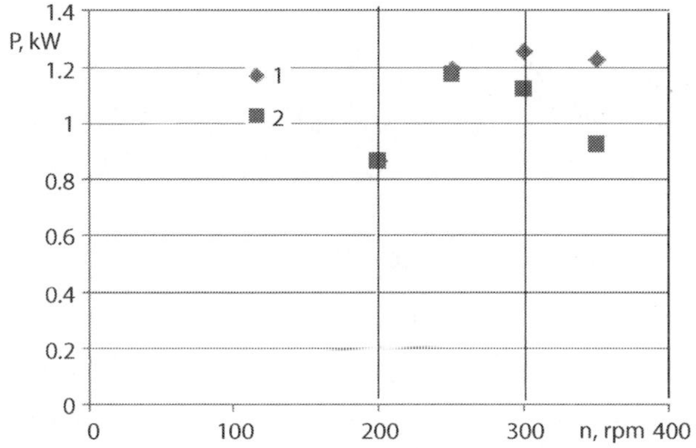

Fig. 4.21 The capacity of a VAWT of diameter D=2m, GAW- 1, 3^0, L=3m, U=10m/s.1-one blade with a chord 0.48 m, 2-three blades with a chord of 0.16 m.

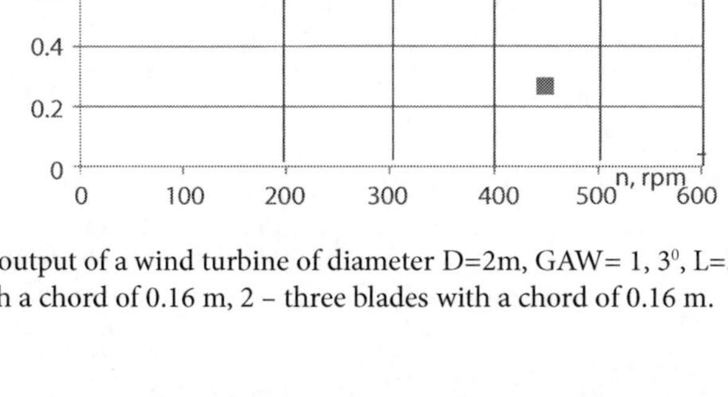

Fig. 4.22 The output of a wind turbine of diameter D=2m, GAW= 1, 3^0, L=3m, U=10m/s.1-a blade with a chord of 0.16 m, 2 – three blades with a chord of 0.16 m.

Already with a diameter of 1.28 m (solidity 0.125 and more), an increase in the number of blades with a chord of 0.16m does not lead to an increase in the maximum power of the unit (Fig. 4.16). All these results are valid for turbines in a limitless flow, where the flow rate inside the turbine varies significantly with the change in speed and number of blades (Fig. 4.23).

Given the decrease in the flow rate inside the turbine, we can try to install a second working blade at a smaller radius. Such a scheme, however, does not lead to an increase in the maximum power (Fig. 4.24).

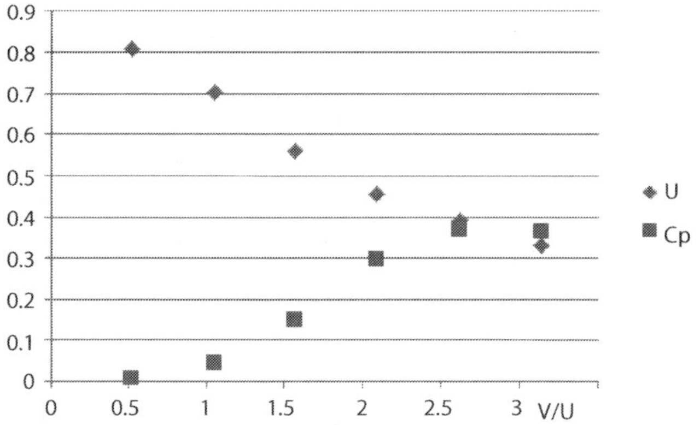

Fig. 4.23 The relative velocity of flow in the turbine (u/U) on the approach to the rear section of the route (x=0.7, y=0) and the turbine efficiency (Cp). GAW-1,3^0, D=10m, i=1, L=20m, b=3m, U=10m/s.

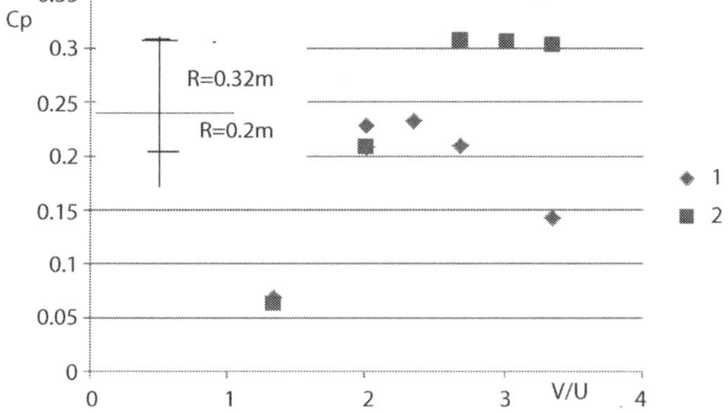

Fig. 4.24 The efficiency of a turbine with two identical blades (b=0.16 m, GAW1, 3^0), located at different radii (0.2 m and 0.32 m)-points1 in comparison with the efficiency of a single –blade turbine with a radius of 0.32 m.-point 2. L=0.9 m, U=10m/s.

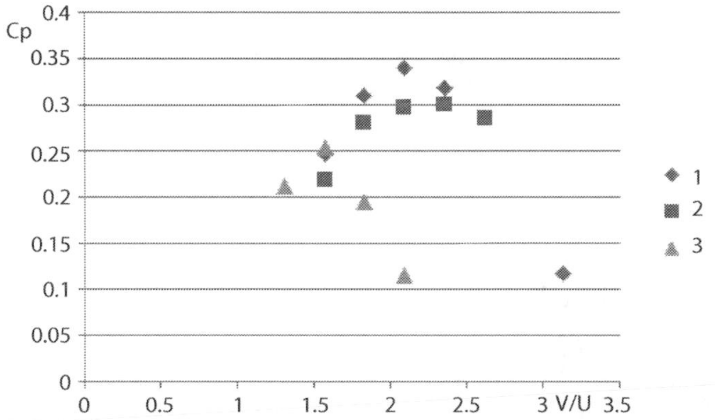

GAW1	R=5m	b=5m	3grad	1tr	btr=5sm	U=10m/s	L=20m	i=1	тол.тр5см

Fig. 4.25 The efficiency of turbines with a diameter of 10 m with blades profile GAW-1, 20 m long, deployed at 3^0. (1) and (2) – the same solidity (0.5), but i=1, b=5m (points 1) or i=2, b=2.5 m (points 2); 3-same chord b=5m, but i=2.

All the experiments and calculations show that a single-blade turbine in a limitless flow is the best. This is obtained by comparing the turbines with both the same solidity and the same length of the chord blades (Fig. 4.25).

The high efficiency of the optimized single-blade turbine is obtained despite the fact that significant torque is observed only on the frontal section of the route. With two or more blades, the back section of the route makes a significant contribution to the average power, although the total maximum power is less.

4.3 One and Two Blades Turbines

Ripple torque (and power) in single-blade turbines which limit the application is resolved in a modern multi-tiered and spiral (without central axis) balanced designs (Fig. 4.26) [5].

These structures do not load the supports with centrifugal forces. They provide a small pulsation of the torque on the shaft of the generator, but do not eliminate the pulsation in the support nodes of the bending moment from the aerodynamic forces, and in variants with straight blades have reduced efficiency due to a large number of traverses. A large number of traverses reduce the efficiency of the turbine, an important trait of multi-tiered schemes. The effect of traverse can be very strong. For example,

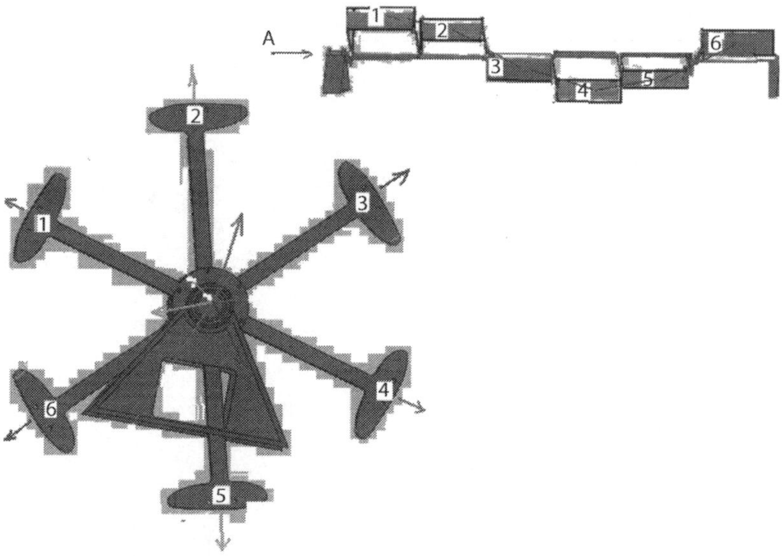

Fig. 4.26 Version of a single-blade six-tier turbine, patented by author.

the use of 4 traverse (blade support at the ends) instead of two (console fastening of blades in the middle) with the aerodynamic profile thickness of the traverse 31% on the optimized double-blade turbine reduced the maximum efficiency of the turbine from Cp=0.3 (at V/U=2.7) to Cp=0.08 (forV/U=2.3) [7]. A similar result was obtained when testing the first variants of single-blade turbines. The turbine with one blade, fixed in disks with balancing counterweights, embedded in disks, showed the highest efficiency (Fig. 4.27).

On the contrary, the fixing of the blade with the profile of the GAW-1 two ordinary cross-arms with a counterweight, adding to the resistance, led to a sharp decrease in the efficiency (Fig. 4.28).

The increased turbulence of the flow in the hydraulic must reduce the effect of Reynolds number. To test this effect (Re effect), rotor tests with a diameter of 400 mm in another tray with a width of 2m (B/D=5) and a depth of up to 1.5 m at a flow rate of up to 1.6 m/s were conducted.

The tests confirmed the qualitative conclusions given here, but the quantitative evaluation of the effectiveness of all units proved to be higher. For example, the maximum efficiency of a two-bladed two-tier rotor on traverses with solidity 0.315 and b_t/b = 44/63 reached C_p = 0.6 at V/U = 3.7. However, even on this large model, the introduction of horizontal

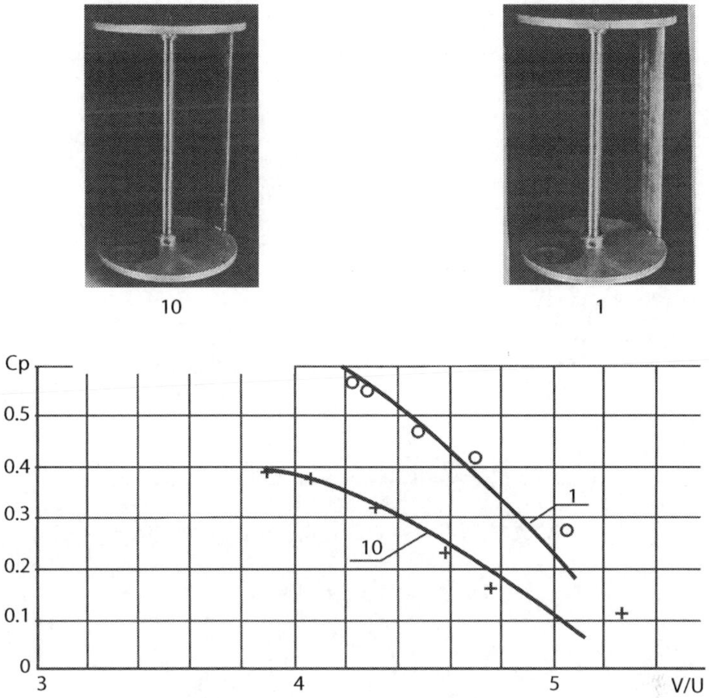

Fig. 4.27 One blade on a disk with a balancer. Chord 50 mm. (right, line 1) and 30 mm (left, line 10). D=200mm H=300mm. Measurements in a channel cross-section of 1x1 m^2. The flow of water at a speed of 1 m/s.

extensions with a diameter of only 0.3 mm in the middle of each tier halved the efficiency of the rotor (Fig. 4.29).

According to the test data in this two-bladed machine tray, the maximum efficiency at solidity 0.315, 0.265 and 0.220 was 0.6, 0.5 and 0.35, respectively. The optimal speed of the blades is from 3.5 to 3.7 flow rates. In this case, the optimized design of the two-tier rotor showed similar results in a small tray (Fig. 4.30).

High efficiency values in both cases may be associated with relatively small flow widths and depths. Full-scale experimental single-blade wind turbine (Fig. 4.31) had a rotor diameter of 7.05 m, balanced by a cylindrical counterweight streamlined at a distance of 1.8 m from the axis of rotation. The rotor via a base-brake unit was connected to a motor reducer MP2-315-15-80 with built-in the asynchronous generator 4AM132M4-U3 11 kW power, the nominal slip 0.0267, critical slide 0.175 nominal torque of 72 NM, the critical torque of 307 NM.

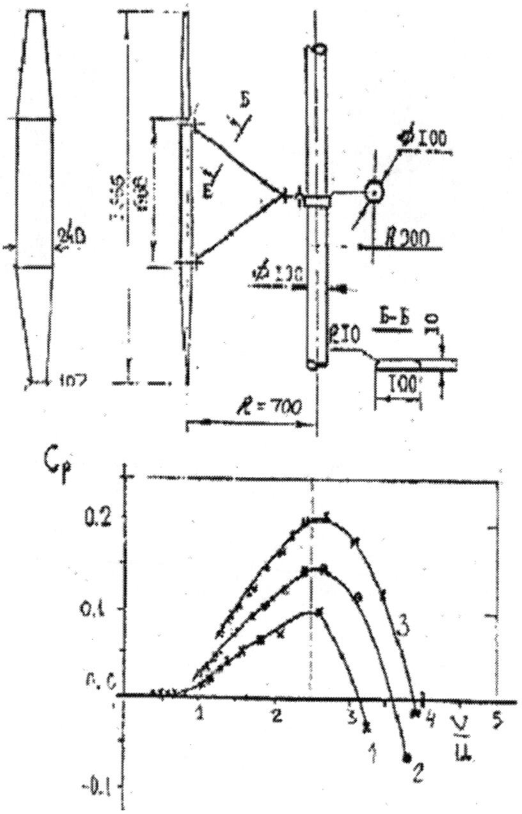

Fig. 4.28 Energy efficiency C_p as a function of blade speed (V/U) at different Reynolds numbers Re= Vb/ν. Test single-blade models with blade trapezoidal shape, the profile of GAW-1, the chord in the middle part b=240mm, the ends – 107 mm, the model radius R=700 mm Different Re. S when Re> $5.8x10^5$.

Mechanical and electrical losses in the generator-reducer system were determined on the basis of preliminary tests of the rotor-less system:

Shaft capacity	kW	0.4	4.58	7.44	10.8	14.1	18.6	24.5
El. Power at the terminals of the generator,	kW	0	3.55	6.40	9.43	12.6	16.0	20.1
Power loss,	kW	0.4	1.03	1.37	1.04	1.5	2.6	4.4

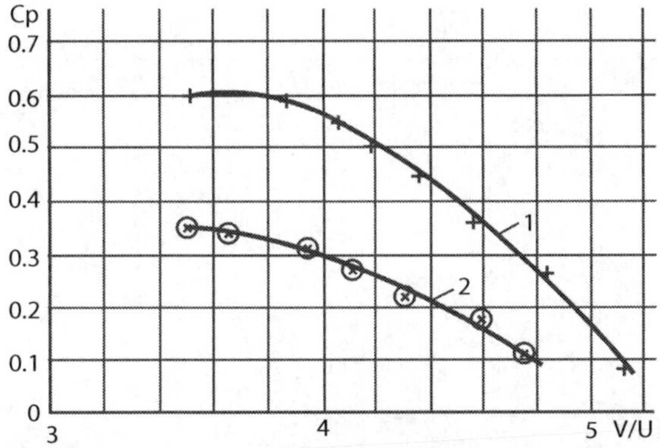

Fig. 4.29 Two-tier, two-blades rotor in a large tray (2x1.5x20 m) D=400 mm, b=63 mm, b_t=44mm, H=800mm, L=400 mm. 1 - no stretch marks 2 stretch marks with a diameter of 0.3 mm. U = 1 m/s.

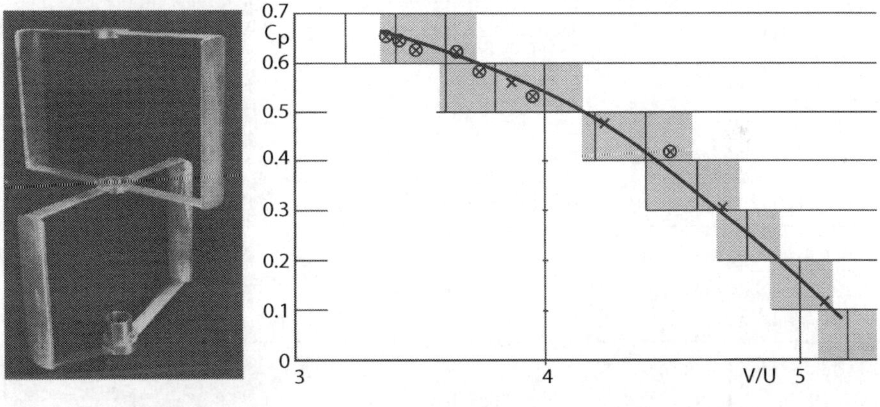

Fig. 4.30 The optimal model, tested in a tray with a width of 1 m. D=200mm, b=30mm, L/b=4.16, b_t/b=0.5, the ends of the blades are rounded along the outline of the traverse. C_p=0.661, V/U=3.35.

The results of the tests of wind turbines in the large wind tunnel of TSAGI, the smallest dimension of the cross section which is almost twice the length of the blade, as shown in Table 4.1.

As can be seen from the Table 4.1 losses on the rotor in the absence of wind reach almost 5 kW. If these losses could be eliminated, the efficiency of the turbine would exceed 40%.

The relatively low efficiency of a single-blade rotor, which at a high Reynolds number.

Fig. 4.31 Orthogonal single-blade wind turbines to the test in the largest pipe TSAGI. The radius of the track of the blade 3.54 m, blade area 6.4 m2, the average chord 0.74 m, length of the blade is 8.6 m, the Profile NASA0021.

Table 4.1 Test results of a single-blade wind turbine in the TsAGI wind tunnel.

The wind speed U, m/s 0 3,1 6,44 7,58 8,76 8,94 10,2 11,0 11,7 11,0 10,2 9,3 8,3 7,3 10,1 10,9 11,5 12,4 13,7
The speed of the blade V, m/s 29 29 29,4 29,6 29,9 29,9 30,1 30,3 30,5 30,3 30,2 30,0 29,8 29,7 30,2 30,3 30,5 30,6 30,6
Power at the terminals, kW -5,8–4,9–3,1–0,6 1,1 2,4 4,8 6,6 8,9 6,9 4,5 2,7 0,5 -1.3 4.8 6.4 8.0 9,8 12,3
The rotor power P, kW -4,8 –3,9 –2,1 0 1,7 3,3 5, 9 7,8 10,3 8,1 5,6 3,6 1,0 -0,6 5,9 7,5 9,2 11,2 13,8
V/U 3,9 3,41 3,34 2,95 2,75 2,61 2,75 2,96 3,22 3,59 4,07 2,99 2,78 2,65 2,47 2,23
Efficiency, CP 0,068,127,153,160,177,167,146,122,05 - ,158,160,168,162,148

Re= 1.9 10^6 was almost the same as that of the GAW-1 wing model with Reynolds number three times lower (Fig. 4.28) may be associated with a low solidity of the rotor ($\sigma = 0.11$), as well as with a large aerodynamic drag made of duralumin with a vertical strut (Fig. 4.31), and the counterweight.

For testing of wind turbines of greater capacity was produced other blades trapezoidal shape (Fig. 4.32) with a length of the central part of 3760 mm and a width of 900 mm, the distance between the axes of the supports 3600 mm, the total length of the blade taking into account the end fillets 8600 mm, excluding fillets-8400 mm.

The working area of the blade-6.4 m2, the average length of the chord profile 0.76 m. Blade profile was adopted symmetrical, close to the profile NASA0021. The blades were made of plastic reinforced with fiberglass, laying out in special forms with subsequent heat treatment. The surface of the blades came out very smooth. Weight -78 kg. Blades after manufacturing at the plant were subjected to static loadings, immaterialism and aerodynamic loading in separated and unseparated flow, and the action of centrifugal forces. At a given calculated total load on the blade 2800 kg plastic (residual) deformation was not observed at loads of 1.5 and 2 times higher than the calculated.

With such blades, a wind turbine was designed, manufactured and tested in the same wind tunnel, allowing the use of two- and four-bladed rotors with a diameter of 9m or more. This traverse of the rotor was made of steel, with no intermediate struts with external 465x88mm^2 section with a round leading edge and smooth narrowing of the bottom edge (the power part of the traverse had a section 234x88mm^2 in cross-sectional area of the metal is 37 cm^2, and the moments of inertia and 284 2389 cm^4). Traverse the axis of rotation was attached to the cross, planted through the support of the rotor on the axis of the multiplier MP2-500-13-80-00 with a ratio

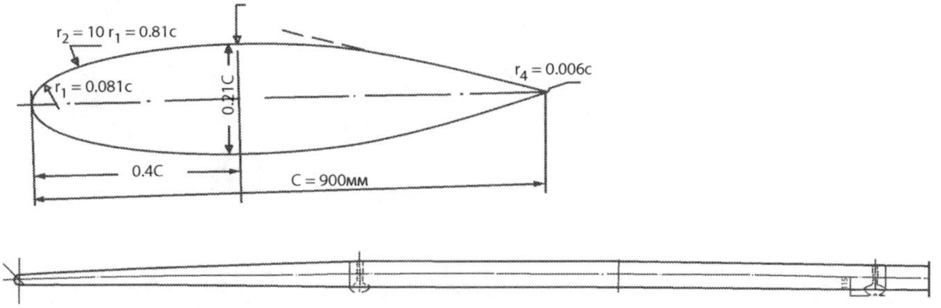

Fig. 4.32 Blade for head samples of large wind turbines tested in a large TsAGI pipe. Profile NASA0021.

of 18.6. As a generator used asynchronous motor 4AM250 8/6 U3 with a rated power of 40 and 50 kW at speeds of 750 and 1000 rpm. Nominal slip 0.0133 and 0.015, critical slip 0.0449 and 0.0515. Critical torques in generator mode 1017 and 966 NM. The weight of the generator, 510kg. Total mass energy of the node along with the braking system 1684кг. The rotor crosshead was arranged so that it could be fixed to another set of two blades shifted in plan by 45 degrees from the previous pair. The two-speed generator is allowed to carry out tests at two speeds of rotation of the rotor– 38.8 +/-0.1 and 51.6+/- 0.1 rpm (speed of blades 18.2-18.3 and 25.3-25.4 m/sec). Previously, it was established that the idling loss of the motor generator is at a low speed of 1.2 to 1.7 kW, at a high speed of 1.4 to 1.9 kW. Power consumption during operation of the motor with reduction were, respectively, 1.79, and 2.52 kW. These values of power losses were added to the power measured at the terminals of the generator when calculating the power of the rotor. The results of measuring the active power at the terminals of the generator, carried out by different methods - the registration of instantaneous values of the current vector and the energy meter, which gave similar results, showed that at low rotational speed, the two-blade machine enters the mode of issuing power at lower flow rates than the four-blade with a rotor of the same diameter at the same rotational speed (Fig. 4.14). Two-bladed machine proved more effective. Increasing the frequency of rotation dramatically increases losses in the power node, so even in a two-bladed machine power output begins only at a wind speed greater than 8 m/s (Fig. 4.33).

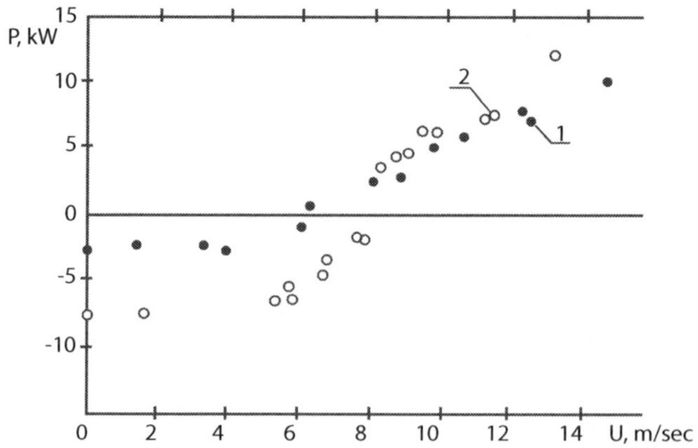

Fig. 4.33 Capacity at the terminals of the generator two-bladed machine with a small (1) and large (2) rotation speeds.

The output power on all tested machines of orthogonal pulses had a frequency of twice the frequency of rotation. The number of blades affects only the shape of the current curves – the four-blade machine changes the current much smoother and noticeably closer to the sine wave (with some constant component!), than, for example, a two-blade machine (Fig. 4.34).

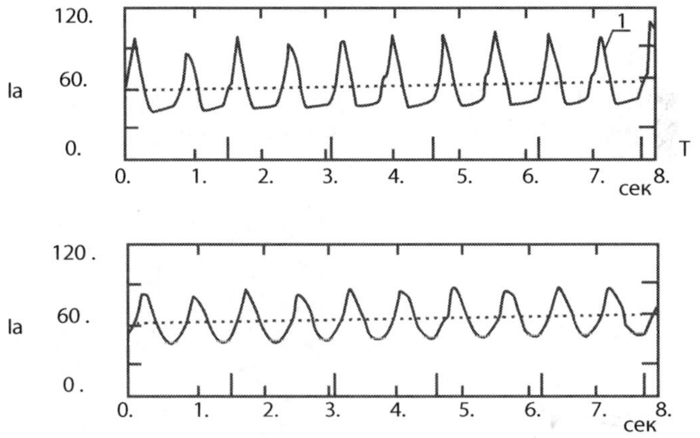

Fig. 4.34 Change of current in one of the generator phases in time at two-blade (1) and four-blade (2) wind turbines at the same rotational speed (38.8-39 rpm) 1-U =14.7 m/s, P = 11.4 kW, 2-U =13.1 m/s, P = 12.9 kW.

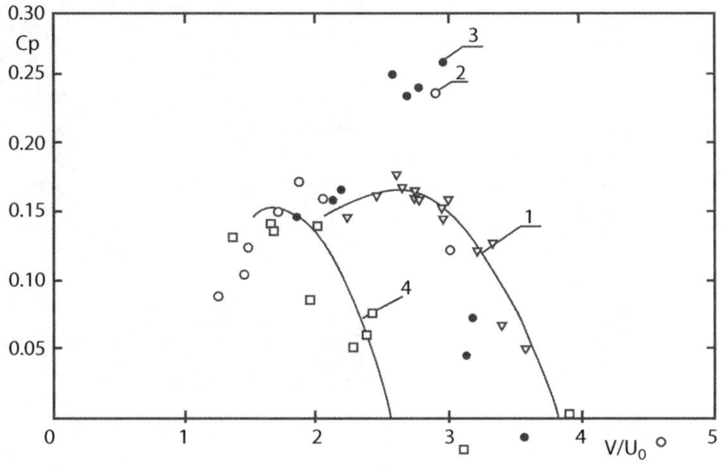

Fig. 4.35 Energy efficiency of the rotors: 1-single-blade rotor, 2, 3-double-blade rotor at low (2) and high (3) rotational speeds, 4 – four-blade rotor at low rotational speed.

Since the solidity of the two-bladed machine σ = 0.164 is closer to the optimal, the efficiency of its C_p rotor was higher than that of the single-bladed machine (Fig. 4.35).

Based on the results obtained, a two-tier wind turbine with two blades in each tier was manufactured with the same power equipment (Fig. 4.36). Its tests were not conducted. After a small reconstruction-optimizing the angle of rotation of the blades and increasing the rigidity of the traverse with improving their flowability, this installation can be used as a demonstration network machine.

In connection with the above, it is important to reduce the number of traverses, using, for example, single-blade balanced helical turbines in cantilever or double-rotor design. The aerodynamic quality of the helical blade

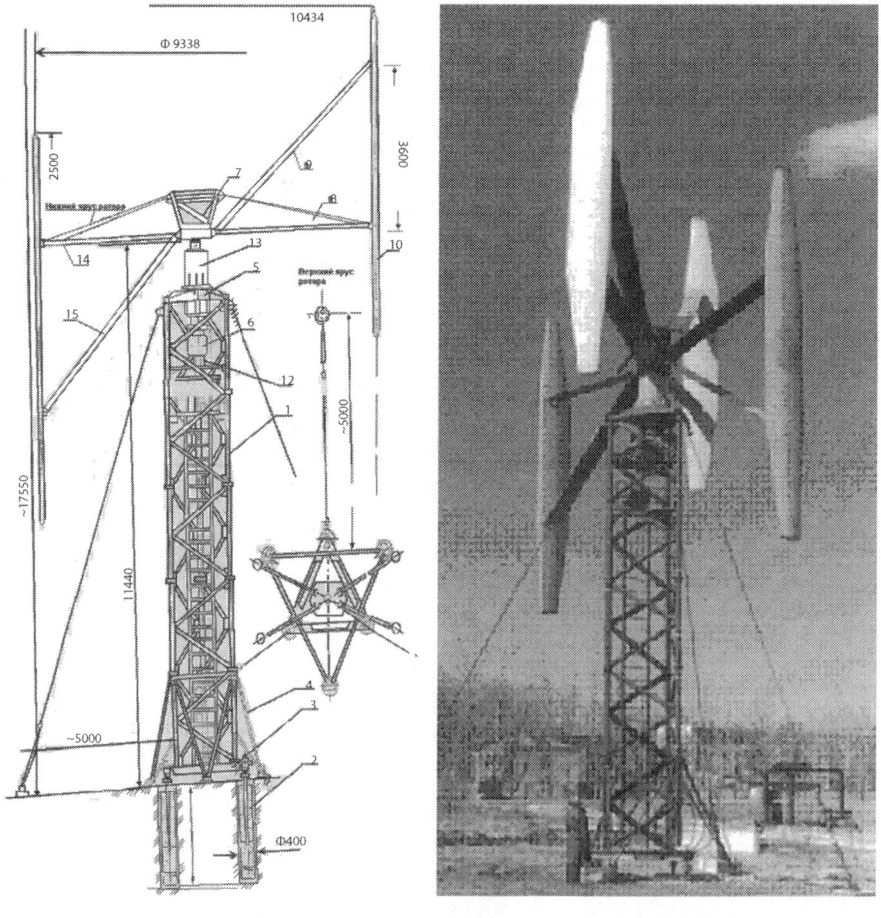

Fig. 4.36 Network wind turbine with a gear-motor 50 kW. Two tiers with two blades each.

may be maintained, for example, by arranging thin protrusions perpendicular to the axis of rotation every 60^0 along the axis of the turbine.

Reducing the number of traverses can be achieved by placing them not at the ends, and against the midpoints of the working blades. This, however, can reduce the aerodynamic quality of the blades. The pulsation of the bending moment in the support is eliminated when the support is located in the center of the turbine (Fig. 4.37).

Of course, six tiers of a single-blade turbine can be replaced by three tiers of a double-blade turbine or two tiers of a three-blade turbine arranged in the same scheme. In this case, the blades of the upper and lower turbines can be oriented in the opposite way, which will ensure the counter rotation of these turbines. This design will allow the use of a generator with double the speed of intersection of magnetic fields and improved economic performance.

According to the scheme of Fig. 4.37 turbines are expected at the lowest cost per unit of capacity and output. The installed capacity is determined by the dimensions of the blades and the diameter of the turbine. At the cheap production of blades available in the Russian Federation with chord of 160mm or 180mm (heavier, stronger and expensive) turbines under the

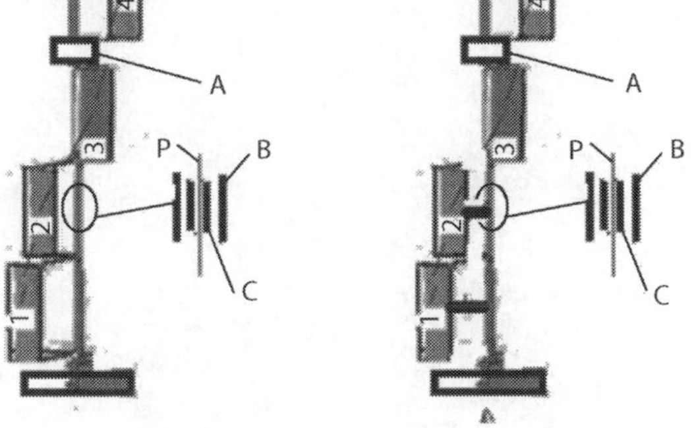

Fig. 4.37 Balanced six-tier single-blade turbine. A-support-generator fixed unit, B - rotating shaft of the turbine, carrying the blades, C-fixed support stand, bearing the support-generator unit a, P-power output cable and turbine control. Right – scheme of the cantilever for supporting the blades.

scheme Fig. 4.36 may have a capacity of several kilowatts. The transition to the dimensions of the wings of modern aircraft allows us to move to large wind turbines, focused on the use in the power system. The power of such machines may be of the order of megawatts.

4.4 Double-Acting Turbine

Any version of a high-speed orthogonal turbine has poor starting characteristics-turbines have a low starting torque, do not overcome the deceleration of rest in the supports and the generator and are not unwound from a stationary state. For machines operating in the power system, it does not matter – they are untwisted by their generator in the engine mode. In small machines with an isolated energy consumer for the promotion of the turbine, a starting system is used in the form of a Savonius unit (Fig. 4.38).

Fig. 4.38 Orthogonal wind turbine with acceleration turbine Savonius on the axis.

In the modern author's design, the acceleration system is performed in the form of a spiral with a radius of 2÷3 times smaller than the radius of the working blade. These two helices allow you to run hard-balanced designs with no central shaft – a console or frame, on two semi-shafts (Fig. 4.39).

A cantilever version of the model of a wind turbine with a double-acting turbine in a small wind tunnel is shown in Fig. 4.40.

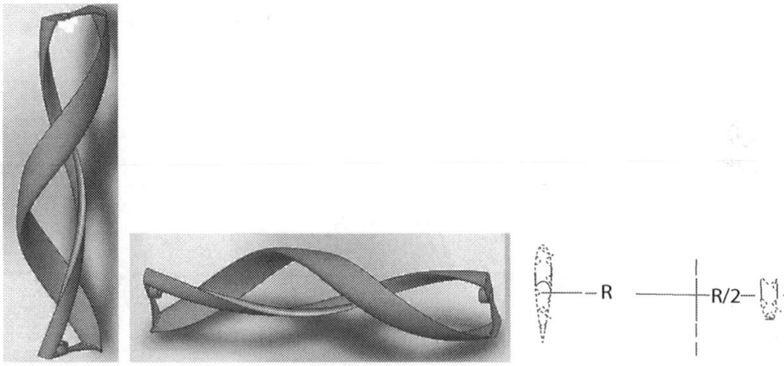

Fig. 4.39 Balanced helical turbine with one working and one acceleration blades.

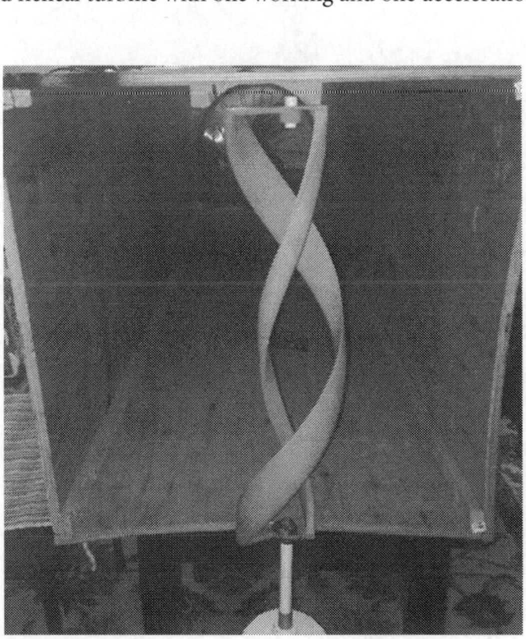

Fig. 4.40 The console model of wind turbine with turbine double action.

Appropriate choice of mass of the upper stage of the blades and installed on each of the blades are structurally associated thin spacer plates to prevent longitudinal (along the axis) flow, the turbine makes tough, durable and, probably, highly effective (Fig. 4.41).

Turbines can be located on a magnetic suspension combined with the generator. When using a turbine in the power system, complete, for example, with an asynchronous motor-generator, when it is possible to stable operation of the turbine at any point of its characteristics, the acceleration blade is not required and the turbine takes a particularly simple form (Fig. 4.42).

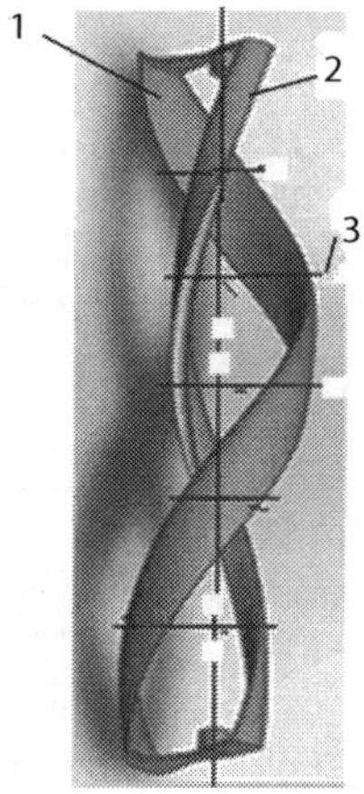

Fig. 4.41 Helical turbine double action with constructive ties between the blades. 1 - working blade, 2 - blade overclocking, 3 - thin flat constructive communication, preventing longitudinal shifting.

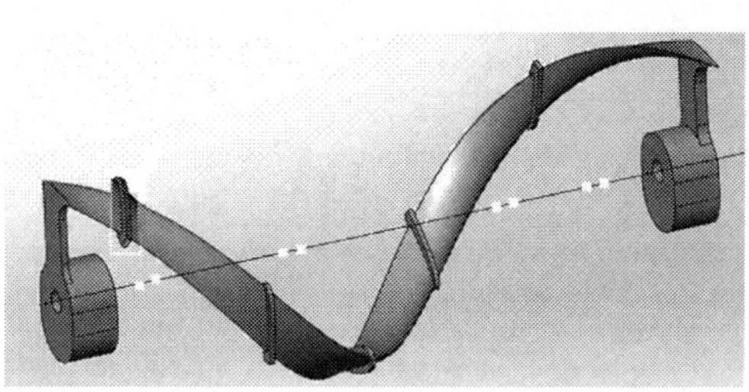

Fig. 4.42 Helical turbine assembly for a network with asynchronous generators. Thin protrusions perpendicular to the turbine axis are visible on the blades, preventing the flow in the boundary layer.

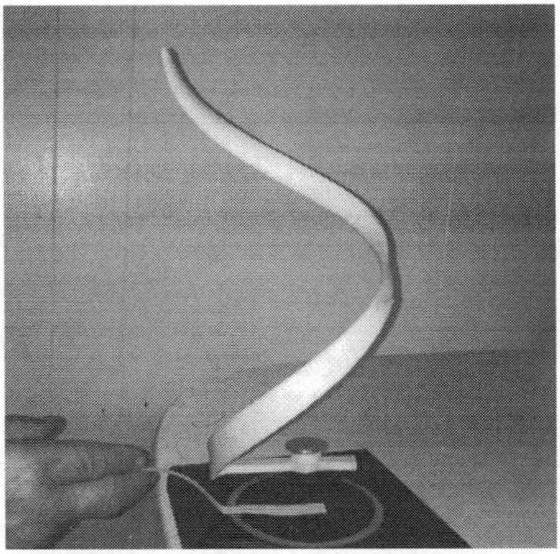

Fig. 4.43 Spiral turbine model on magnetic suspension.

In this case, the inertial only require a balancing of the traverse. Balanced cantilever turbine without an acceleration blade (for power system), located on a magnetic suspension (Fig. 4.43) can be highly effective.

However, under the action of the flow it will vibrate-precession. Therefore, it may be advisable to do large power units in the frame according to the double-bearing scheme (Fig. 4.1 to the right, Fig. 4.2).

4.5 Many Blades Turbines with Large Diameter and Control Position of Blades

More promising for units of large capacity seem to be a scheme with many blades (more than 2-3), or placed on a separate traverse or (better) on the single ring, suspended on a central pole or supported on a platform via magnetic suspension. As a result of the vertical turbulent exchange, the wind speed in front of the rear structure of the blades turns out to be the same as in front of the front (Fig. 4.44).

Since the velocity of the incoming flow at different points of the blade route is different, it is an idea to change the position (reversal) of the blades at different points of the route [8] or to jet into the boundary layer of the blade. It is estimated [8], the rotation of the blade to a different angle at different points of the track allows you to raise the efficiency of turbines (Fig. 4.45).

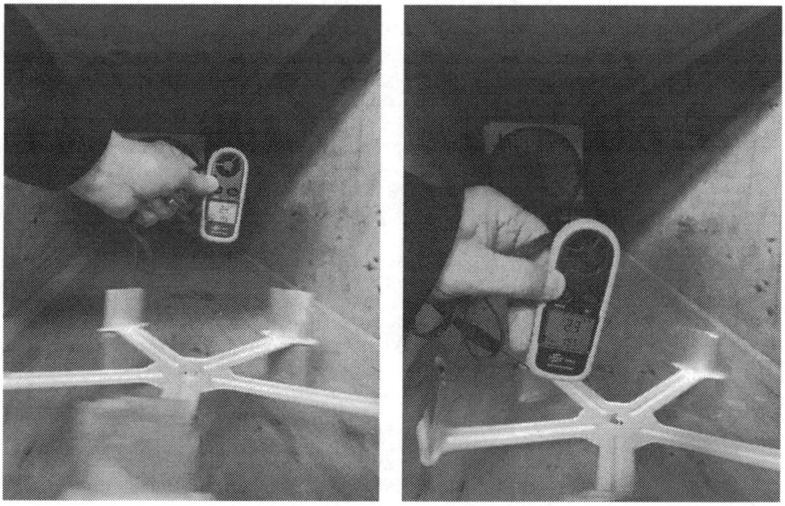

Fig. 4.44 The wind speed outside the multi-blade turbine (left U=2.4 m/s) is almost the same as in front of the rear line of the blades (right U=2.3 m/s) in the turbine with 5 blades whose height (from the support to the end) is 10 times smaller than the diameter of the turbine.

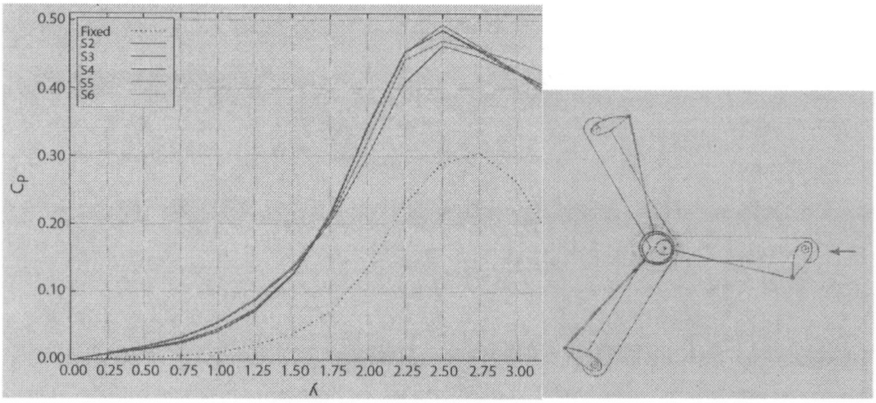

Fig. 4.45 Efficiency of an orthogonal turbine with shading 0.3÷0.4 at fixed blades (dotted line) and at various schemes of cyclic control of blades rotation (B. Kirke [8]).

According to the data published in 2017 [15], the increase in the maximum efficiency of the turbine due to the optimal rotation of the blades can reach about 7%. There was a proposal of the author (A. S. USSR 1325186 from 05.03.85) to do a straight rotatable blade such that its center of gravity was located slightly farther from the toe of the profile than its aerodynamic center. At the same point is the axis of rotation of the blade (Fig. 4.46).

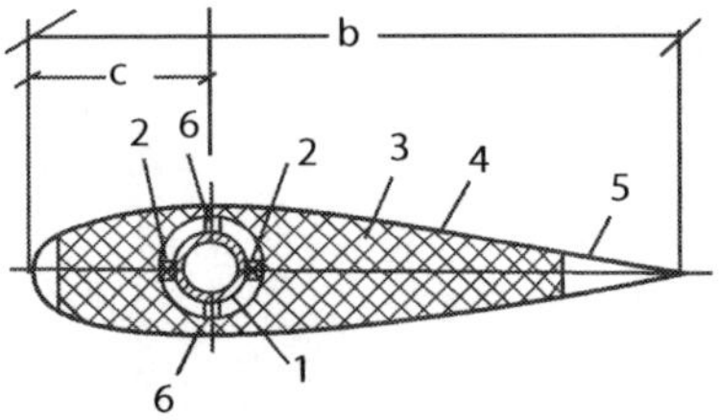

Fig. 4.46 A cross-section of the blade is optimal. The rotary pen (3) of the blade with the frame (4) and the skin (5) can rotate relative to the spar (1), with the center combined with the center of gravity of the blade, located at a distance ¼ b<c<1/2bot of the blade wear, on bearings (2) and elastic elements (6). Elastic elements begin to work if the profile deviates from the nominal direction - 50 socks outward from the tangential to the route of the aerodynamic center.

The centrifugal force acting on the blade does not affect the position of the blade, which unfolds until the aerodynamic center falls into the center of the spar, which in the design is chosen to meet the maximum pulling force of the profile (the critical angle of attack). The aerodynamic center of the NACA 0018 profile at subcritical angles of attack is almost exactly at a distance of 0.25 b from the tip of the blade. So the center turns of a blade should be positioned at a distance of, for example, 0.26 to 0.27 b from the tip of the blade. The action of the resultant aerodynamic (hydrodynamic) force on this point at an angle of attack less critical, a little kompensaci elastic reactions in the supports (6). As the angle of attack of the incoming flow increases above the critical center, the force moves to the middle of the profile. The blade rotates under the influence of the rising moment and the angle of attack decreases with the movement of the aerodynamic center closer to the profile wear and the wing tends to return to the previous position. When moving the blade on the highway Aero (hydro)dynamic force will rotate the blade – it will "scour" about the position corresponding to the maximum pulling force.

This situation occurs only in those sections of the route where the initial angle of rotation of the blade with respect to the tangent to the track of the blade is less than the angle between the vector of the relative velocity of the blade flow and the tangent to the track of the blade (Fig. 4.47). For rice.46 it can be seen that in the overclocked state, when the speed of the blade is much higher than the speed of the oncoming flow, a significant part of the blade route passes with a negative angle of attack (the entire route in the right figure and the entire rear part of the route in the left figure), regardless of the initial. It is on these sections of the route that normalna occurs to the chord of the blade force, the force to deploy the blade in the direction of

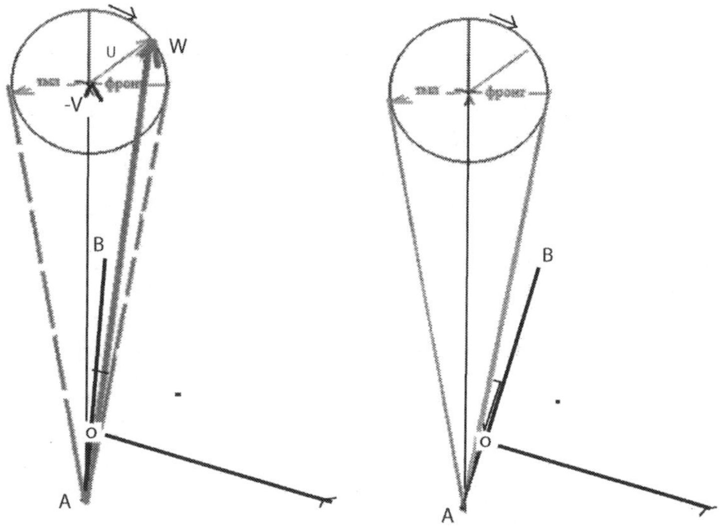

Fig. 4.47 The scheme of flow around the working blade AB, the sock of which (a) moves at a speed **V** in the flow at a speed **U**. the relative speed of the flow around the sock of the blade **W=U-V**. on the Left – the picture with a small deviation from the chord of the blade tangential to the track of the sock, on the right – with a significant deviation.

increasing the modulus of the angle of attack until the occurrence of separation. After the separation occurs, the normal force moves to the center of the blade and leads it to the initial state. The left Figs 4.46 shows that on the front section of the route the equal forces from the stream at small positive angles of attack, located at a distance of b/4 from the wear of the blade (closer to the wear than the support of the blade O), tends to expand the blade in the direction of reducing the angle of attack. This turning moment is small (the shoulder of the force equal to the length of the OA-1/4b segment is small) and is perceived by the elastic element of the support. Thus, the pulsation of the blade and the corresponding increase in the angles of attack and pulling forces is expected, mainly, on the rear section of the route.

This proposal, of course, requires experimental verification and optimization. Jet circulation control also allows you to raise the efficiency of turbines [9, 10]. All these techniques in the conditions of unlimited two-dimensional (flat) flow seem doubtful – the increase in the energy consumption of the flow causes its additional braking – the flow leaves the turbine, without significantly increasing its overall energy efficiency. Available experimental "proofs" can be connected with the inevitable distorting influence of limitation of cross-section of the experimental channel. A different situation may occur in large-diameter turbines (D/L>>1), where the role of transverse turbulent transport, which restores the energy of the flow in front of the rear structure of the blades, is essential [14]. Under these conditions, and under conditions of limited flow, control of the flow around the blades can give quite a noticeable result. In conjunction with the proposal for Fig. 4.45 such a turbine can be highly effective.

Systematic studies of orthogonal power units were conducted in Siberian research institutes with experiments in the hydro channel of the Novosibirsk Academy of water transport. The channel is 70 m long, 6 m wide, and up to 3.5 m deep. In the experiments, the water depth was 1.3 m. the Models were fixed in a frame with a cross section of 0.7x0. 6 m^2, which moved along the channel on a cart in standing water at a speed of 0.76 m/s. An electric drive is mounted on the upper part of the frame, which ensures stable rotation of the shaft at a frequency of 12 to 72 rpm. The measured power consumed by the engine at a fixed speed characterizes the torque acting on the turbine. The diameter of the turbine shaft was 2 cm, three blades 0.45 m long each were fixed between disks with a diameter of 0.65m. The profile of the blades is symmetrical with a relative thickness of 0.18. Set of blades was selected with the chords b=3, 4.5, 6, 8 cm, which corresponds to a solidity of 0.15; 0.225; 0.3; 0.4. Before testing rotors with blades, rotors without blades were tested, and the power consumption ("parasitic") was determined. The power of the system, which was added to the power output of the turbine during tests with blades. An example of the test results of a two-tier rotor with 3 blades in each tier with a real traverse d effect and with the exception of this effect is shown in Fig. 4.48 (b=8.4 cm, L/b=4.05, solidity 0.42, 4^0).

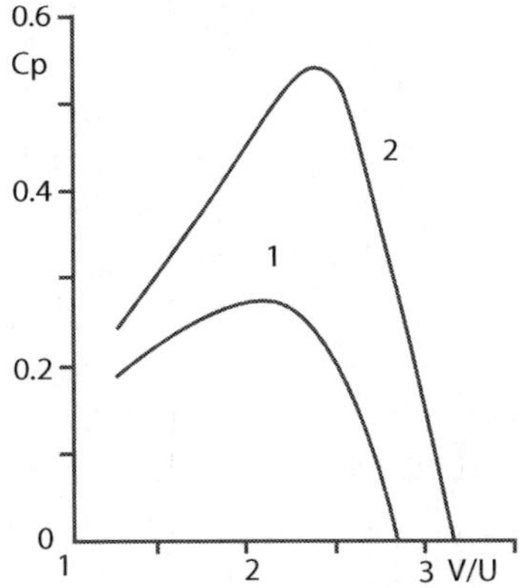

Fig. 4.48 Efficiency of the 2 stories rotor with 3 blades in each stories (1). 2-recalculate without traverces.

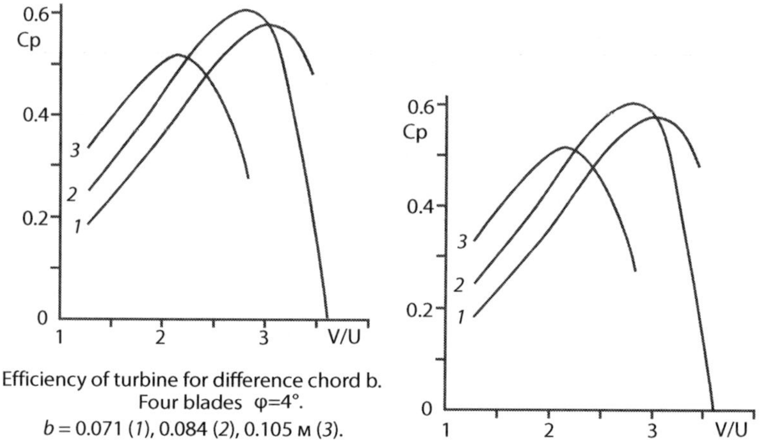

Fig. 4.49 Efficiency of 2 stories rotor with 2 blades in each stories. b=7.1 (1), 8.4 (2), 10.5 cm (3).

E-+xperimental values of the efficiency of two-tier turbines with two blades in each tier as a function of the speed of the rotor with different blade chords and a 40-degree turn (toe out) are shown in Fig. 4.49 (excluding losses on the traverse).

The maximum efficiency value obtained in experiments reached C_P = 0.72 and exceeds the limit value of 0.593 for propeller-type turbines [ ]. The results of systematic experiments presented in the reviewed publications allow optimizing the turbine for various operating conditions.

4.6 General

The General conclusion is that in the limitless flow among the turbines streamlined by a flat flow (the length of the turbine is much larger than the diameter), the most effective balanced turbines are with one blade and a shading of about 0.5 (the chord of the blade is equal to the radius of the turbine). However, such turbines for individual use (in the absence of a powerful electrical network) require the use of acceleration blades or a special motor generator. The use of multiple blades in one tier and (or) the use of multi-tier turbines partially eliminates this problem. At present, turbines with a capacity of about 1 kW (when the wind is 10m/s) with one, two or three blades in one tier are ready for mass production. The height of the turbine is 2.7 m, diameter 1.8 m. These units can advertise as real objects and as model units of megawatt power. In 1988, the author put forward and implemented the idea of using the wings of aircraft in the design of wind turbines that have worked their flight life. Detailed tests of different models of such units were carried out in TSAGI [2a, 3, 13]. The first units with capacity 130 and 1000 kW was constructed in the USSR in 1988 and 1991 (Figs. 1.23, 1.24). Experiments and calculations have revealed an important property of such a layout: at a constant speed of rotation, selected, for example, from the condition of maximum efficiency of the unit at an average annual flow of wind energy, the maximum torque (and maximum power) on the turbine shaft is observed at a wind speed exceeding the nominal one, no more than 2 times (at an unchanged, nominal speed). For high-power turbines, multi-blade machines with a diameter much larger than the height (length) of the blades are the most promising [14]. The project of such a machine with a capacity of 20 MW, performed in 2002, showed its high economic efficiency.

References

1. Darrieus, G., Turbine Having its Rotating Shaft Transverse to the Flow of Current, U.S.Patent 1834018, December 1931.
2. Victor Lyatkher. *WindPower. Turbine Design, Selection, and Optimization*, Wiley, 2014, 311p.
2a. Лятхер В.М. Испытания головных образцов ортогональных ветроагрегатов, ж.Гидротехническое Строительство, №3, 2002, ср.31-39.
3. Баклушин П.Г., Вашкевич К.П., Самсонов В.В., Экспериментальное исследование аэродинамических характеристик ортогональных крыльчатых ветроколес, Сборник Научных Трудов Гидропроекта, вып.129, Ветроэнергетические станции, М. 1988, стр.98-105.
4. McGbee R., Reasley W. Low Speed Aerodinamic Characteristics of a 17-percent-thic Airfoil Section Designed for General Aviation Applications, NASA TN D-7428, 1973.
5. Lyatkher V. Orthogonal Power Unit, Patent US 8007235 B1, Int.Cl. F03B 3/12(2006.01), Aug. 30, 2011.
6. Lyatkher V. Orthogonal turbine having a balanced blade. Patent US 9687715 B1, Int.Cl F03D 3/00, May 23, 2017.
7. Rawlings, G.W. Parametric Characterization of an experimental vertical axis hydro turbine. BASc, University of British Columbia, 2008.
8. Brian Kirke. Tests on two small variable pitch cross flow hydrokinetic turbines *Energy for Sustainable Development*, 31 (2016) 185–193.
9. Application of Circulation Controlled Blades for Vertical Axis Wind ... M. Wilhelm J.P. Angle G.M. Smith J.E ... McGrain D. Angle G.M. Wilhelm J.P.

Pertl E.D ...Article (PDF Available) in *Energies* 6(8):3744-3763, August 2013.
10. Andrew Shires and Velissarios Kourkoulis, Application of Circulation Controlled Blades for Vertical Axis Wind Turbines,*Energies***2013**, *6*(8), 3744-3763. ww.mdpi.com/journal/energies Article Application of Circulation Controlled Blades for Vertical Axis Wind Turbines.
11. В.М.Лятхер, Ю.Б..Шполянский, .В.Самсонов и др. Влияние турбуленции ветра на ветротурбины. ИзвестияАкадемииНаукСССР, Энергетика и транспорт, N°2, 1990.
12. В.В. Самсонов, П.Г.Баклушин, А.И.Дмитриев, А.В.Зосимов Вертикально осевая ветротурбина со струйным закрылком. ПромышленнаяАэроди нвмика, N°4 (36), 1990.
13. V. Samsonov, P.Baklushin, Comparison of different ways for VAWT aerodynamic control. *Journal of Wind Engineering and Industrial Aerodynamics*. Vol..39, N°1-3, May 1992.
14. V.Lyatkher,.Orthogonal units to use energy of flows with the highest efficiency and power. *International Scientific Journal for Alternative Energy and Ecology (ISJAEE)*, 2017.
15. A.Rezaeiha, I.Kalkman, B.Blocken, *Applied Energy*, 2017, (132-150)
16. Д.Н.Горелов. Аэродинамика ветроколес с вертикальной осью вращения, Омск, 2012. 66 стр.

5

Turbines with Transverse Turbulent Energy Transfer

Orthogonal windmills and hydro units of big capacity and high power efficiency are offered. Units can have rather big individual capacity and at the solution of a number of technical problems should provide high efficiency of power. In orthogonal power installations, at which diameter of the route of movement of blades much more than their height (length), the flow speed before the rear blades may be the same as in front of the unit. The test and calculations confirm it.

5.1 Introduction

Working elements of the systems transforming energy of currents (flow kinetic turbines) usually move on the closed routes [1]:

- some surfaces whose axis of symmetry can be parallel (collinear) to flow,
- or is perpendicular (orthogonally) to a stream.

Respectively we will allocate two groups of power units –

- Collinear at which the axis of symmetry is approximately parallel to a stream (HAWT – named in USA practice); and
- Orthogonal at which the axis of symmetry lies to the direction of a stream (VAWT – named in USA practice).

Elements of units (blades) can move more slowly or quicker than a stream on the way to the system. Accordingly, then, the units are called low-speed (active) or high-speed (reactive) turbines. Collinear, high-speed

Victor M. Lyatkher and Ziaur Rahman. Wave, Wind and Current Power Generation, (195–216) © 2022 Scrivener Publishing LLC

units in the turbines of today have the greatest distribution. Orthogonal units can have a horizontal or vertical axis.

When comparing various types of power aggregates – even in one class – certain difficulties emerge. This is owing to the fact that it is essential to take into account the differing criteria that are to be measured in different units. Such criteria consist of the following: specific material capacity (per unit of capacity and development), quality of the developed energy, degree of simplicity and factory readiness, convenience and reliability in operation, labor input of a construction and operation, and ecological purity. Quality comparison is especially difficult for machines of various classes. Therefore, the choice of each kind of scheme is considerably conditional. However, indicators of the economic, energy and ecological factor that are presently taken into consideration may well remain relevant even in the years to come.

Factoring in the speed of sound waves in the approach to measuring the speeds of bodies in motion yields a sharp increase in the values for the aerodynamic resistance and acoustic radiation. Consequently, the elements of windmills are designed in such a manner as to allow them to spin in a circular motion and produce significantly lower sound speeds (with low Mach numbers). It is to be noted here that air can be considered to be almost incompressible; this leads to results of aerodynamic calculations derived by accounting for traffic conditions of incompressible liquids qualifying for recalculation against incompressible liquids of varying densities. In this manner, all results can be directly used for the analysis of river or oceanic power stations with similar configurations. At the same time, it must be noted that, for quiet water flows with small numbers of Froude, the role of a free surface of a flow is somewhat close to a role of a hard smooth wall [2]. Therefore an analogy between land wind and river (oceanic) installations can be doubl in case of an arrangement of aggregates on a water surface or at the bottom of a reservoir. Of course, designs of nodes, the applied materials, strength and technical and economic estimates for power aggregates in air and water can be significantly varied.

In any scheme of selection of power at a stream, it makes sense to replace as a first approximation the unit which is carrying out this process by some hypothetical flat permeable figure, the contours of which are projections of borders of a body, swept around by the rotor, on the figure plane, perpendicular to the direction of a stream on the way to the unit (Fig. 5.1).

In case of traditional, collinear units, the circle or a ring will be such a figure. For orthogonal units, it can be a strip (as in the considered model), a rectangle, a trapeze, or even an ellipse figure (unit section to Darrius) or

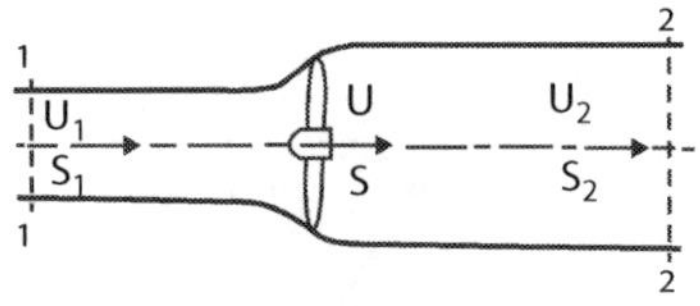

Fig. 5.1 Scheme of currents in a turbine zone.

a triangle. Whatever the shape, at a distance before it, the speed of a stream of U and pressure of p do not maintain indignant values –

$$U = U_1, p = p_0, \tag{5.1}$$

and at some distance from her downstream where distribution of pressure across a stream is leveled and comes back to a reference value of p_0, speed in the tube of current leaning on the allocated figure has a little smaller size $U_2 = (U_1 - v)$. If current speed through the allocated figure to designate U, and pressure with frontal and the back of a figure through p_1 and p_2 respectively, then Bernoulli's equations for frontal and back sites of a tube of current will have the following appearance –

$$P_1 + \rho U^2/2 = p_0 + \rho U_1^2/2 \tag{5.2}$$

$$p_2 + \rho U^2/2 = p_0 + \rho(U_1 - v)^2/2 \tag{5.3}$$

If the speed of a stream and pressure upon surfaces of the allocated tube of current were to accept constants, then from the equation e, we receive the difference of pressure on the allocated flat figure replacing the power unit

$$\Delta p = p_1 - p_2 = \rho 2(U_1 - U)U \tag{5.4}$$

here

$$U = U_1 - v/2 \text{ or } v = 2(U_1 - U) \tag{5.5}$$

The power lost by a stream and transferred to the turbine is equal to

$$P = U\Delta p\, S = \rho u(U_1 - v/2)^2\, S = \rho 2(U_1 - U)U^2\, S \tag{5.6}$$

Resistance coefficient

$$C_D = 2 \quad p/\rho U_1^2 \tag{5.7}$$

And power coefficient

$$C_P = 2N/\rho U_1^3 S \tag{5.8}$$

Is the function of only one relative value U/U_1.

$$C_D = 4\,(1 - U/U_1)U/U_1 \tag{5.9}$$

$$C_P = 4\,(1 - U/U_1)(U/U_1)^2 \tag{5.10}$$

Maximum of power will be at $U/U_1 = v/U_1 = 2/3$ and equal

$$P_{max} = 16/27 S\rho U_1^3/2, \quad (C_P)_{max} = 16/27 = 0.593 \tag{5.11}$$

The received ratios (5.4), (5.6), (5.11) usually connect with the works of Zhukovsky (1912) and A.Betz (1920), giving to these ratios a sense of certain limit laws [3]. However t the initial equations aren't exact and reflect only some model of the phenomenon. In particular, equation (5.3) is obviously not fair at $U \to 0$. In this real limit case (the impenetrable unit) p_2 pressure behind the unit will be obviously less than pressure of p_0 from the unit at a distance (5.3). However the idea of replacement of the aggregate with one or several permeable flats or other figures on which there is a gap of pressure and concentrated selection of capacity was very productive.

Let's review a characteristic example in which on the permeable flat figure modeling the power unit, the constant in all points of a figure U including speed, perpendicular plate surfaces is set. The non-stationary equations of hydrodynamics (Euler) by method of final differences are solved. The decision for a figure in the form of a strip plate (a two-dimensional non-stationary task) is various at the introduction of a condition of symmetry of a current and without introduction of such condition (Fig. 5.2). This distinction is shown in a picture of currents and distribution of pressure only behind a plate. It is essential to the definition of loadings. It is curious that though the difference in loads of a permeable plate in different

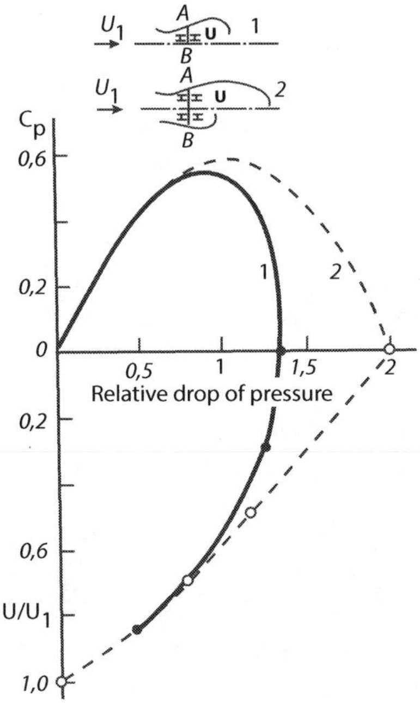

Fig. 5.2 Efficiency of the power unit C_P modelled by a flat permeable plate, depending on the relative speed of a flow U/U_1 through the plate to a hypothesis of a symmetric flow (line 1) and without any additional hypotheses (2). In calculation, the relative speed of a flow through the U/U_1 plate is set, with relative pressure difference being $2\Delta p/\rho U_1^2$ and power factor (efficiency) of being $C_P = 2\Delta pU/\rho U_1^3$ as a result.

statements is enough (especially at small U), the maximum power factor differs slightly, remaining in limits.

$$(C_P)_{max} = 0.55 \div 0.59 \tag{5.12}$$

Contrary to the Betts-Zhukovsky theory, here, the resistance coefficient of a flat, impenetrable plate - discounting the additional condition of the symmetry of current flowing behind the plate - was precisely equal to the experimental value $C_D = 2$ along the infinite length of a plate. U At a symmetric flow of a plate, the pressure declining behind it was small in number (due to the influence of the return stream); the result of the settlement also corresponds to that which was derived from rectangular plate having a long party that fell short by 15 times ($C_D = 1.38$). A pressure difference on the permeable plate modeling the power unit can be found, generalizing, for example, a task about a jet flow of a plate on a case of existence of a filtration through a plate the GGS model [4]. Coefficient of the pressure upon a permeable plate here was significantly smaller than the real experimental value. The maximum power factor 0.301 yielded a low value in the GGS model only because real pressure decline behind a plate wasn't taken into account. The filtration condition, too, formulated by the authors while isolating the problem definition through a plate actually was not used in the derivation. Models giving other estimates and the correct results at the known limit points are possible ($U = 0$ and $U = U_1$) [1].

If, for example, instead of equations (3) and (4), simply parametric representation of the coefficient of resistance of a plate were used, then

$$C_D = 2(1 - U/U_1)^m \tag{5.13}$$

The maximum power factor $(C_P)_{max} = 2m^m(1+m)^{-(m+1)}$ turns out at $U/U_1 = 1/(1+m)$ which, at

$$C_D^0 = 2 \text{ and } m = 0.5, 1.0 \text{ or } 2.0$$

yields

$$(C_P)_{max} = 0.577, 0.500 \text{ or } 0.296 \text{ for } U/U_1 = 0.667, 0.5 \text{ or } 0.333 \tag{5.14}$$

It is clear that the widely applied parameters of Betts-Zhukovsky do not result in a reliable assessment. Primitive representation (13) can be closer to reality if, instead of coefficient 2, the empirical values of coefficients of resistance of impervious plates were used, depending on a ratio of the parties:

Relation of the parties =	20:1	10:1	4:1	1:1
$(C_D)_{max}$ =	1.45	1.29	1.19	1.10

When accounting these ratios instead of estimates (14) for the turbine with cross section in the form of a rectangle with a ratio of the parties 4:1, we will receive maximum efficiency about 0.458 ÷ 0.343 ($C_{Pmax} = 0.298$) that is close to the empirically achieved results. If the same reasons to approach assessment of maximum efficiency of the collinear aggregates were replaced in the zero-dimensional consideration by a permeable round disk then, in case of resistance of an impervious disk equalling 1.25, maximum efficiency of the collinear turbine (or any axisymmetric system) can be expected to be within 0.48÷0.36. According to the GGS model mentioned above, the most achievable efficiency of free line turbines makes 0.301 at a current speed, producing in the turbine 0.613 speed from the running flow.

Neither the elementary theory, nor the more difficult calculations reflect specifics of the power unit that can be disclosed by consideration of hydro aerodynamics of the rotor which flows round with U speed. Using this, or the mathematical model of a rotor, it is possible to construct the impact of pressure difference on a rotor from U speed. Uniting this decision with the decision modeling unit, and with a permeable figure along with the explosive field of pressure, the relative speed of a stream in a zone of the U/U_1 unit and all its characteristics can be found. Such an approach was the cornerstone of many works dedicated to the research for optimum designs of power units. There are now mathematical models that produce a complex hydro aerodynamic calculation of power units by taking into account currents beyond its limits even without the use of a model of a permeable plate.

5.2 Efficiency of Ordinal VAWT – Brake of Flow Within the Aggregate

The actual efficiency of aggregates largely depends on their design and can differ considerably from the given estimates. This is evidenced by the smaller values that are also explained by the non-optimality of a design or mode and energy losses, and by the great values that can be connected with attraction of energy of the part of a flow which isn't passing through the conditional flat section of the aggregate. Such a positive effect takes place in the offered orthogonal aggregates at which the diameter of the route of movement of blades considerably exceeds the blade length. In regular orthogonal aggregates, braking of a flow within the aggregate is large, and the more the aggregate shading, the more it leads to there being a larger number of blades. Even in the single-blade optimized orthogonal wind-driven generator, the torque on a back section of the route is 2.5 times less than it is on the frontal site (Fig. 5.3).

Increase in the number of blades (i) leads to a bigger reorganization and braking of a flow. As a result the maximum power of an orthogonal hydraulic unit in the same overall dimensions and in case of the same flow rate doesn't increase, and optimum frequency of rotation decreases (Fig. 5.4), causing additional weighting (and rise in price) the generator.

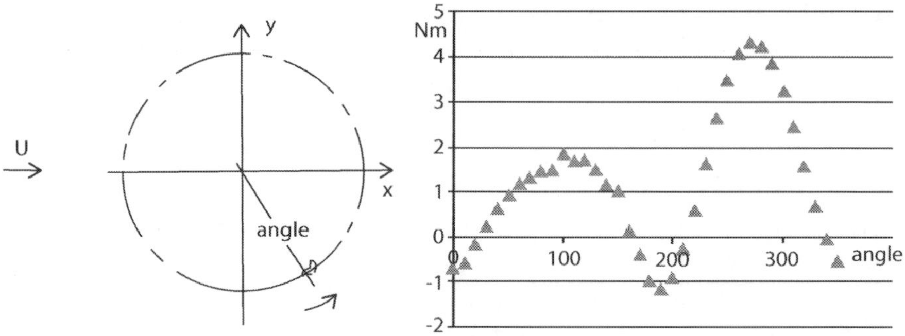

Fig. 5.3 Torque moment from one blade orthogonal wind turbine. D=1.4m, chord b = 0.16m, angle of blade 3 grad, GAW1, length L = 0.45m, two traverses (hands) b_t = 0.16m, thickness 0.01m. Wind speed U=10 m/s, frequency of rotation 500 rpm. Tip speed V/U=3.66. Calculation by method of discrete vortices [5] with experimental results for profile GAW1 [6].

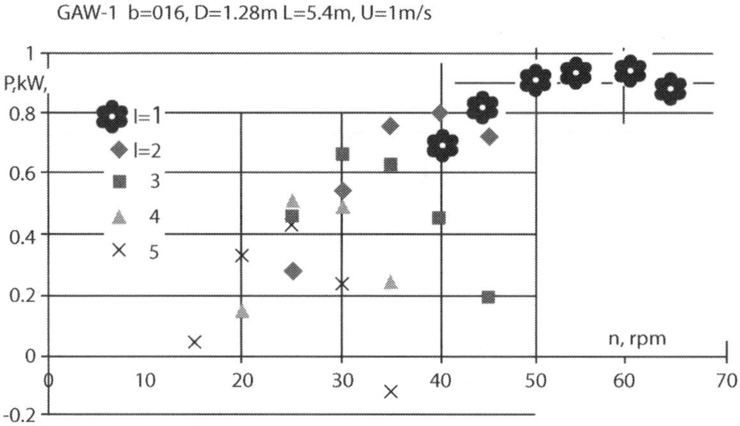

Fig. 5.4 Power of straight blade hydro turbines with length L = 5.4m, Blades GAW-1, b = 0.16m, $\alpha = 0 \div 2^0$. D = 1.28m, Water speed 1 m/s.

5.3 New Design with Turbulent Vertical Mixing of Streams

The idea of the offered installations consists of the effect of the turbulent vertical mixing of streams providing restoration of energy of a stream on the way to a back system of blades (Fig. 5.5). If the supporting framework is lifted above the ground, then the arrangement of units one under another with oncoming traffic of blades and the two sides bilateral generator

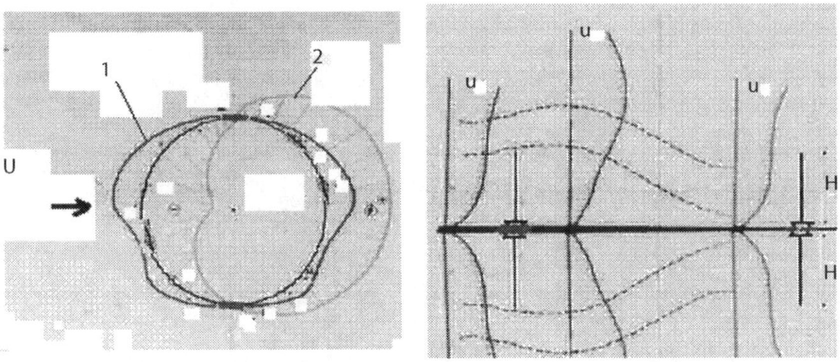

Fig. 5.5 Distribution of loadings on the route of blades (at the left) and the scheme of restoration of a stream before a back system of blades due to vertical turbulent exchange. 1) the pulling force, 2) normal pressure. V = 3U.

(as shown in Fig. 5.5 right) preferably linear (arc) type with a magnetic levitation is possible.

Strict equality of speeds of wind on all sections of the route in the presence of selection of a wind energy is impossible. However, the more the relative diameter of the aggregate D/L, the higher possible efficiency of the aggregate. In case of the aspiration to increase the number of blades in such aggregate that in proportion it shall increase its capacity, it is necessary to consider the possible mutual influence of the neighbor blades connected with length of a chord of the blade "b" and distance between blades "B". The distance between blades "B" should be such that zones "E", where relative indignations of the initial field of speed there is more set small size weren't crossed (for example, 2%). Borders of such zones depend on a local angle of attack under which the aggregate wing is doomed (Fig. 5.6).

For good aerodynamic profiles (for example, NACA 0015) the maximum pulling force at the large numbers of Reynolds characteristic of future large units, will take place at angles of attack about 15^0. For these conditions distance **B** between blades in the offered units has to be not less **11b**, and solidity

$$\sigma = ib/D < 0.28. \qquad (5.15)$$

This assessment can be clarified by considering the problem of a flat flow system of the blades, as was done to obtain the results of Fig. 5.4.

Program with discrete vortices was a calculated plane (two-dimensional) unsteady flow around a rotor **diameter of D = 50m with blades of length**

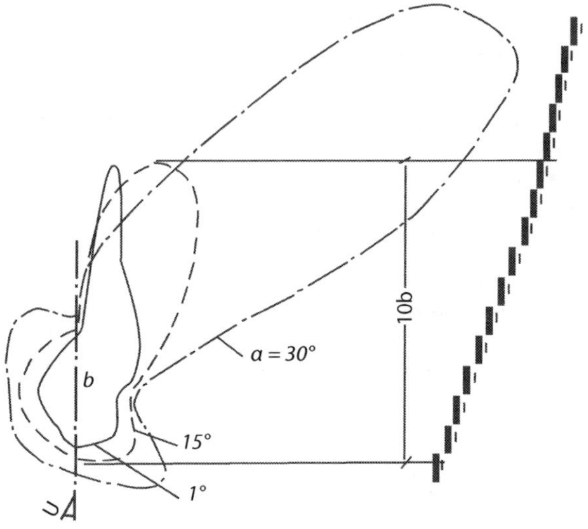

Fig. 5.6 Zones of 2% of distortion of the field of speeds at different angle of attack on plate.

L = 3m, which had a profile NASA-0021, chord **b = 1 m**. In practice may be used two the same rotors, moving in opposite direction, with one linear generator between its (Fig. 5.7).

The wind speed was adopted at 10 m/s, friction losses in the bearing nodes were not considered. The power of the rotor with one blade at the location of maximum chord tangent to the track of the aerodynamic center and increases monotonically with increase of speed of the blade (Fig. 5.7). Under the terms of the limitations of acoustic radiation to increase the speed of the blade in excess of 60-70 m/s is not reasonable. Therefore, to analyze the selected speed of the blade 68.1 m/s, which in real terms (taking into account the losses in the support system) is likely to be close to optimal. At this speed the power system with one blade may give power 12.8 kW (efficiency $C_p = 0.137$, tip speed V/U = 6.81) - Fig. 5.8.

Unit with two symmetrically arranged blades in conjunction with the increase in deceleration of flow in front of the rear part of the track, does not give a doubling of power, but in this case more efficient: the power P = 20.13 kW, $C_p = 0.215$. The variation of the radius of the track of the second blade 1m is almost not changing the characteristics of the system. It is important that the convergence of the blades with their location on the angular distance of 24^0 (1/15 part of the track), that is B = 10.5 m, and even up to 18^0 (**B = 7.8 m** between the blades) did not cause significant reduction of the average power (P = 20.08 kW, $C_p = 0.214$ at 24^0, P = 20.04 kW and the same $C_p = 0.214$ at 18^0). Noticeable mutual influence appears at an angular distance

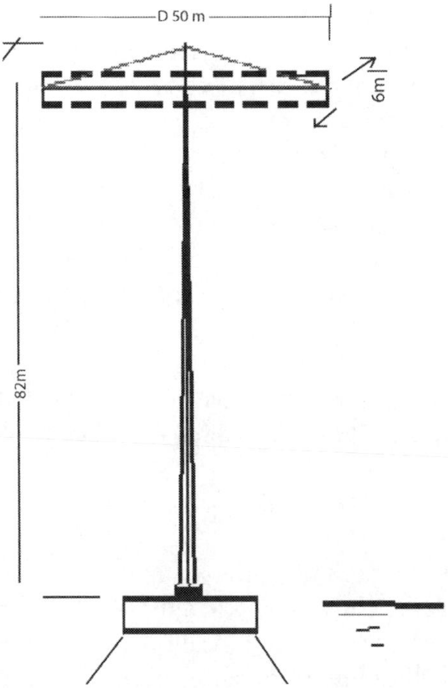

Fig. 5.7 Multiblades wind power system on the tower from usually HAWT.

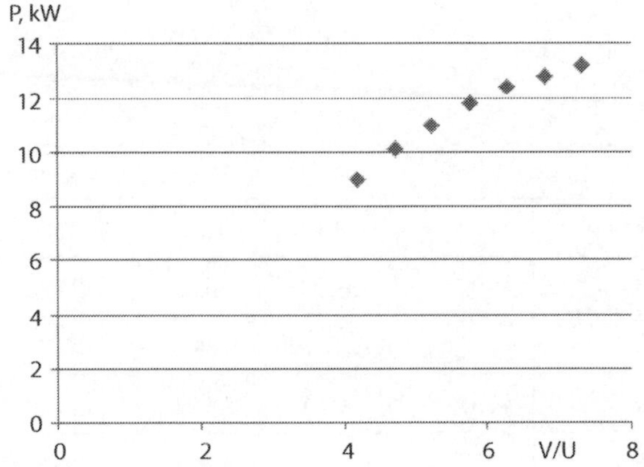

Fig. 5.8 Averaged power of windmill D=50m with one blade NASA0021, b=1m, L=3m as function of tip- speed. Wind 10 m/s.

of 9^0– P= 19.8 kW, C_p =0.21. Importantly, these differences are only on the back part of the track of the blades. Typical load (N) for a first and second blades following each other at distances of 24^0, 18^0 and 9^0, obtained in the above described example, are shown in Table 5.1 (averaged, max, min).

Based on this example, it can be assumed that when fully restored velocity field of the wind on the back part of the track (due to the vertical turbulent exchange) in this case it would be possible to place 20 or may be 40 of the blades on the rotor and to increase the power up to at least 200 (**20 blades**) or 400 (**40 blades**) kW for wind speed 10 m/s. Positioning one layer under another street and general linear generator, will receive a wind turbine with a capacity of up to 400-800 kW at a wind of 10 m/s. At wind 12 m/s (nominal as usual) nominal power may be **700-1400 kW**. As for the condition (15), then it can probably be replaced with the requirement

$$B> 10b \text{ or even } \mathbf{B>7.8b}, \text{ or may be } B>5b, \tag{5.15a}$$

The most important question of the optimal ratio between the diameter D of the track of the blades and a height (length) of the blades L remains the least certain. Hydro or aerodynamic optimization is determined by the local natural conditions(stratification, turbulence) of flow around the power system, design and mode of operation of it. For preliminary estimate s it is possible to use empirical data on the length of the vortex in the scheme of sudden expansion of the flow [1, p. 130; 2, p. 158], or flow around normal obstacles [7]. According to these data it turns out that the ratio of the diameter of the unit and height (length) of the blade should be around 20 -30 or more:

$$D/L >30 \ (?) \tag{5.16}$$

Table 5.1

	First blade			**Second blade**		
24^0 F_t	169	427	-61	130	336	-61
F_r	-436	4095	-5928	-418	3786	-5554
18^0 F_t	175	441	-61	123	320	-61
F_r	-442	4157	-6014	-442	4157	-6014
9^0 F_t	198	495	-57	98	261	-59
F_r	-469	4418	-6366	-389	3489	-5069

The first one-sided version of such a machine was realized with use of the helicopter blades and asynchronous generators located on axes of wheels on which the ring bearing blades slid (Fig. 5.9).

For similar multi-blade cars with long routes crucial importance has friction resistance in oporno - running knots and aero (hydra) - the dynamic resistance of structural elements. The role of these resistance is so big that in separate designs of large vetro agregat options of almost full refusal of mechanical losses due to use of centrifugal forces (Fig. 5.10) were considered.

Multi-blade single-tier units can be realized or with the blades installed on the separate carts connected in the ring moving on a platform (the train curtailed into a ring – Fig. 5.11), or, on the contrary, with the blades which

Fig. 5.9 The first one-sided windmill with large relative diameter. Doubki, USSR, 1987y.

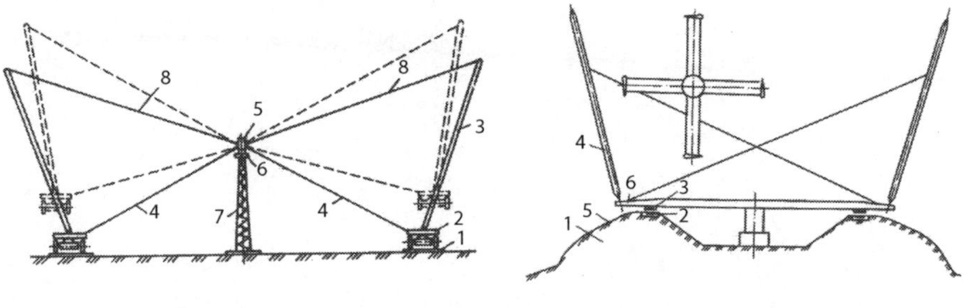

Fig. 5.10 Radical lowering of friction in support. The node (5) - at the left is mounted on a stand (7) at height exceeding heights of layout of the center of gravity of blades and carts. Centrifugal force lifts them above the ground. Author Certificate (A.C) USSR 1539380, 11.04.88. (I. I. Ilinykh, O. B. Borsky, O.N. Korolkov and V. M. Lyatkher) the Wind turbine using a land relief (1) with a low roughness (on the right) for concentration of a flow on the blades (4) leaning through traverses (6) on the support (3) combined with the linear (arc) generator (2) (V. M. Lyatkher, A.C. USSR 1156445, 04.06.82; 1240950, 07.05.84).

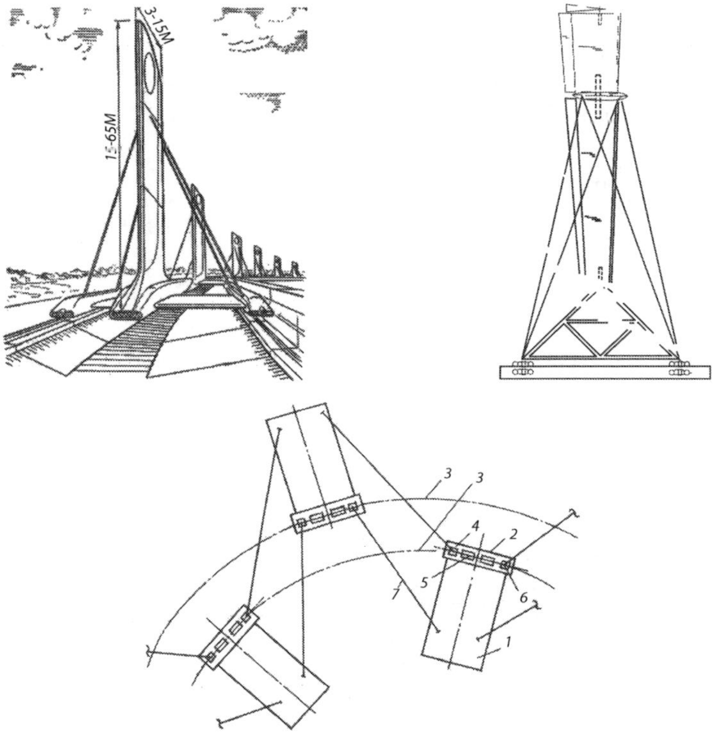

Fig. 5.11 Carts with blades move on the ring route. Above - a fragment of the patent of the USA, 4302684, November 24, 1981 (Laird B. Gogins). Below - separate carts are flexibly connected among themselves by A.C. USSR 1209919, 20.04.84 (V. M. Lyatkher).

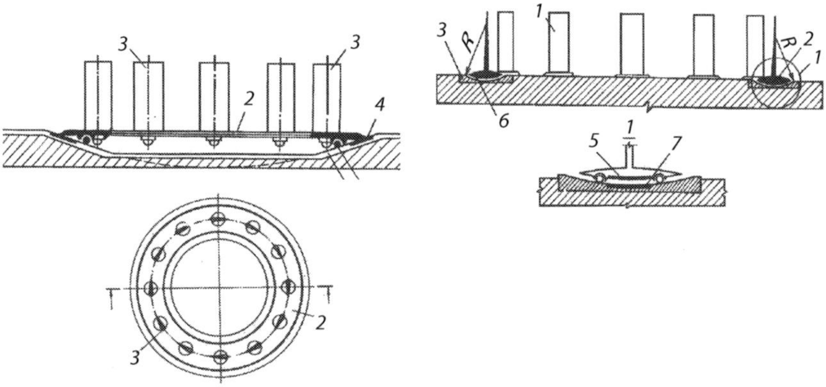

Fig. 5.12 Blades on a rigid ring, but the ring itself chooses a position of dynamic balance. A.C. USSR, 1188364, 15.02.84. Each cart is connected with neighbors, but independently chooses a balance position in a trench of the route - on the right. A.C. USSR, 1182196, 15.02.84 (V. M. Lyatkher, M. S. Brook, O.P. Bortsov).

are rigidly closed up in the uniform ring moving on motionless support (Fig. 5.12).

The large models of the same turbines were tested in aero and hydro dynamical tubes with good results – increase the number of blades in two times increase C_p in 1.5 times (Fig. 5.14)- compare with Fig. 5.4 – big difference!

The capacity of wind and hydro units of this kind can be essential more than at units of traditional execution, and specific capital investments and operational expenses for it are less. For example, according to the project executed by the author, the windmill with rated power of 10 MW has a mass of 748 tons (75 kg/kW) and relative average capacity from 19 to 43 % at mid-annual speed of a wind from 5.5 to 8.5 m/s. Development of energy of such cars can be considerably increased at the expense of jet management of an interface on blades.

The offered power installation protected by new patents of the USA and Russia (Fig. 5.15) has two established the friend over the friend with rotation possibility to the opposite sides blades turbines with a vertical axis of rotation, the electro generator and fastening system to a basic design. The unit can lean on the motionless or floating base (V. Lyatkher, patents USA 8047785 B2, Nov. 1, 2011; 7586209 B1, Sep. 8, 2009; 7566983, Jul. 28, 2009).

Active zones of the motionless inductors located between circles of turbines, have opposite directed fields running in opposite directions with identical speed, defined in the distance between phases of windings and adjustable frequency of a current in windings. The motionless inductor between turbines can be one for both short-circuited rotors of the top and bottom turbine (Figs. 5.13, 5.15) and have a uniform three-phase winding, the general for both active zones.

The scheme of an arrangement of the windings forming magnetic fields, running in opposite directions on the different parties of the inductor (below at the left on Fig. 5.16). And, In, With – entrances of windings of phases on the face-to-face party of the inductor, the X-th, Y, Z - entrances of windings of phases on the back party of the inductor, 1) the direction of movement of short-circuited rotors, 2) the direction of distribution of a magnetic field, 3) an inductor body. The scheme of a design of a basis of the inductor prepared for placement of windings (below on the right). 1, 2) the magnetic circuit motionless plates closing magnetic fields, 3, 4) aluminum plates (flat rings), moving in opposite directions, 5) an inductor body from sheets of a magnetic circuit material with grooves for laying of windings.

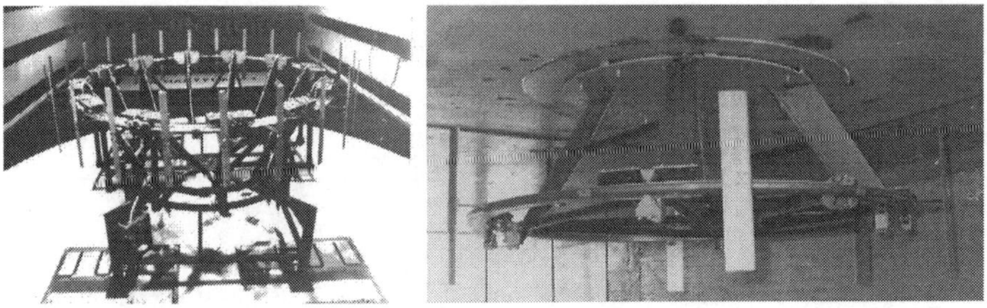

Fig. 5.13 Models of multi-blade rotors with linear generators on tests in big wind tunnel of the Dmitrovsk auto testing Company (Moscow area, Russia). The rotor with a diameter of 2330 mm, different speed of air (at the left) and in the hydraulic pressure head channel section 6x2.5m^2 the Siberian VNIIG branch (Krasnoyarsk, Russia)with a speed of water flow up to 2.2 m/s. Model diameter of the route of blades was 2.7- 3.98 m, length of blades was 0.72m, from ceiling to blades 0.733m, a chord of 0.12 m and the GAW-1 profile, diameter of ring 2.33m.

The design of the basic knot of such a unit can be the most various. On Fig. 5.17 the carried-out conservative decision with a system of basic rollers is presented. And effective use of modern bearings of sliding with air greasing can be more progressive. Basic knots of the model intended for tests in a wind tunnel, are executed in the form of cones on internal ball quality bearings with a low friction. The conicity of a basic contour was expected so that to avoid sliding in contact between a rotating ring and a support. This decision yields positive results – noise and model vibration decreases, power consumption of inductors of the linear motor-generator in dispersal mode decreases. In production tests inductors were connected according to the scheme "star" and запитаны by an alternating three-phase current to tension of 380 volts, frequency of 50 Hz. Rotors of the model rotated thus in different directions with identical frequency of 70-72 rpm. The consumed current on each phase made 3,2 amperes. Duration of rotation of rotors after food shutdown to a full stop made 22-24 seconds. For improvement of a supply of a stream to workers the blade constructive knots between the turbines rotating in opposite directions, were closed by a streamline casing of a toroidal look (Fig. 5.17).

Installation is completely prepared for complex tests and demonstrations in a wind tunnel. The main idea of units consisting in use of energy of layers of a stream, not passing directly through cross-section section of the turbine was confirmed with experiences in the largest hydrodynamic test canal on Krasnoyarsk hydroelectric power station and the mathematical modeling which is not containing empirical constants. A numerical

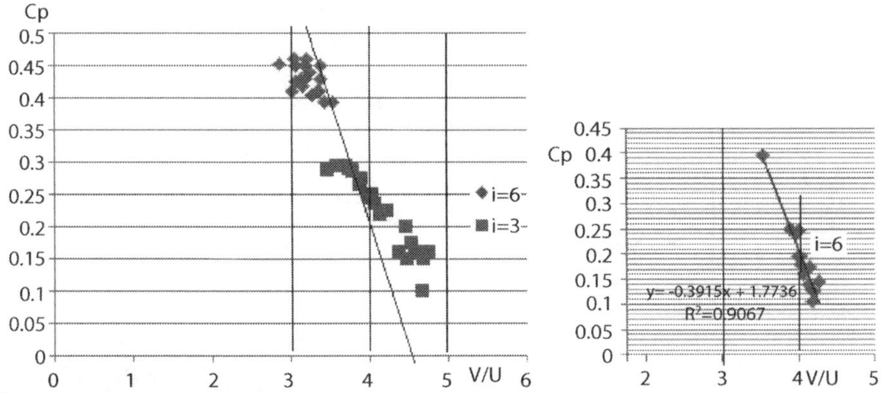

Fig. 5.14 Results of tests of hydro turbines. D=2.7m, b=0.12m, L=0.36m, U=2.30-2.15 m/s. D/L=7.5 (only).

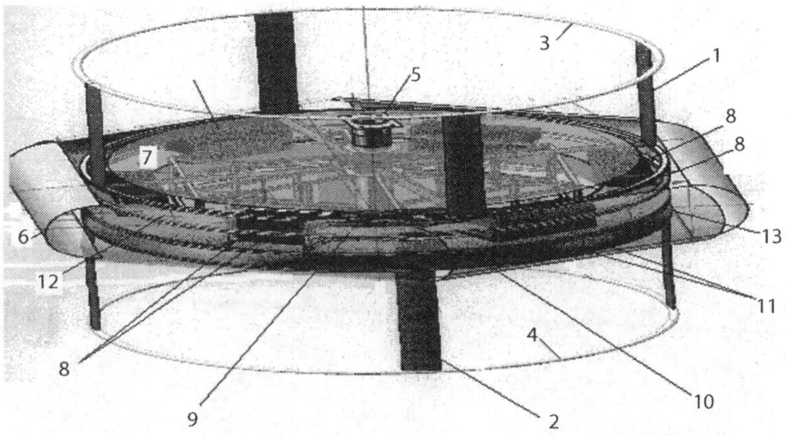

Fig. 5.15 General view of the power unit 2.5. 1, 2) the blades focused in opposite directions, 3, 4) the streamline rings distributing loadings between blades and creating spatial rigidity of turbines, 5) the central basic hub perceiving horizontal loads of turbines from a stream, 6) the toroidal stream-lining forming a smooth approach of a stream to blades of turbines, 7) a cover of a casing of the turbine, closing the motionless bridge and the rotating traverses of the turbine transferring radial loads, 8) motionless inductors of the top turbine, 9) a short-circuited rotor of the top turbine - the flat copper ring connected with the top number of blades, 10) a short-circuited rotor of the bottom turbine, 11) motionless inductors of the bottom turbine, 12) a bearing ring of the top turbine, 13) a bearing ring of the bottom turbine.

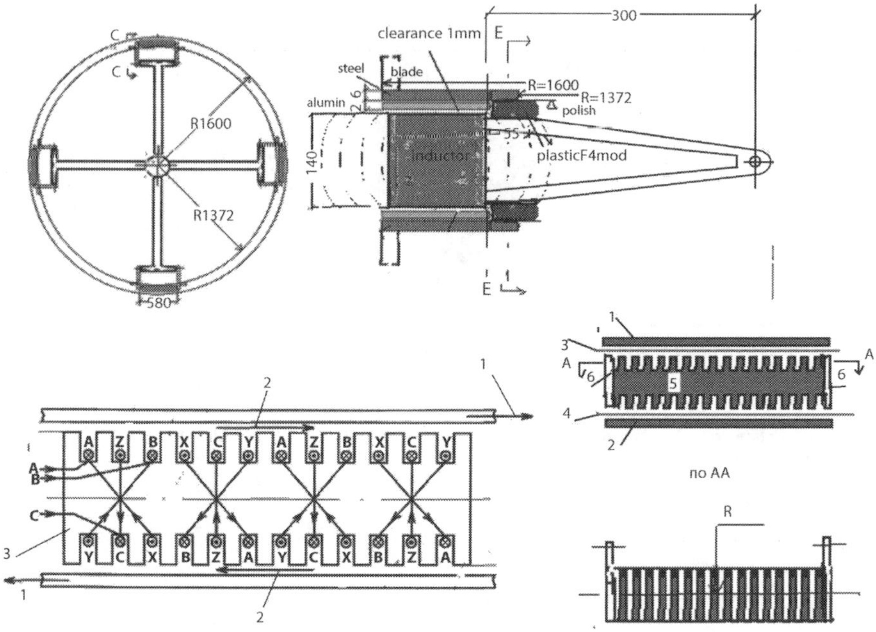

Fig. 5.16 The linear generator with opposite moving of rings.

Fig. 5.17 A general view of the model prepared to test in a wind tunnel.

calculation model has been developed and implemented for the flow space system pattern - Fig. 5.18.

Fig. 5.19 shows the results of calculations of the model with a diameter of 2.4 m for three different lengths of blades - 0.3 m, 0.6 m and 0.9 m. The calculations were performed for the same water flow rate of 3.5 m/s and the blade linear speed of 7 m/s. The load (radial force) acting on one of the 6 blades with a chord of 160 mm was determined. The top turbine rotates counter-clockwise, the bottom in a clockwise direction. The origin

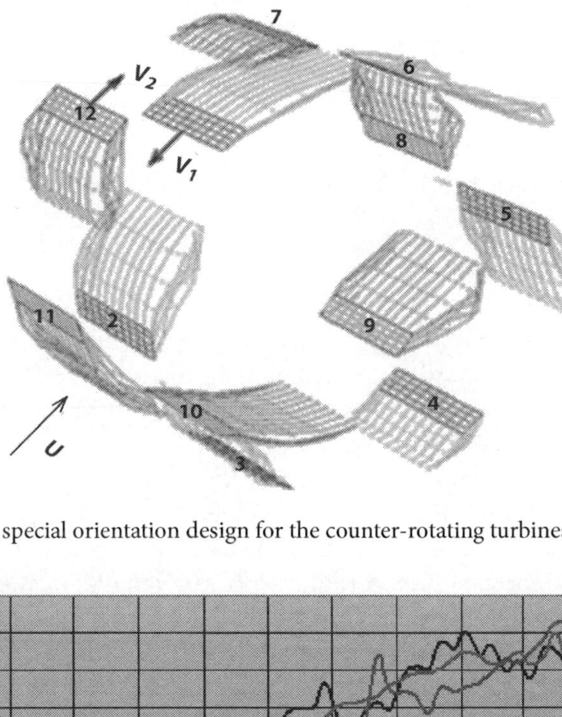

Fig. 5.18 The special orientation design for the counter-rotating turbines.

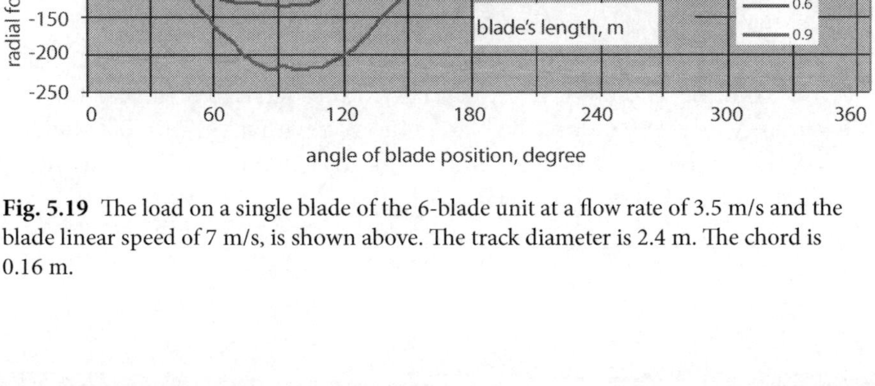

Fig. 5.19 The load on a single blade of the 6-blade unit at a flow rate of 3.5 m/s and the blade linear speed of 7 m/s, is shown above. The track diameter is 2.4 m. The chord is 0.16 m.

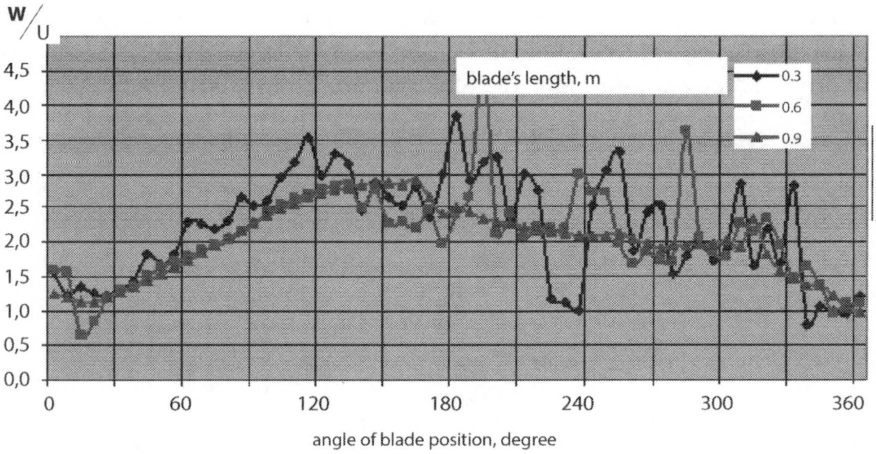

Fig. 5.20 The flow velocity module at the blade track points that are distanced apart by the chord's length before the blade front. The speed is measured in the fractions of the turbine incoming flow rate.

corresponds to the point where the blade goes against the flow. It can be observed that when the blade is at the front, the load exerted on it is negative (force directed towards the center) and the maximum force on the short blade (0.3 m) is 3 times smaller than the force directed on the longest blade (0.9 m) and the maximum points are not congruent. The maximum loads on the blades, with lengths of 0.6 m and 0.9 m, are observed almost in the same place and defer in magnitude by approximately 1.6 - 1.7 times, which roughly corresponds to the ratio of the lengths of the blades. The most interesting thing is observed on the blade rears. The load on the rear side of the track, of course, is directed away from the center of rotation (positive). It pulsates (vortex effect, cascading from the blades at the front), but most importantly - in the area of the first peak (240 - 275^0), the load on the short blade is almost the same as the load applied on the long blade. The flow rate on the back side of the blade track with short blades is much higher than that with the long blades (Fig. 5.19).

Our new windmills may be located on the surface of water (Figs. 5.7, 5.21).

The windmill and large watermill may be associated. It is clear that the flow energy recovery before the blade rears is very active, and that the proposed high efficiency large diameter machines will have the highest energy production rates!

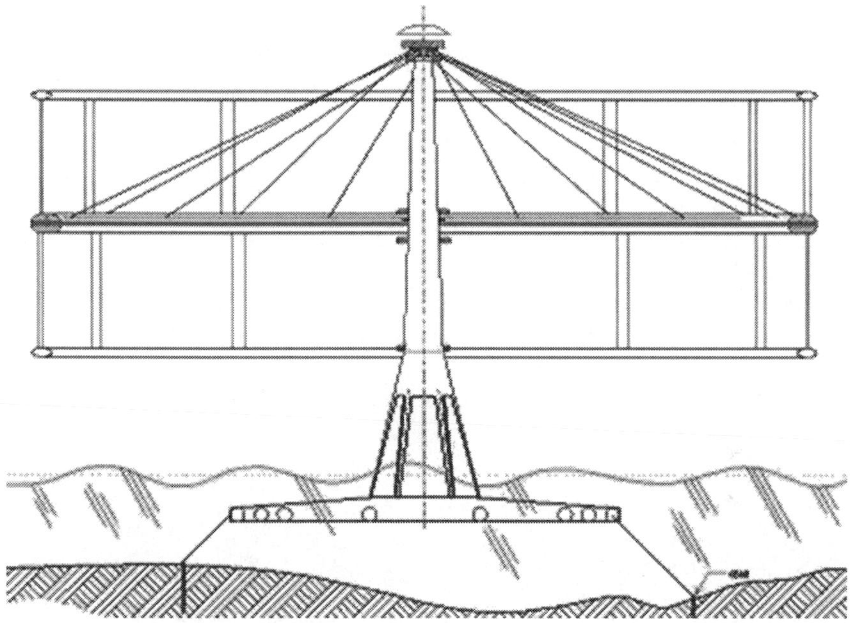

Fig. 5.21 The windmill is going on the pontoon in dry dock and afloat delivered to the job site in the open sea.

5.4 Conclusion

The described system shows the high efficiency of converting wind or water flow energy, achieved through the use of turbulent transport of current energy in the vertical direction and the restoration due to this velocity field in front of the back action of the blades. The consumption of this system per unit of power generation is approximately 2 times lower than that of the best modern traditional versions with high installed capacity.

In the area of any East or West beaches of the Americas and other countries, possibly more effective will be the location of the windmills in the seas without the use of expensive land in these areas.

References

1. Victor Lyatkher, *Wind Power: Turbine Design, Selection, and Optimization*, Wiley, ISBN: 978-1-118-72092-9, 328 pages, December 2013.

2. Victor M. Lyatkher, Alexander M. Proudovsky, *Hydraulic Modeling*, ISBN-13: 978-1118946190-ISBN-10: 1118946197, 594 pages, 2016.
3. Gijs A.M. van Kuik, The Lanchester-Betz-Joukowsky Limit, *Wind Energy*, 2007;10:289-291
4. Gorban' A.N., Gorlov A.M., Silantyev V.M. Limits of the Turbine Efficiency for Free Fluid Flow, *Journal of Energy Resources Technology*, ASME, 2001, v. 123, pp. 311-317.
5. Belotserkovsky S.M., Lifanov I.K. *Method of discrete vortices*. CRC Press. Roca Raton, USA, 1993, 451 p.
6. McGbee R., Reaslev W., Low Speed Aerodynamic Characteristics of a 17-percent – thick Airfoil Section Designed for General Aviation Applications, NASA TN D-7428, 1973.
7. Arie, Rouse H., Experiments on two-dimensional flow over a normal wall. *J. Fl. Mech.*, 1956, v1, N1.

6

Damless Hydropower and Tidal Power Plants

Balanced orthogonal hydraulic units can be installed in the lowland rivers or regions of high tidal flows, not necessarily as a part of an electric grid system. The lowest consumption of materials has the single-blade machine (Fig. 6.1).

Currently freely flow hydropower station with one orthogonal turbine (Figs. 6.1, 6.2), to position perpendicular to a stream, and the generator DC (or AC) is ready for a batch production. Connection of two coaxial turbines with an opposite direction of rotation and power knot between turbines is possible. To get hydro forces balanced, two turbines locate on one shaft and rotate as one unit (Fig. 6.3).

Units settle down under water (under ice) on a pontoon or are fixed by ground piles. On the Fig. 6.2 top picture is the turbine's general view, on bottom – view on an A arrow. Chord of the blade of 0.16 m, with profile GAW-1, will turn on 3^0 outside from a tangent. Three pairs of identical blades 0.45 m length, each is fixed on the 70-100mm shaft with traverses. Planes of blades pairs are turned from each other at 120 degrees. The system is counterbalanced both from inertial, and in hydrodynamic respects. Vectors of the bending moments of the inertial forces arising from blades are colored in corresponding colors on the image. Radius of a line of blades is 0.32 m (diameter of the turbine 0.64 m), length of the one turbine is 2728 mm (between supports). The optimum frequency of rotation from 100 to 200 rpm is reached at speed of a stream from 1 to 2 m/s. The highest frequency of rotation (without a load) in 1.5 times is more. Peak efficiency C_p in the formula for definition of capacity P:

$$P = C_p \rho U^3 D L/2$$

Victor M. Lyatkher and Ziaur Rahman. Wave, Wind and Current Power Generation, (217–230) © 2022 Scrivener Publishing LLC

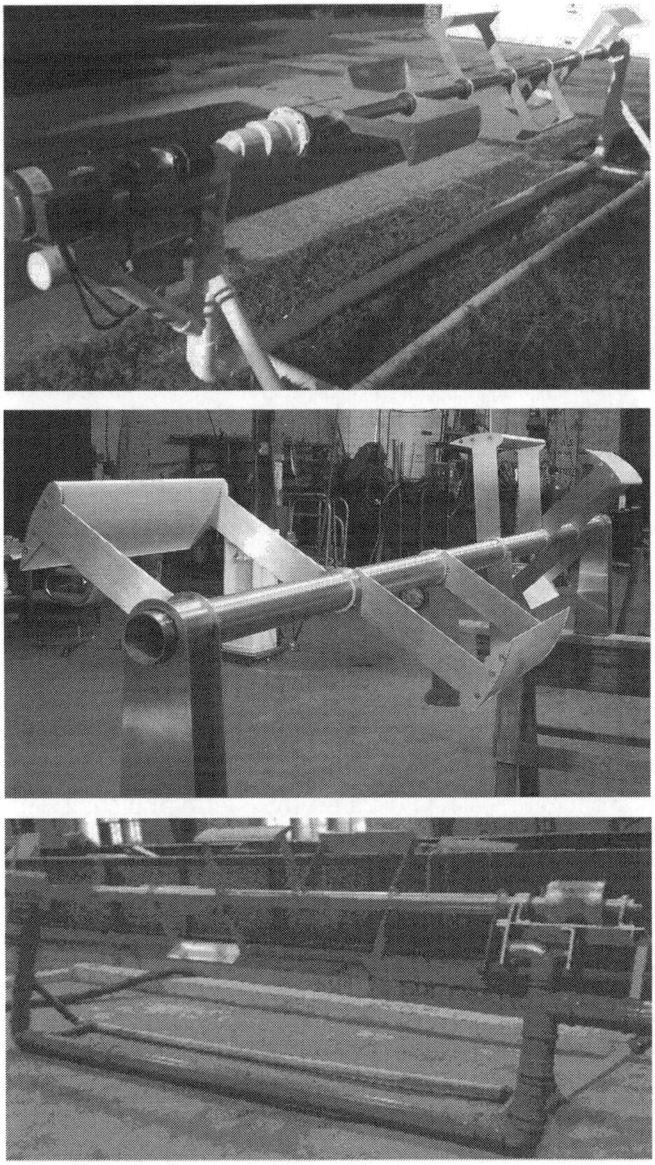

Fig. 6.1 Six tier balanced turbine with one blade in each tier.

at optimum rapidity V/U=4.2 reaches 60 %. Fig. 6.4 shows the results of experiments in a channel with width of 1 m and a depth of 0.5 m using models of the diameter 0.2 m at a flow velocity of water U = 1m/s. Balancing was achieved by the lead disks having internal smaller heavy disks in models.

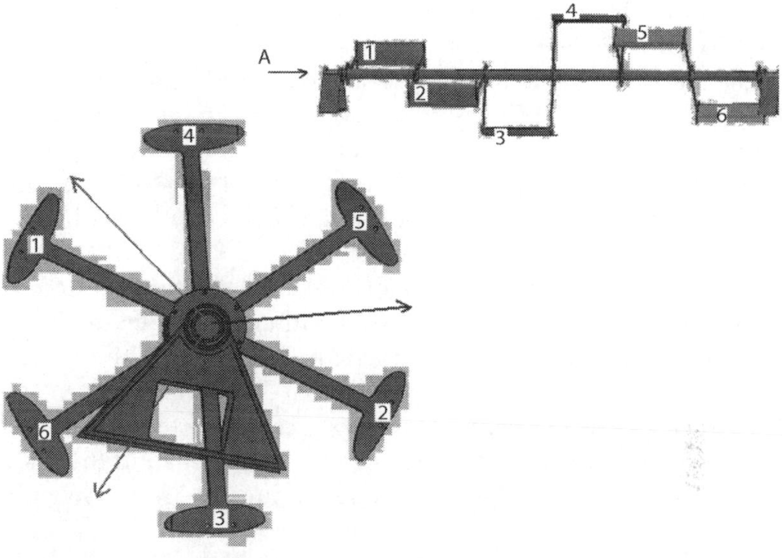

Fig. 6.2 Turbine common views.

Fig. 6.3 Completely balanced turbine concerning the central point of contact.

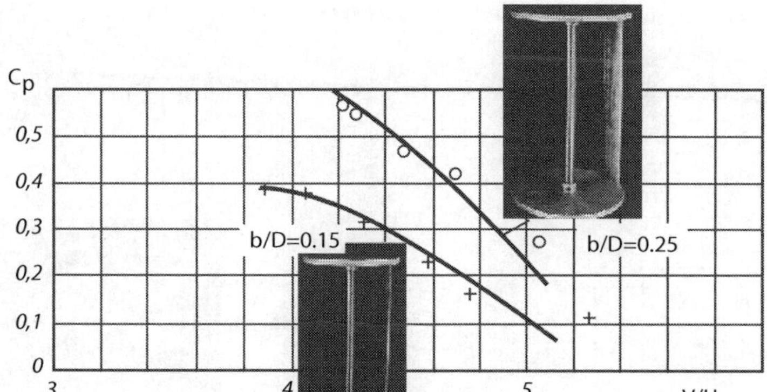

Fig. 6.4 One blade on a disk with the balance weight.

Chord of the blade of 50 mm. (top line) and 30 mm (bottom line), profile NACA0015. D=200mm, L=300mm.

In offered dimensions (0.64m x 2.7m) the maximum capacity of one turbine will make about 0.5 kW at speed of a stream of 1 m/s or 5 kW at speed of a stream of 2 m/s. Weight of the turbine (without support units and the generator) when using the aluminum alloys – 15.7 kg. Turbine blades have improved hydrodynamic profile GAW-1 with very high coefficients of lift and pulling force at low resistance.

The constructive profile of the blade (Fig. 6.5) is optimized from a condition of the minimum weight at the maximum durability and comprehensible manufacturability for manufacturing of blades by an extrusion method from alloy AD31-T1 or AD-33.

Tests have shown that with a length of free supported (on two pillars) blades of 450 mm, the concentrated load 322 kg, applied in the middle of the blade, causes only an elastic deflection limited to 3mm. Weight of the blade of 1.65 kg/m.

The ends of blades were closed by plastic to keep the inside of blades from the water.

The turbine shaft is made from a hollow pipe diameter of 100mm with the wall 2.5mm, having the edges strengthened with inserts of thickness is 1.6 mm. The pipe is made from the same material as the blades. Tube's weight is 2.82 kg/m. From the power prospect of the turbine as smallest steel pipe diameters are the most preferable. Let's see the results of characteristic loads on one blade with length of 0.45 m, a chord of 0.16 m, diameter of a line 0.64 m, turn of the blade 3^0, speed of a stream 2 km/s, speed of the blade 8.3 km/s. Weight of blades 1.65x2.7 = 4.46 kg.

The maximum rotation frequency under load is 226 rpm, acceleration-339 rpm.

Circular frequency Ω = 23.65 1/s gives an acceleration→ 18.2g→

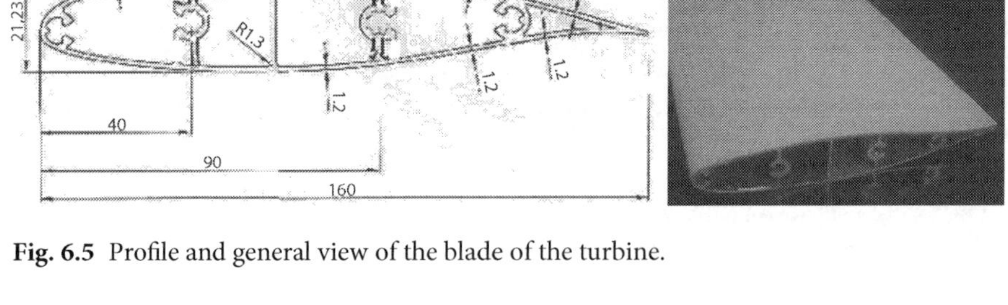

Fig. 6.5 Profile and general view of the blade of the turbine.

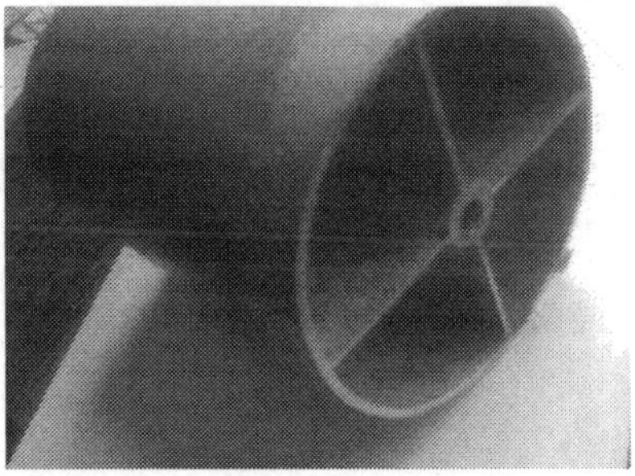

Fig. 6.6 Fragment of an aluminum shaft.

Centrifugal force on one blade 1.65 × 0.45 × 18.2 = 13.5kg, during acceleration-30.4кг

Mass with tube length of 3 m of diameter 0.1 × 0.0025m from AD 31-T1 2.82 × 3 = 8.46 kg

Weight of transverse section 0.1 × 0.003m from AD31-T1 $3 \times 10^{-4} \times 2.7 \times 0.29 \times 12 = 2.82$ kg

The total mass of the turbine without supports, transmission and generator 15.74 kg

Weight of the generator of 22 kg

The maximum load on the blade (from a rotation axis) 74 + 13.5 = 87.8 kg (without water)

The maximum load on the blade (to a rotation axis) 224 -13.5 = 210.5 kg

Average load on the blade - 60 kg. The maximum deflection of the blade – 4 mm.

The maximum deformation of a shaft of the turbine of 23.5 mm

The maximum effort in aluminum shaft of 1247 kg/sm^2

Diameter of a shaft of 100 mm is too great for the turbine in diameter of 640 mm and that can worsen efficiency of the turbine. If to use a steel pipe 70x3 mm the maximum pressure in it will be 2240 kg/sm^2, and the maximum deflection 20.5mm. Weight of such a pipe of length of 3 m - 14.8 kg.

Capacity of the turbine is calculated by the formula

$$P\ (W) = C_P \rho\, U^3\, D\, L/2 = 1000\, C_P\, (U/V)^3\, V^3\, D\, L/2$$

U – speed of a stream on the way to system, D = 0.64 m – diameter of the turbine, L = 2.7 m – length of the turbine, V – linear speed of the blade. On the basis of the skilled data the relative characteristic of the turbine can be presented in a following kind (Fig. 6.7):

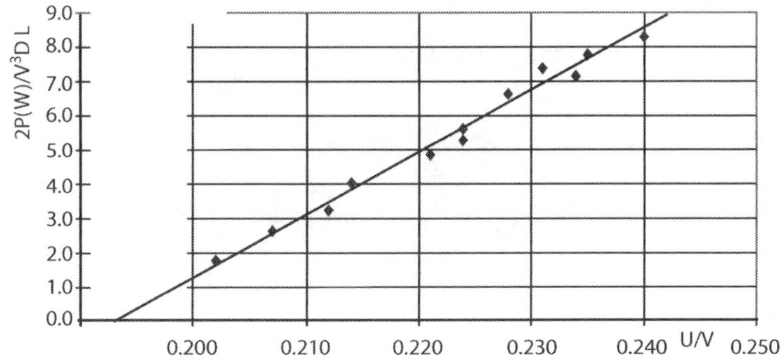

Fig. 6.7 Turbine power P opposite speed relation.

$$Y = 178(X - 0.193), \quad 0.193 < X < 0.7$$

$$P = 89(U/V - 0.193)V^3 DL = 153.8(U/V - 0.193)V^3 =$$

$$= 153.8U\,V^2 - 29.68V^3 = 0.1727U\,n_t^2 - 0.00111n_t^3$$

The generator with the constant magnets, applied in a complex with the turbine, has a weight of 22 kg (Fig. 6.8).

Turbine on Fig. 6.1 with sleeve bearings has been tested in the water channels of the Moscow branch of TsAGI. Turbine with a generator mounted on a depth of 0.5 m from surface water under a cart that can move along the channel with a given speed. We measured the speed of the turbine, voltage and current in each phase at the generator terminals with different resistive loads. At Fig. 6.9 shows the characteristics of the generator, issued by the manufacturer and obtained in the tests.

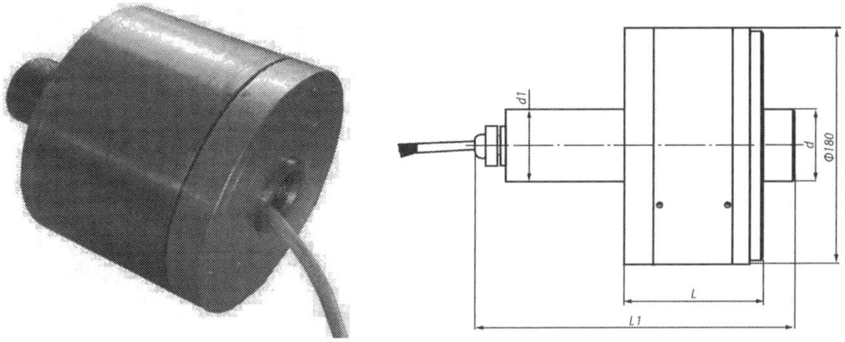

Fig. 6.8 The generator with the constant magnets ГВ-2/650-110-12Г («Erga», Kaluga).

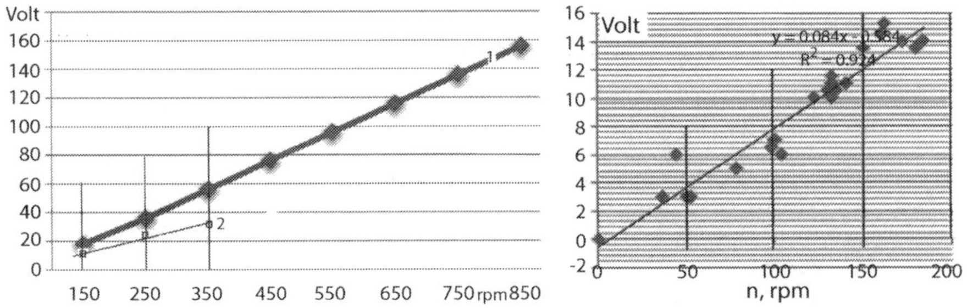

Fig. 6.9 From left - a comparison of passport (1) and experimental (2) data. Right-experiments with different load (∞, 2, 4 ohm).

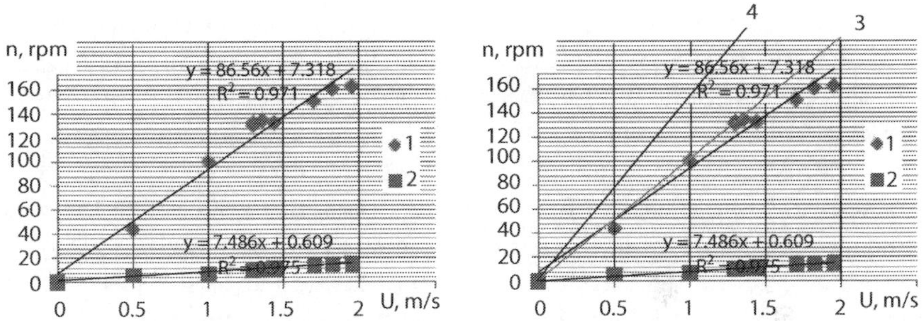

Fig. 6.10 Rotational speed of the turbine (1) and the voltage on the generator (2) as a function of feeding speed turbine Flumes TsAGI without load. 3 - the maximum power line, 4 - Overclocking frequency of laboratory tests.

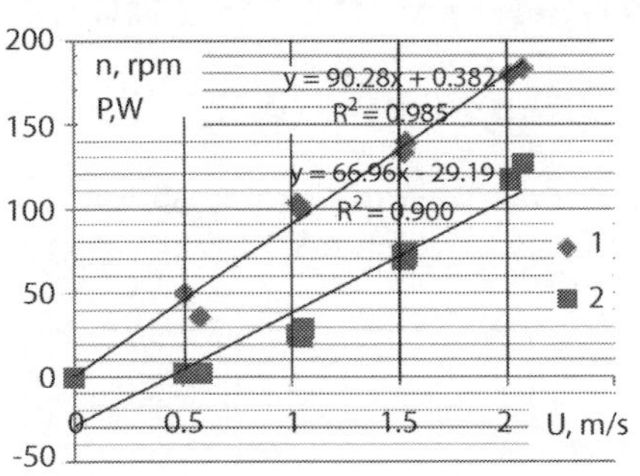

Fig. 6.11 Rotational speed of the turbine (1) and power at the generator terminals (2) under a load of 2 ohms.

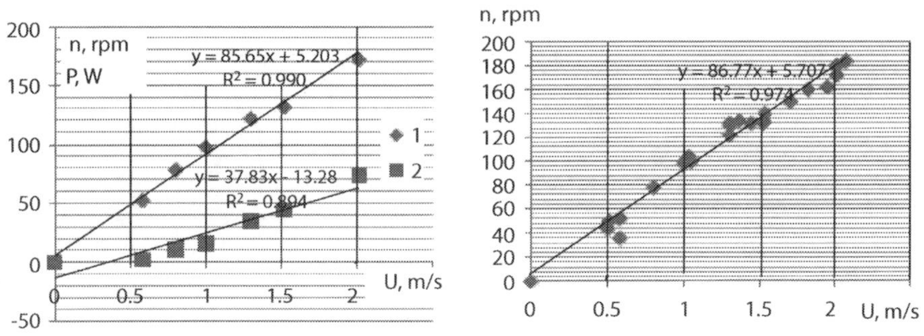

Fig. 6.12 Rotational speed of the turbine (1) and power at the generator terminals (2) at 4 ohms (left), right-all experiments (under different loads).

Results of tests in the channel with different electric loads of the generator shown in Figs. 6.10-6.12.

The general conclusion from these experiments is that the scheme is workable and effective. However, due to the large friction in the bearings, the turbine, even without electrical load is at the limit of the stability, meaning at the point of maximum efficiency. This is confirmed by the fact that the turbine, for example at a speed of 1 m/s would not start (it had to be pushed). Adding electrical load actually almost did not change the rotation frequency, meaning the turbine remained in the zone of maximum efficiency - at the stability limit (the zone to the left of the peak on the graph C_p (V/U)-unstable). According to test results support units were redesigned, providing a dramatic reduction of friction in the bearings. To determine the capacity of the complex «turbine-generator ГВ-2/650-110-12Г» we use representation of capacity as a function of the parameters of the turbine and frequency of rotation:

$$\mathbf{P = (89\ UV^2 - 17.18\ V^3)DL = (0.244U\ n^2D^2 - n^3D^3)DL}$$

$$\mathbf{P = 3.18\ n_t i\ \text{ипи}\ 3.18 n_t i = 0.1725U\ n_t^2 - 0.00111 n_t^3}$$

Solving these equations, we will receive values of capacity on generator outputs at the set speed of a stream and the chosen gear ratio **i** of the installed

Stream speed, m/s		1		2		2.5
Optimum gear ratio		2		8		12
Frequency of rotation of turbine, rpm	102		204		265	
Maximum capacity, Wt		**649**		**5186**		**10100**

gear connecting the turbine and the generator. That result defined the optimum gear ratio that will provide a maximum of capacity of the system.

The generator with constant magnets should be connected with the turbine through the gearbox, raising the frequency of rotation of a shaft of the generator. The optimum gear ratio of gearbox *i* for the specified generator depends on the speed of the stream U. If the gear ratio is equal 1, the frequency of rotation of the turbine giving the maximum capacity is presented by the formula

$$n_t = 77.415U + /\text{-}\sqrt{5993U^2} \text{ -}2864.86$$

Value of capacity of the generator (Wt) corresponding to this frequency calculated by the formula:

$$P = 3.18\ n_t$$

The terminal voltage of the generator (V)

$$0.195\ n_t - 11.25$$

Maximum overclocking speed without load on the generator (rpm):

$$n_{max} = 155.4\ U\ (m/s)$$

The table below shows the results for the above formulas, displaying that without the use of a multiplier the power output is significantly lower than possible. For example, at a speed of a stream of 2 m/s without the use of the gearbox, instead of a possible 5 kW (at the gear ratio 7-8) it received only 0.7 kW.

The usual three-phase motor can be used with the short-circuited rotor, united with a planetary reducer (motor-reducer). Fig. 6.13 is the connection diagram of an asynchronous generator as the engine requires controlled capacitors.

The next feature of the turbine is necessary to control loads – at excess of some particular maximum the turbine can stop. The turbine – in the process of increasing loads, frequency of rotation decreases. Control value of the minimum frequency of rotation makes 2/3 from maximum overclocking frequencies of the turbine at the given speed of a stream.

Pair of units with an opposite direction of rotation of turbines can be united in micro-hydroelectric power station (Fig. 6.14).

Table 6.1

Flow rate, m/s		1	1.5	2	2.5
Maximum of frequency turbine rotation, rpm		155	233	311	388
Frequency of rotation at the maximum loading, rpm	133	180	223	263	
Minimum admissible frequency of rotation, rpm	104	156	208	260	
Maximum capacity, Watt	**424**	**574**	**708**	**838**	
Voltage at maximum load, Volt		15	24	32	40
Amperage, Ampere		28	24	22	21
The terminal voltage without load, Volt		19	34	49	64
Loading (ballast resistance), Ohm	0.5	0.7	1.4	1.9	

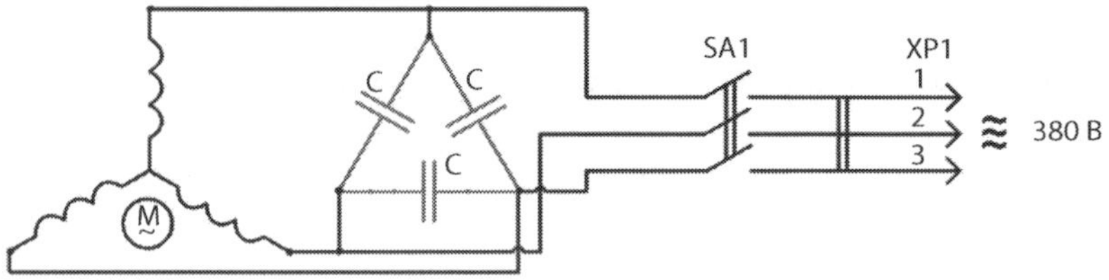

Fig. 6.13

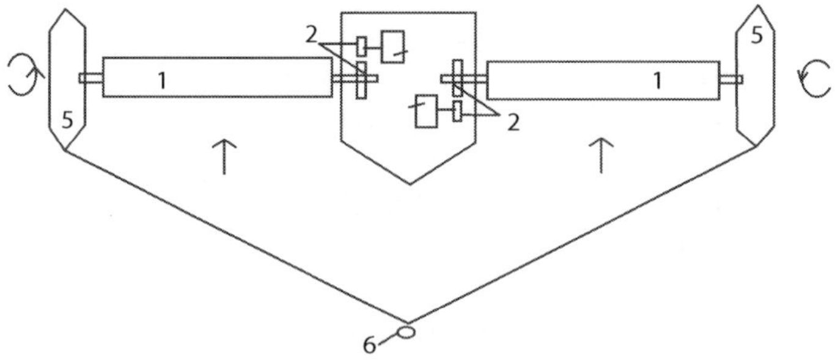

Fig. 6.14 Scheme of micro-hydroelectric power station 1) turbine (2.7 × 0.64 m2), 2) shift pinion, 3) generator with a rectifier and voltage relay, 4) energy pontoon, 5) ballast pontoon, 6) anchor.

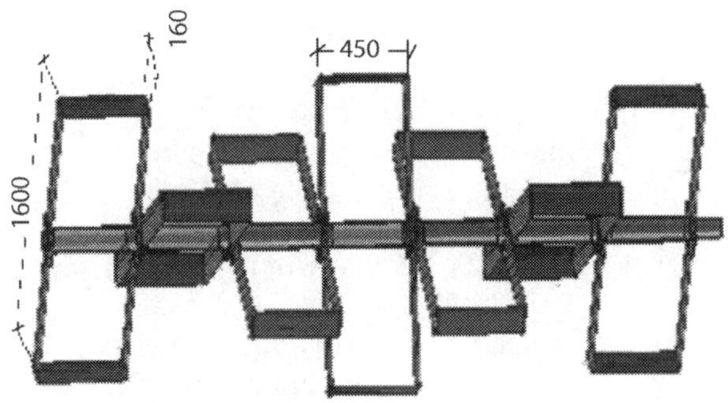

Fig. 6.15 Completely balanced turbine with 2 blades on each tier.

At a deep water of a stream with use of the same constructive elements (the blade, the shaft, the generator) turbines can be effective with two or three blades in one tier. The two bladed turbines with diameter from 1.3 to 1.6 m with 7 tiers (Fig. 6.15) is completely balanced with regards to inertial and hydrodynamic forces.

Its efficiency concedes to the one bladed machine (Fig. 6.16) slightly, but the capacity of the almost identical turbine's length can be twice more. If V = 2.8U,

$$\mathbf{n_t(rpm) = 33.4U(m/s)}$$

$$\mathbf{P\ (kW) = 0.36\ U^3\ 7x0.45x1.6\ /2 = 0.91\ U^3}$$

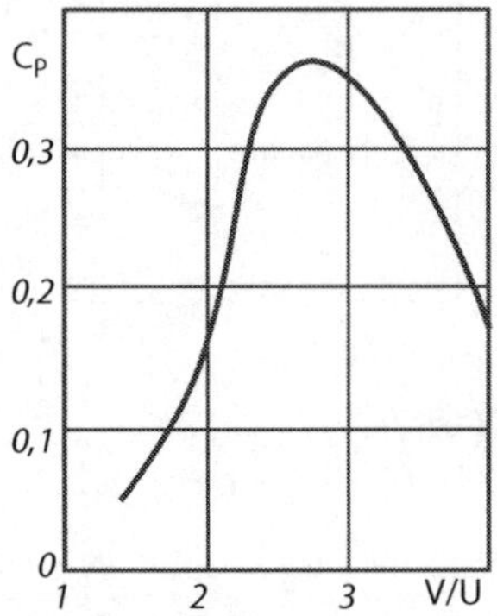

Fig. 6.16 Efficiency of two blades machine. Solidity 0.2.

At speed of a stream 1, 1.5 or 2 m/s possible capacity of the turbine **0.9, 3 or 7 kW**.

If the diameter of the turbine will be smaller (D=1.08m) and solidity will be ib/D= 0.3, the optimal turbine will have a higher efficiency (C_p =0.43) for the slower blade's speed (V/U=2.2 – Fig. 6.17).

$$\mathbf{P\ (kW) = 0.73\ U^3,\ n_t(rpm)=38.9U(m/s)}$$

In the presence of variability of current's directions, it is suitable to increase the diameter of the turbine and quantity of blades, reducing length of the turbine and making it vertical (Fig. 6.18).

The maximum capacity of this turbine:

$$\mathbf{P(kW) = 0.42\ U^3\ 0.9x1.6/2 = 0.3\ U^3}$$

At optimum frequency of rotation $\mathbf{n_t}$ **(rpm) = 45.4 U**. In any case, the length of blades can be increased according to the durability limits with corresponding increase in capacity of the unit if necessary. Tests of blades for a bend, spent on a press with the blade's length of 300 mm, have shown that the blade does not receive residual deformations at the bending moment reaching 78.8 kgm. Taking into account real loads at speeds of a stream of 2-2.5 m/s the length of blades can be increased approximately by

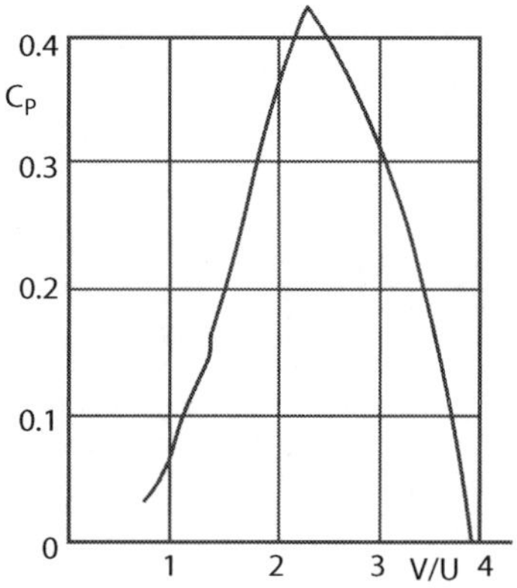

Fig. 6.17 Efficiency curve. Solidity 0.3.

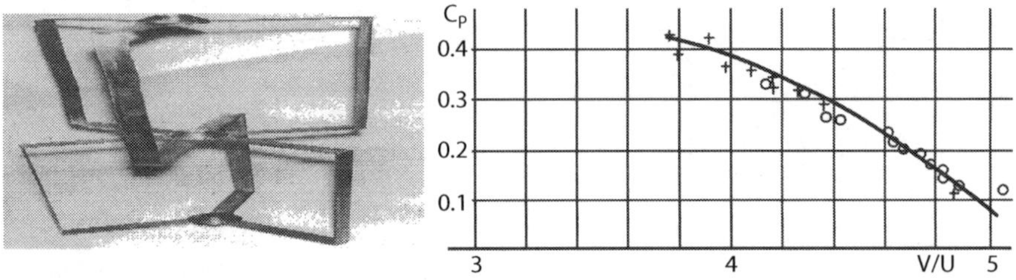

Fig. 6.18 Short hydro-turbine diameter of 1.6 m and a height of 0.9 m with 3 blades in the tier (at the left) and efficiency (on the right).

2.2 times with proportional increase in capacity of the unit – in the basic variant, with one blade turbines – will receive up to 2-3 kW in the simple scheme without the gearbox.

All the described schemes of turbines include the use of blades with the same cross-section (chord 160 mm). Various schemes can be optimized for different design speeds and different flows to receive the particular capacity. The greater the flow rate the greater the danger of the cavitation since the speed of blades is much higher than the speed of the current. The increase of blades in same level will decrease the danger of the cavitations but the efficiency will slightly decrease.

The smaller blades in a single tier of the turbine, the higher its energy and economic efficiency, but the possible deterioration of performance characteristics - surge of support reactions, the vibration of the machine. It is very important the required scheme of the power output - a powerful network (which is always preferable) or on the individual consumer, in the single-phase AC power or battery charging direct current with controlled loads.

Developed, tested and ready for mass production several types of orthogonal turbines to generate electricity from rivers without dams. Turbine blades of the optimal profile with chord of 160mm make a basis for the serial production and manufactured by extrusion. Unit capacity of these turbines in the flow rate of 2 m/s can be from 5 to 10 kW. Another effective design may be two coaxial turbines rotating in the opposite direction. Turbines can be protected by the cylindrical mesh, secured on the turbine shaft. The same turbines may be proposed, for example, in Nepal rivers (Table 6.2).

Table 6.2

Name	Location	Elevation	Discharges, m3/ sec		
			Dry season year		
			Minimum		
			Mean dry mean		
Taktor Khola 668.4	27 31 45. 86 33 30.	2350	1.68	0.98	5.97
Melamchi Khola 627.5	28 02 30. 85 32 00.	2134			
Rosi Khola 640	27 34 50. 85 30 50.	1480	0.83	0.40	2.79
Nakhu Khola 540	27 34 30. 85 18 50.	1400	0.15	0.01	1.11
Bagmati River 530	27 42 30. 85 21 00	1300	0.50	0.01	3.11
Surnagad River 170	29 27 30. 80 33 10.	1110	0.88	0.17	7.03
Khudi Khola 439.3	28 17 15. 84 21 45.	990	4.14	1.60	8.60
Mardi Khola 428	28 18 30. 84 20 50	915	3.03	1.01	16.4
Puwa Khola 730	26 55 00. 87 54 40.	802	0.66	0.17	8.40
Phalandu Khola 446.8	27 58 25. 85 11 15.	630	1.60	0.14	13.2
Rapti River 460	27 26 30. 84 58 15.	332	7.10	1.10	28.5

7

Tidal Power as Basis for Hydrogen Energetic

It is shown the relations to estimate maximum capacity and power that may be transformed from the tidal in closed basin. Using these relations is estimated the world power from tidal and volume of the annual producing hydrogen liquid or gas. It is enough to change the power situation in the world. The new technology of the tidal power and hydrogen manufactures construction, produced liquid hydrogen with low cost and low initial investment, is described. To convert the power from tidal flow the new floating orthogonal turbine may be used. The tidal power station with these power units will not be as expensive as other modern kinds because we do not need to cross the entrance of a gulf – the power unit converts flow power without a dam. We do not need to invest too much money, because the power output and hydrogen manufacturing may go up step by step, from one block to another, but we will refund the investment very fast—immediately after the first block will go on. The produced hydrogen will keep in storage and delivery to the industrial centers by the ships or by the refrigerated glass pipe lines. Tidal power stations with these units have no environmental problems.To produce the hydrogen as fuel for cars and electric power plants the best source may be the tidal power plants (TPP) with new technology and equipment. The traditional tidal power plants, which provide for the building of dams, complete cutting from the sea of the bay being used and creation of one or several energy basins, have four fundamental drawbacks restraining until now the development of this branch of power engineering. All projects provide for the building of the closed head front of TPP, the dams and other hydraulic structures which cut off the TPP basin from the sea. The existence of such a front noticeably changes the ecological situation in the basin. The maximum power, which can be issued by TPP, is determined by the flow regime, the basin configuration and the TPP design. In any event it varies in the time from maximum to zero attainable twice in a 24-hour period. The heads at

Victor M. Lyatkher and Ziaur Rahman. Wave, Wind and Current Power Generation, (231–256) © 2022 Scrivener Publishing LLC

TPP are not large, the traditional hydropower equipment is expensive and production is relatively small. The cost of TPP building and head front constructions designed for gale oceanic waves and heavy ice load is great. This determines large building terms, high capital investments per unit of the installed capacity and relatively high prime cost of energy.The existence of head front determines the unfavorable regime of object financing – it is necessary to completely finish the building (completely pay for all the works) and only after this the station begins to generate the energy and return capital investments.Another approach to TPP designing, which is free from mentioned drawbacks, is proposed [1]. The main distinctive feature of the proposed approach consists in the use of free tidal flows in those parts of the water area, where the absolute values of these flows speeds are maximum. TPP in our proposal consists of the stationary or floating, fixed at the bottom, underwater (in the regions with heavy ices) hydropower units, which convert tidal flow energy into the electric power and further into gaseous or liquid hydrogen, in the case if the direct issuing of electric power is impossible or economically inexpedient. *There is no need to completely cut off the estuary of the basin and create the head front for complete utilization of tidal flows energy.* **The highest possible power and power production is obtained in the case of well-defined hydraulic resistance brought in by the turbines which are installed at the entrance into the TPP basin (in its narrowest part) [2]. The selection of the installed capacity of TPP is a technical and economic task.** *The distinctive feature of our proposal consists in the fact that TPP power can have any value, that corresponds to the financial possibilities of the investor but is smaller than the limiting maximum value determined by hydrological and topographic conditions.* Hydropower units can have various construction. They can be located freely in the form of the simplest complex "hydro turbine + generator" or as a part of power unit. It is important that there is no need to completely cut off the estuary of the basin for complete utilization of tidal flows energy. In the general case the energy efficiency of tidal basin of the area Ω at the rise of tide A and period T is determined from the formula

$$P = C\,\Omega A^2/T \tag{7.1}$$

Here by symbol P the installed capacity of TPP or its average annual value, which determines the production of the station, can be denoted. In [2] it is shown that maximum power and maximum power production in the case of the fixed basin area Ω and fixed height of tidal wave A at the approach to the basin has place in the presence of particular value of dimensionless parameter

$$X \equiv \alpha = [\Omega/\Omega_p]^2 \xi A/4gT^2 \tag{7.2}$$

Here Ωp is the cross-sectional area of the channel or the strait, where power units are installed (or the summary cross-sectional area of turbine paths in case of traditional TPP composition), ξ is the coefficient of the hydraulic resistance of turbine path, $g = 9.8\ m/s^2$ is the acceleration of gravity, T is the period of tidal wave. The maximum power and power production of tidal power station (Fig. 7.1) is obtained with

$$\alpha = 0.03 \div 0.04 \tag{7.3}$$

In this case the coefficient C in (1) for the limiting maximum power is equal to

$$C = C_{max} = 7.27 \tag{7.4}$$

For the average power, which determines maximum production,

$$C = C_{average} = 2.82 \tag{7.5}$$

The relative time of usage of the installed capacity τ = 0.388.

At that the amplitude of the water level fluctuations in the basin amounts approximately 78% of the amplitude of the sea level fluctuations and phase shift φ is determined by the equality

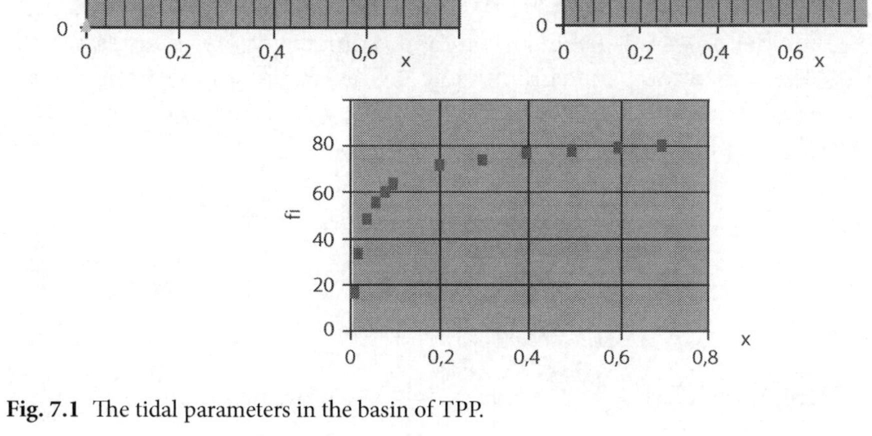

Fig. 7.1 The tidal parameters in the basin of TPP.

$$\text{Cos } \varphi = 0.67 \tag{7.6}$$

The tidal regimes in the sea $z_0(t)$ and in TPP basin z(t) are periodical:

$$Z_0(t) = A \text{ Cos } (2\pi t/T) /2 \text{ , } z(t) = A \text{ } z_m \text{ Cos } (2\pi t/T - \varphi)$$

Here: A – tidal range, T- tidal period.

Depending on the adopted lay-out diagram of TPP and the selected type of hydropower units the parameters, which are included in the formula (7.2), can change; however, for ensuring the maximum power and power production their combination must give the value defined by the relation (7.3). If the area of the basin is measured in km^2 and the period of the flow is taken such that in the course of twenty-four hours there would be 3.87 half cycles (tidal period T=12.4 hours), then in case of measurement of rise of tide A in meters the formula (7.1) takes the following form.

The maximum (installed) capacity of TPP

$$\mathbf{P_{max} (kW) = 163 \ \Omega(km^2) \ A^2(m^2)} \tag{7.7}$$

The maximum average power of TPP

$$P_{average} (kW) = 63 \ \Omega(km^2) \ A^2(m^2) \tag{7.8}$$

The maximally possible average annual power production

$$E \text{ (MWh)} = 552 \ \Omega(km^2) \ A^2(m^2) \tag{7.9}$$

The maximum value of rise of tide should be inserted in the formula (7.7) and in the formulas (7.8) and (7.9) – the average root-mean-square value.

The tidal process is irregular, but approximately it may be presented as the sum of the moon and sun influences. The relation of the amplitudes of these two components is 2.17 about. Because maximum and minimum of the tidal amplitudes may be estimate as follow

$$A_{max} = A_0(1 + 1 /2.17) = 1.46 \ A_0 \tag{7.10}$$

$$A_{min} = A_0(1 - 1 /2.17) = 0.54 \ A_0 \tag{7.11}$$

Here $A_0 = 0.685 \ A_{max}$ – the tidal height acted by the moon.

The root mean square of the tidal's height, determined the possible power output of a tidal power plant must be calculated by real information of the tidal regime. Approximately from (7.10) and (7.11) follow:

$$A_{averaged} = \sqrt{(A_{max}^2 + A_{min}^2)/2} = 1.1A_0 = 0.754A_{max} \quad (7.12)$$

or

$$A_{max} \approx 1.33\ A_{av} \quad (7.13)$$

Using (7.13) it may be rewritten (7.7) and (7.9) in the same forms included averaged (mean) tidal range A_{av}:

$$P_{max}(kW) = 288\Omega(km^2)A_{av}^2(m^2) \quad (7.14)$$

$$E(MWh) = 552\Omega(km^2)A_{av}^2(m^2) \quad (7.15)$$

In reality the energy from TPP can be received only in the case, when the water level in the basin differs from the sea level. Our results reflect this condition. These estimations are thru, if the power basins have a regular form about a circle with one narrow water entrance-exit.

In other limited situations the basin would be as a channel (for example, Strait of Uldolmok in Pyongyang Channel, South Korea). The flow in this channel is the function of difference of the water elevation between the inlet and outlet of the channel. The power N_0 lost by the flow under everyday conditions at the moment of maximum of flow speeds (when $\partial U_0/\partial t = 0$) is determined by the following expression (per unit of water discharge):

$$N_0 = \rho g \Delta H = (\lambda L / R + 1)\rho U_0^2 / 2 \quad (7.16)$$

Here λ is the reduced factor of the hydraulic resistances of the channel, L is the length of the channel, R is the hydraulic radius of the channel.

After the building of the power station in one of the sections of the channel the water discharge through the channel will change and the flow speed at the approach to the power site will take on a value U less than U_0. Assuming that the dropping between the ends of the channel ΔH caused by external causes remains constant, let us rewrite the equality (7.16) with the introduction of the concentrated pressure drop Δp (power loss of the flow) at the power station:

$$(\lambda L/R+1)\rho U_0^2/2=\Delta\rho+(\lambda L/R+1)\rho U^2/2 \qquad (7.17)$$

The pressure drop and the change of flow strength in the channel before and after the building of power station are connected as following:

$$\Delta p=(\lambda L/R+1)\rho(U_0^2-U^2)/2 \qquad (7.18)$$

The power P withdrawn in the entire water flow by section Ω passing through the power station comprises:

$$P=\Delta pU\Omega=(\lambda L/R+1)\rho U(U_0^2-U^2)\Omega/2 \qquad (7.19)$$

In order to find the maximum value let us equate to zero the derivative of the right side of expression (7.19) by the parameter U:

$$\partial P/\partial U=(lL/R+1)\rho(U_0^2-3U^2)\Omega/2=0 \qquad (7.20)$$

Thus the withdrawing power is maximum, if

$$U = U_0/\sqrt{3} \qquad (7.21)$$

Under this condition the value of the maximum withdrawing power is equal to:

$$P_{max}=(\lambda L/R+1)\rho U_0^3\Omega/3\sqrt{3} \qquad (7.22)$$

The relative value of maximum power which can be withdrawn in the flow in the channel (in fractions of initial maximum power) comprises

$$P_{max}/N_0\,U_0\Omega = 2/3\sqrt{3} = 0.385 \qquad (7.23)$$

or

$$P_{max} = 0.385\ \rho g\Delta H\ U_0\Omega \qquad (7.24)$$

This result is fundamental estimation free from specific hypotheses.

The stated principle of TPP designing and building can be efficiently realized, in particular, with the use of the orthogonal turbines according to the Russian patent of the author No. 2242634 (13) C1 with the priority of

05 May, 2003.The idea of this turbine consists in the use of pull force of the wing flowed around with the attack angle smaller than the critical one (that is common for orthogonal turbines), but at that owing to the relatively large diameter of blades route the recovery of flow energy at the approach to the rear order of the blades takes place and the effectiveness of turbine turns out to be above usual one. The turbine consists of two (or more) ring hollow rotors (Fig. 7.2) in the form of regular polygons or tores located one above another and moving towards each other, on which the blades of aerodynamic profile, oriented in opposite directions, are vertically fixed. The upper ring rotor, which carries the inductors of the linear generator, is connected by the stretchings with radial thrust bearing, located at the top of the hollow central pylon. The lower hollow ring rotor, which carries a strip from current-conducting material and magnetic core and has the excessive buoyancy, is pressed to the upper ring rotor owing to the excessive buoyancy and the electromagnetic interaction of the inductors and the magnetic core. In other variants the motionless inductor of a linear generator may be located between the rings. *The use of such very turbine in the proposed approach is not the only possible one, however, the installation of type under study has the following advantages over other known schemes:*

- The counter-rotation of rotors releases the central pylon from the torque, developed by the blades, and ensures the mutual compensation of transverse forces, acting on the rotors; that reduces the summary load on the central pylon, decreases

Fig. 7.2 General view of the orthogonal multi blade power unit. The top of the central pylon is located at the level of the upper rim but under the lower edge of ice.

the specific consumption of materials and increases the reliability of the installation.

- The counter-rotation of rotors, each of which bears the blades, oriented in opposite directions, ensures high relative speeds between the inductors of the linear generator and the conducting plate; that reduces the mass of inductors and increases the efficiency of the installation.
- The counter-rotation of rotors releases the central pylon from the torque, developed by the blades, and ensures the mutual compensation of transverse forces, acting on the rotors; that reduces the summary load on the central pylon, decreases the specific consumption of materials and increases the reliability of the installation.
- The combining of the inductors into the units with different rated speeds of armature allows to begin to use the energy of flows from relatively small values of flows speed and ensures high energy effectiveness of installation in the entire significant range of changes of flows speed due to the optimum selection of rotational speed of rotors for each range of changes of flows speed.

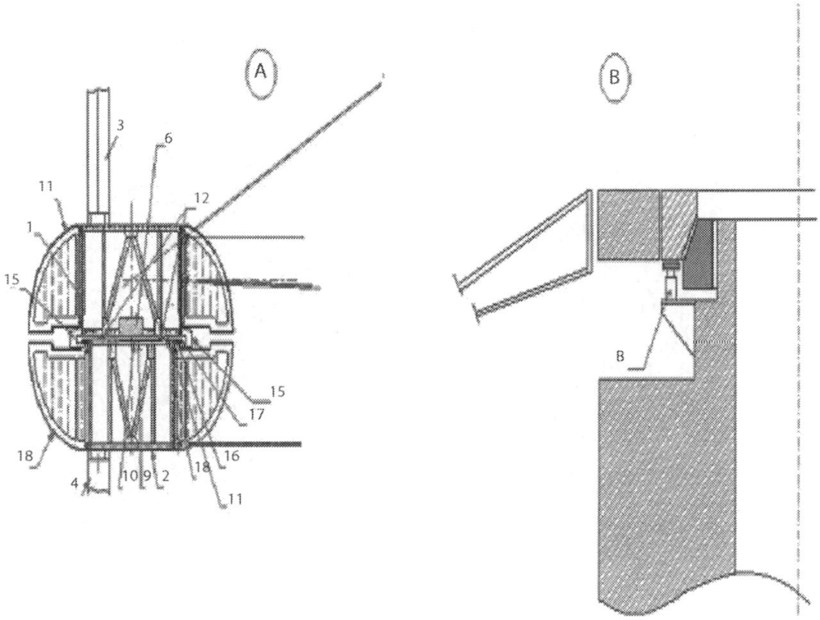

Fig. 7.3 Cross section of the rings.

- The control of circulation on the blades allows to enlarge the range of flow speeds, which correspond to the maximum efficiency, and owing to that enhance the power production and increase the effectiveness of the installation.
- The installation does not influence the ecological situation in the basin – the distance between the blades is great and their speed of motion is small (only 2-3.5 times more than the flow speed). Fish have the opportunity to pass freely through the installation or go around it.
- In case of need the spatial arrangement of installation can be changed both by the depth and in the plan. The installation can be manufactured in the dry dock, delivered or towed to any given place and installed there with minimum installation works.
- The power production of the installation is determined by the regime of the flows speeds regardless of their direction. Under the conditions of stable flows typical, for example, for the straits of Cook Inlet, Gulf Stream or Southern Kuriles ridge, the total hours of utilization of the installed capacity per year can reach 6,000-8,000 hours, while under the conditions of tidal flows in case of optimum project it can reach 3,400 hours per year.
- In connection with exceptional simplicity of construction the cost of the installed kilowatt in case of calculated flow speed of 3 m/s for the units of average capacity of about 2-5 MW or more is evaluated within the limits of 500 – 600 dollars/kW in case of manufacturing of 10 and more sets. The expedient capacity of one unit amounts from 2 to 75 MW.
- The installation practically does not have the wear parts. Its operation life is determined only by the processes of corrosion and/or fouling by microorganisms. Both these processes can be controlled by contemporary technologies.

The torque developed by the blades on the upper and lower rings of the power unit during the work must be identical. Practically this is achieved by the fact that the absolute speed of the upper ring increases and of lower ring decreases (with the maintaining of the relative speed of rings) with the corresponding decrease of the torque on the upper ring and increase of the torque on the lower ring till their alignment.

The turbine power P depending on the speed of the incident flow U and the linear speed of blades V can be represented in two forms:

$$P = C_p \rho U^3 A_t /2 \tag{7.25}$$

$$P = C_N \rho V^3 A_t \sigma /2 \tag{7.26}$$

Here A_t = D H is the area of the axial section of the figure swept by the blades, (A_t σ) is the area of the median (chord) surface of working blades, σ = ib/D is the turbine solidity, i is the quantity of the blades in one tier, b is the blade chord. Presentation (7.25) is a traditional one, presentation (7.26) is introduced by the author for describing the asymptotic properties of orthogonal turbine characteristics. The efficiency factor C_p and the power factor C_N, which depend on the outlines of the turbine and the correlation of speeds V/U, are connected by the identity

$$C_N = C_p (U/V)^3/\sigma \tag{7.27}$$

The overall sizes of installation are determined by the design flow speed. At flow speed of 2 m/s the energy flow density amounts 4 kW/m^2 and in case of hydropower unit effectiveness of 50% the necessary cross section amounts 0.625 m^2 per kW of required power. For example, for obtaining 1000 kW the cross-sectional area of installation A_t= 625 m^2 is necessary. The blade's height (length) is determined according to the minimum depth of the flow taking into account the reliable lowering of the upper point of installation under ice. All basic units of the proposed installation were tested experimentally. Under one and the same boundary conditions the form of the turbine – blades and traverses quantity and form of section, flow speed and turbulence – has a great impact.

The energy effectiveness determined on the large hydraulic model (Fig. 7.4), which reproduced both tiers of blades but without the opposing motion, proved to be noticeably higher than 50% (Fig. 7.5).

According to the calculation and test data the power factor C_N is approximately described by the linear function (Fig. 7.6)

$$C_N = B\,(U/V - B_0), \text{ if } B_0 < U/V < 1 \tag{7.28}$$

Thus the energy characteristic of the turbine can be approximately specified only by two parameters B and B_0. Parameter 1/B_0 is the maximum relative speed of the blades, reached by the turbine without braking,

$$n_{max} = 60U/\pi DB_0 \,(\text{rpm}) \tag{7.29}$$

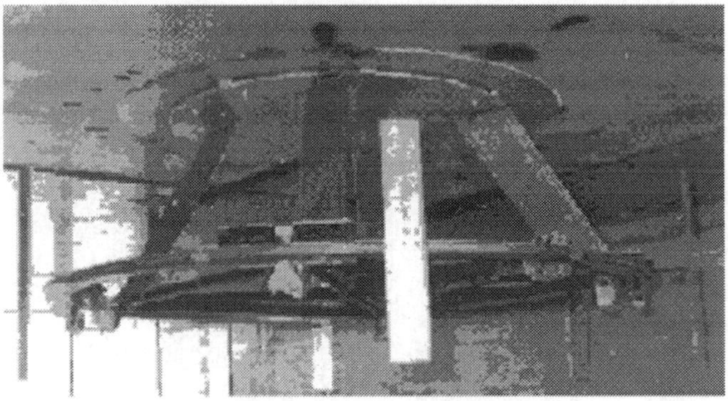

Fig. 7.4 The model of multi-blade rotor in the hydraulic channel.

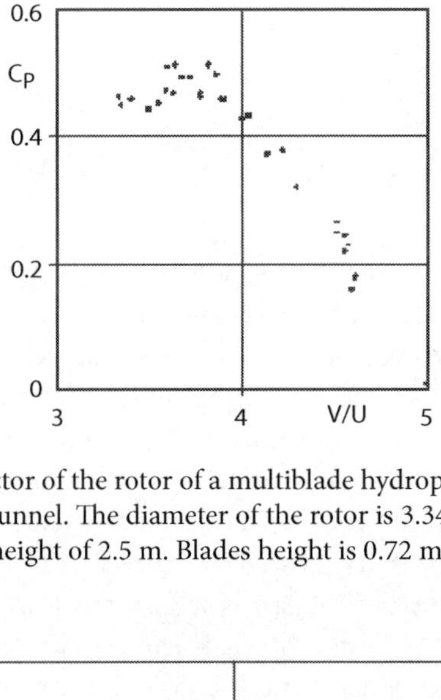

Fig. 7.5 The efficiency factor of the rotor of a multiblade hydropower unit according to the test data in the water tunnel. The diameter of the rotor is 3.34 m in the channel with the width of 6 m and the height of 2.5 m. Blades height is 0.72 m, chord is 0.12 m, solidity 0.22.

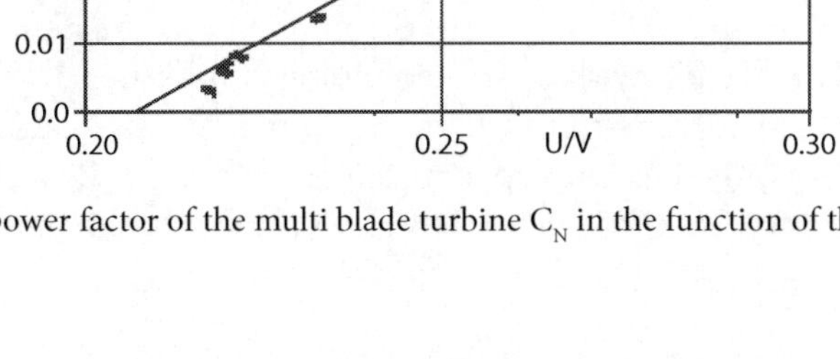

Fig. 7.6 The power factor of the multi blade turbine C_N in the function of the relative flow speed.

According to the data of Fig. 7.5 for the model of the proposed hydro turbine

$$B = 0.55, B_0 = 0.205 \tag{7.30}$$

The power output of the turbine depends on the rotational speed and the flow speed

$$P = B\rho\pi^2 D^3 H\sigma n^2 (U - B_0\pi Dn/60)/ 7200 \tag{7.31}$$

The maximum power P_{max} will be in case of optimum rotational speed n_{opt}

$$n_{opt} = 2/3\, n_{max} = 40U/\pi DB_0 \text{ or } V_{optimum} = 2U/3B_0 \tag{7.32}$$

$$P_{max} = C_{Pmax}\rho U^3 A_t / 2, \; C_{Pmax} = 4B\sigma / 27B_0^2 \tag{7.33}$$

All basic design features of the power unit except the counter rotation of different tiers were reproduced at the hydraulic model. In particular, the linear generator for power output was reproduced.

The described version of an orthogonal turbine is the best, but not the only possible one for implementation of the general idea of the use of tidal flows energy. Within certain limits the construction schemes of «helicoid turbines» (V.M. Lyatkher, I.V. Semenov – certificate of authorship of USSR №1150395 with the priority from 17.01.1983, A.M. Gorlov, Patent of USA 5451137, 19.09.1995), which ensures self-starting and the torque constancy, Davis turbines and any other the turbine blocks can be used. The drawbacks of these solutions are noted above. However in principle all these solutions are applicable, but according to the proposed idea these turbines or turbine blocks should not necessarily completely cut off the estuary of the used basin! To day any professionals begin to understand these general ideas and to construct the new kind tidal power plants without the dams using the usual propeller turbines – generators on the pillars, fixed on the bottom, or swimming (Hammerfest Strom Co., Marine Current Turbines Co., SMD Hydrovision with TidEI technology and other).

The proposed turbines as well as any other orthogonal machines are highly effective only in case of specific correlation of the flow speed and the blades speed. The speed of the tidal flow continuously changes. Therefore it is necessary either to change the turbine rotation speed or select this speed as a constant one, which ensures the maximum production.

It may be shown [1], that for periodical tidal with maximum speed of flow $U_{m.}$

The maximum value of power from one speed our new orthogonal turbine will be with optimum speed of the blades motion

$$V_{optim} = 0.57\ U_m/B_0 \tag{7.34}$$

For our experimentally studied turbine $B_0 = 0.205$ and

$$V_{optim} = 2.78\ U_m \tag{7.35}$$

For a helicoid three-bladed turbine with the solidity $\sigma = 0.41$ and the blades profile NACA 0020 in the best layout the parameters have the following values (our processing of A.M. Gorlov's experiments) $B = 0.60$, $B_0 = 0.34$, $Cp\ max = 0.35$.

In this case the optimum speed of blades motion of one-speed turbine amounts

$$V_{optim} = 1.7\ U_m \tag{7.36}$$

At that in any version the average power will have the maximum value equal to 0.378 from the maximum power of the turbine, which is possible in case of the given maximum flow speed. If the rotational speed of the turbine was not constant, but varied in proportion to the flow speed with retention of the maximum effectiveness of the turbine, then its average power would amount to 0.42 from the maximum value. Thus in case of accepting the constant rotational speed we lose approximately 10% of possible power production. With the use of a two-speed generator this loss can be decreased approximately 2 times.

The power production of the installation is determined by the regime of the flow speeds regardless of their direction. Under the conditions of stable flows typical, for example, for the straits of Southern Kuriles ridge or Shantarskiye Islands, the total hours of utilization of the installed capacity per year can reach 6,000-8,000 hours, while under the conditions of tidal flows in case of optimum project it can reach 3,400 hours per year.

In connection with exceptional simplicity of construction the cost of the installed kilowatt in case of calculated flow speed of 3 m/s for the units of average capacity of about 2-5 MW is evaluated within the limits of 500 –

600 dollars/kW in case of mounting of 10 and more sets. The installation practically does not have the wear parts. Its operation life is determined only by the processes of corrosion and/or fouling by microorganisms. Both these processes can be controlled by contemporary technologies.

Such hydropower units are installed in the strait, which connects the basin with the sea, where the flow speeds are sufficiently great. The optimum flow speeds for the proposed units vary from 2 to 4 m/s. If such speeds are not reached in the natural conditions, the preliminary or subsequent contraction of the strait by the coast dams or the ground enrockment is possible.

The proposed units can be produced in the dry docks of industrially developed centers and transported afloat as a whole or in partly assembled form to the site of installation. The expedient capacity of one unit amounts from 2 to 20 MW. The maximum quantity of units in one strait, which determines the maximum total power of TPP, is found from condition (3), (14) or (24). If the section of the strait is completely occupied with the proposed turbines, then according to the tests the coefficient of hydraulic resistance can be taken as $\xi = 1.3$. The proposed floating hydropower units installed in one row across the strait in many instances do not ensure the optimum value of parameter α and do not enable to obtain the maximum energy-conversion efficiency from the basin under development. In such a situation the turbines should be installed in several rows located from each other at a distance sufficient for restoring the normal velocity field. This solution is also useful for retaining the ecological clearness of the object. Comparatively slowly moving blades of the turbines, mounted in the machines sufficiently sparsely, are not dangerous for the fish, but the presence of tranquil flow zones between the rows of the turbines can be useful for the rest of the fish and does not require additional material expenditures. The proposed solution can be used for any basins.

For example, the Bay of Fundy is one of the best places in the world for using the energy of flows, whose height here reaches the record values of 18 m. Many projects of tidal power plants (TPP) worked out for this region during the past 100 years are known. One of them for the capacity of 20 MW in the region of Annapolis Royal was even implemented.

For Cobequid Bay, whose surface area is 264 km^2, the average rise of tide is 11.8 m, the width of the strait between the Ekonomi and Tenni capes is 8 km, the depth (maximum) is 42 m, the parameter α approximately amounts 1.6 10^{-4}. The maximum flow speed in the natural conditions should be about 0.75 m/s. In order to approach the optimum value of α, it is possible, for example, to decrease the width of the channel up to 2 km and install the turbines in 12 rows. In this case α = 0.031, the possible capacity of TPP approaches to the maximum value equal to P_{max} (kW) = 163x264x 11.8^2 = 6 GW, and the annual output will approximately amount 20 TWhr. In Canadian project of 1982 the capacity of this TPS in traditional assembling was determined as 4 GW and the production as 12.26 TWhr.

The proposed hydropower units in case of flow speed of 3 m/s and overall sizes of 48x24 m^2 can have capacity of about 8 MW each. 60 – 70 such units can be installed in one row. Their total capacity will amount 5760 – 6720 MW. It would be expedient to begin the implementation of the proposed approach in Canada not from such major projects, which require specific construction work (contraction of strait by the dams), but from more modest scales. It is proposed to arrange the first experimental-industrial units in the bay near Saint John. The assembling of serial production can also be arranged there. The following serious step is the arrangement of the units with capacity of 20 MW, the diameter of 48 m and the calculated flow speed of 4 m/s in the Split Cod region, opposite Parrsboro city in the strait, which connects Minas Basin with the Bay of Fundy (Fig. 7.7). In this site, where in everyday conditions the maximum flow speeds reach 4 m/s, 100 power units with a total capacity of 2000 MW can be installed

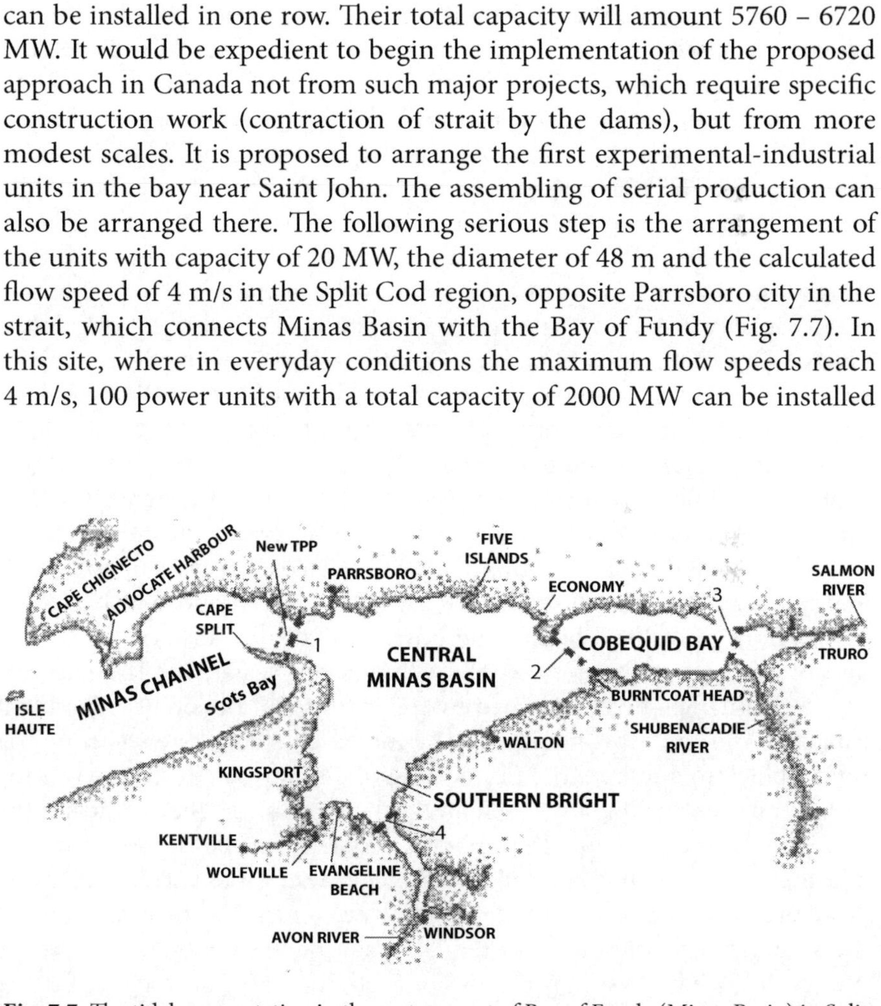

Fig. 7.7 The tidal power station in the eastern part of Bay of Fundy (Minas Basin) in Split cape site. The width of the strait in TPP site is 6 km (between Partridge island opposite Parrsboro city and Blomidon cape – 5 km), the area of Minas basin is 1150 km^2, the maximum rise of tide is 16 (18) m, the average rise of tide is 11.8 m. The maximum power of TPP is 48 GW, the maximum average annual production is 89 TWhr/yr. 1 – the site of the units of the proposed large TPP. 2-4 possible TPP to produce base power.

in one row practically without the additional construction work. Specific capital investments into this TPP will not exceed 600 US dollars per kW of the installed capacity in case of 5.5 thousand hours of usage per year. In succeeding years it is possible to increase the capacity of TPP up to the maximum value of 48 GW, determined according to (7), with a certain decrease of total hours of utilization of this capacity – the production up to 89 TWhr per year.

Favorable conditions for the building of TPP according to the proposed scheme exist in the bays of northern and west gulfs of Australia. For example, in the straits of Van Dienen Gulf, located near Darwin town, may be constructed TPP with installed capacity 84.7 GW and annual output about 163 TWhr per year. In Derby Bay capacity TPP may be up to 14.4 GW and output 27.6 TWhr per year. Good conditions for the building of TPP according to the proposed scheme exist in the bays of the West seas of India. In the Gulf of Kutch with a square of basin 170 km^2 and mean tidal range 5 m (maximum 6.6 m) the tidal power plant can have the maximum capacity of 1.2 GW with the production of 2.3 TWhr per year. In the Gulf of Khambat (area 1970 km^2, mean tidal range 7 m) maximum capacity of TPP may be 27.7 GW, annual power output – 53.3 TWhr per year. The tidal range in New Zealand varies from place to place; for example, at Wellington the maximum range is 1.35 m only, while at Auckland it is more 3.6 m (mean 2.7m). On the Manukau Harbor of Auckland (Fig. 7.7) are good conditions to construct the tidal power plant with new technology. The installed capacity of this plant without a dam may be up to 1.3GW and annual power output about 2.6 TWhours. We propose to begin the installation in this bay the floating power units with capacity 20-30 MW and power output about .50÷100 GWhours per year each. The numeral of these units will be determined by the investment possibility. The plant capacity (and the investment) may be increased step by step – from one power block to another. The first testing plant may be constructed in the narrow channel leading into Tasman Bay between D'Urville Island and the mainland of South Island. The currents here reach speeds up to 3.5 m/sec, and the cost of the power would be the cheapest. Tidal currents in Cook Strait are particularly interesting. High water on the North Island side of the strait occurs five hours before high water on the opposite side, which means that when it is high tide on one side it is almost low tide on the other.

This difference in sea level across the strait gives rise to strong tidal currents. The speed and duration of these currents are made quite variable by the heavy gales which are often encountered in this area and by other meteorological conditions. Particularly strong tidal currents which reach velocities of 2.5 to 3 m/sec are found off Cape Terawhiti. In the other side, the difference of the tidal regime may be used to receive the basis component in the total tidal power – when one plant will stay, the other will generates maximum capacity. Favorable conditions for the building of TPP according to the proposed scheme exist in the bays of northern and eastern seas of Russia. In Tugurskaya Bay (Figs. 7.8, 8a, 90), where the rise of tide reaches 10.1 m, the tidal power station can have the maximum capacity of 19.4 GW with the production of 48 TWhr per year. Hydrological conditions there are sufficiently favorable – everyday speeds exceed 2 m/s. The first experimental units with high economic effectiveness can be arranged in Lindholm Strait (Fig. 7.9), where the measured natural flow speeds exceed 4 m/s. Very tempting conditions exist in the northern part of Penzhinskaya Bay with the area of 5650 km^2 (Fig. 7.10), where the large and very effective TPP of the proposed type can be arranged between Elistratov and Mametchinskiy peninsulas without the changes of the ecological regime of the sea and conditions of navigation. Penzhinskaya Bay is relatively narrow water basin, which juts into the continent by 306 km, its average width is about 63 km. The narrowest place of the bay – narrow entrance to bay – stretches in the length on 7 km with the width of approximately 26.5 km. The tidal wave entering into Penzhinskaya Bay from the

Fig. 7.8 TPP without the dam near Auckland. 1- TPP general, capacity up to 1.3GW, 2-3 – possible TPP to produce the basis power.

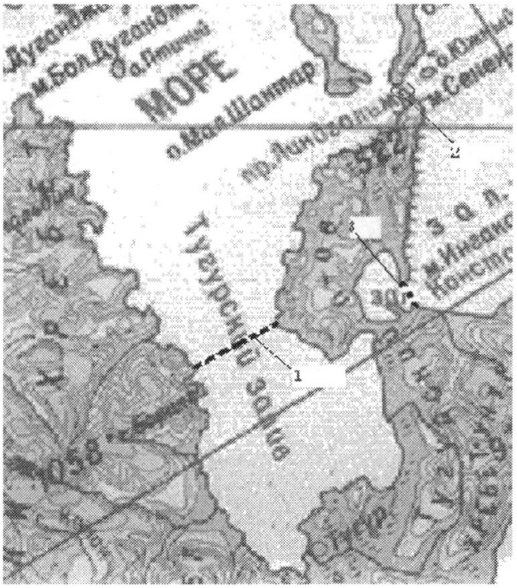

Fig. 7.9 The tidal power station in the narrowness of Tugursky Bay. The width of the strait is 15 km.

Sea of Okhotsk considerably increases. In the apex of the bay the magnitude of tide reaches 13.4 m, that is the maximum value in Russia and the second largest value in the world after the Bay of Fundy. In the quadrature the maximum flow amounts about 5 m, thus the root-mean-square value of rise of tide approximately amounts 10 m.

The maximum rise of tide is 10.1 m, the average rise of tide is 8.6 m. The area of the exploitable part of the bay is 1400 km^2. The maximum possible power of TPS is 19.4 GW. The maximum average annual production is 48 billion kwhr/yr, 1) the site of the units of the proposed TPP, 2) zone of arrangement of the first pilot unit of large capacity, 3) possible TPP to produce basis power.

With a calculated flow speed of 4.8 m/s the power of one turbine is about 75 MW, the power of one conditional block, which occupies the total depth of the flow amounts from 300 to 75 MW for the depths from 120 to 40 m. The coast sections of the strait with the depth less than 40 m remain free for navigation and are used for energy purposes in the last turn. In one site by the width of the strait up to 100 units of TPP of total capacity near 20 GW are installed. Eight or nine such sites are installed along the length of the strait at the distance of approximately 1 km one from another, forming the total capacity of TPP of approximately 160-180 GW. The average

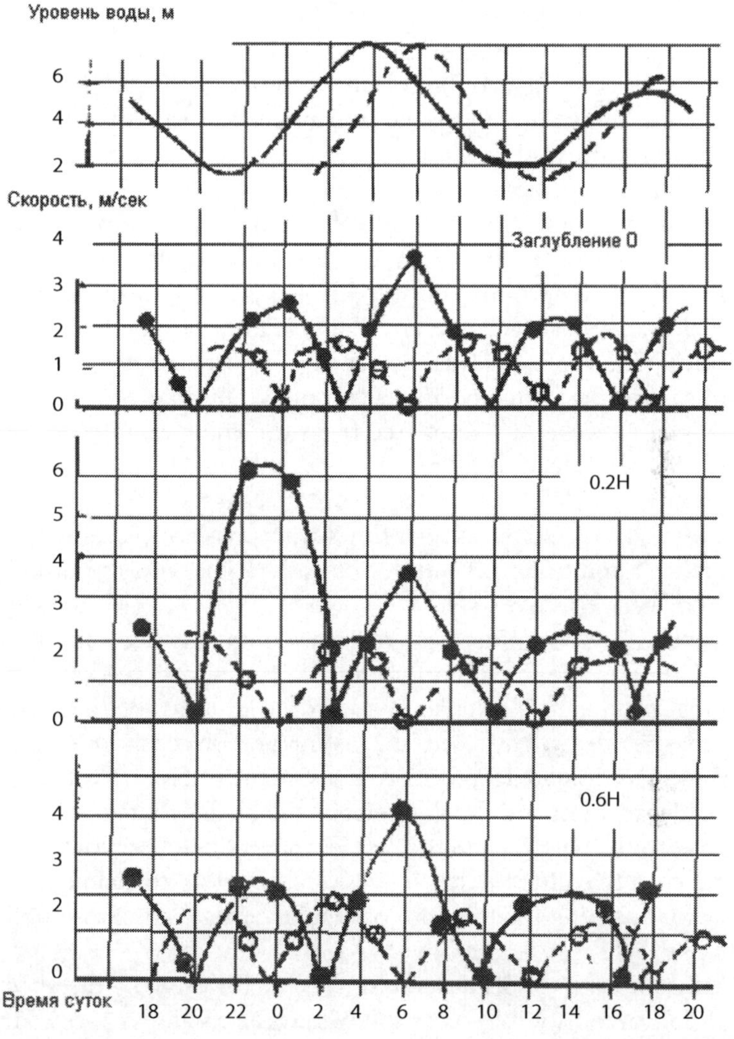

Fig. 7.10 The water levels and flow speeds in Tugur and Lindholm (solid lines) straights.

annual production of this TPP is expected not less than 312 TWhr/yr. The boundaries of turbines location are marked by the pneumatic screens creating the local inclinations of free surface of water, which impede penetration of ships and large fishes into the turbines zone.

The prime cost of energy of TPP under the conditions in question will not exceed 0.002-0.005 $/kwhr. For the filling of power dips of off-system individual tidal power stations the integration of tidal power stations of the proposed type with the tidal power station of the traditional type, whose power is proportional to rise of tide at the given moment, can also be used [3]. All elements of the proposed unit are individually studied and tested in the experiments. The designing and manufacturing of demonstration experimental-industrial models with real capacity, for example, of 10-50 MW is of current importance.

Hydropower units are installed in the strait, which connects the basin with the sea, where the flow speeds are sufficiently great. The optimum flows speeds for the proposed units vary from 2 to 4 m/s. Such speeds in the tidal flows are observed in many places of the straits of many Countries.

The proposed scheme of tidal power stations with the new units is highly economical and very effective in respect of execution of works and capital investments. However, it is not free from the changeability of the issued power in the time depending on the regime of tidal flows. This drawback can be removed by electrical integration of separate TPP of the proposed type installed in the zones of tidal flows with phase displacement. In spite of severe ice conditions TPP under consideration with the units of new type can be very effective.

According to the formula (7.7) the maximum installed capacity of TPP corresponding to the maximum rising tide can be 165 GW, and the average annual production, calculated according to (7.9) for the root-mean-square rise of tide, will amount 312 TWhr/yr. For conditions in question the turbine was drafted according to the scheme of Fig. 7.1 with the length of the blades of 12 m and the diameter of the blades route of 100 m. These turbines with the height of approximately 30 m (including lower pontoon) are installed at the different depth of flow with the shifting along and across the flow for ensuring the freedom of maneuvering during assembling and repairs (Fig. 7.12).

In nature this phase displacement. may be for the long distance points only. More general solution is as follows: at the small bay locates the TPP with maximum capacity and the power's phase late 4.65 hour from the tidal level maximum. The nearest large bay locates the other TPP with the same capacity, but relatively smaller possible capacity in this place. The phase of the power of this plant will be equal to the phase of flow (3.1 h from maximum tidal level) about.

The total power of these plants will have a large constant component. These two different TPP may be located in the narrows of one large basin (Figs. 7.7 & 7.8). It will be efficient because the most difference of the tidal range in the sea and basin would not be more than 22%, but phase difference will be 48 degree and more (Fig. 7.3). The total power of these two TPP with described new multi blade orthogonal units is

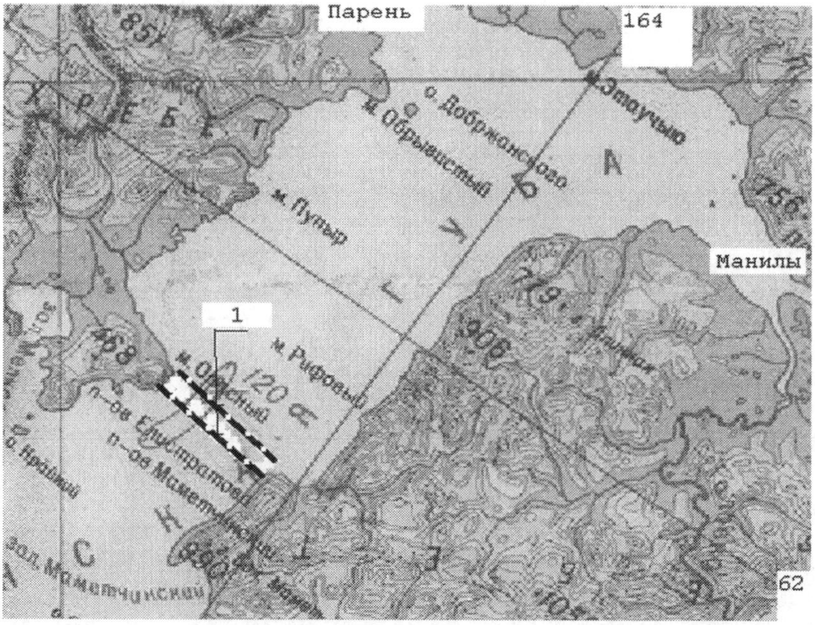

Fig. 7.11 The tidal power station in the northern part of Penzhenskaya Bay. The area of the basin is 5650 km^2, the width of the strait in TPP site is 26.5 km, the maximum rise of tide range is 13.4 m, the average range of tide is 10 m, the maximum power of TPP is 165 GW, the maximum averaged annual power output is 312 TWh/yr. 1 – the site of arrangement of the units of the proposed TPP.

$$P_{\Sigma} = P_0(t) + P_i(t), \tag{7.37}$$

here $P_0(t) = 0.55\ (0.36\,|\mathrm{Sin}\ (2\pi t - \varphi_0)| - 0.205)\{A_t\ \sigma\ \rho\ V^3\}_0\ /2$, если $0.57 < |\mathrm{Sin}(2\pi t - \varphi_0)|$ (7.38)

$$P_0 = 0, \text{ if } 0.57 > |\mathrm{Sin}\ (2\pi t - \varphi_0)|$$

$P_i(t) = 0.55\ (0.36\,|\mathrm{Sin}\ (2\pi t - \varphi_i)| - 0.205)\{A_t\ \sigma\ \rho\ V^3\}_i/2$, если $0.57 < |\mathrm{Sin}(2\pi t - \varphi_i)|$ (7.39)

$$P_i = 0, \text{ if } 0.57 > |\mathrm{Sin}(2\pi t - \varphi_i)|$$

The long time the sum (7.37) may be more the fixed numeral for different conditions.

For example, a distance about 10 km from Reykjavik (Iceland – Fig. 7.14) locates the bay Hvalfjordur (square35 km^2). The tidal range in this

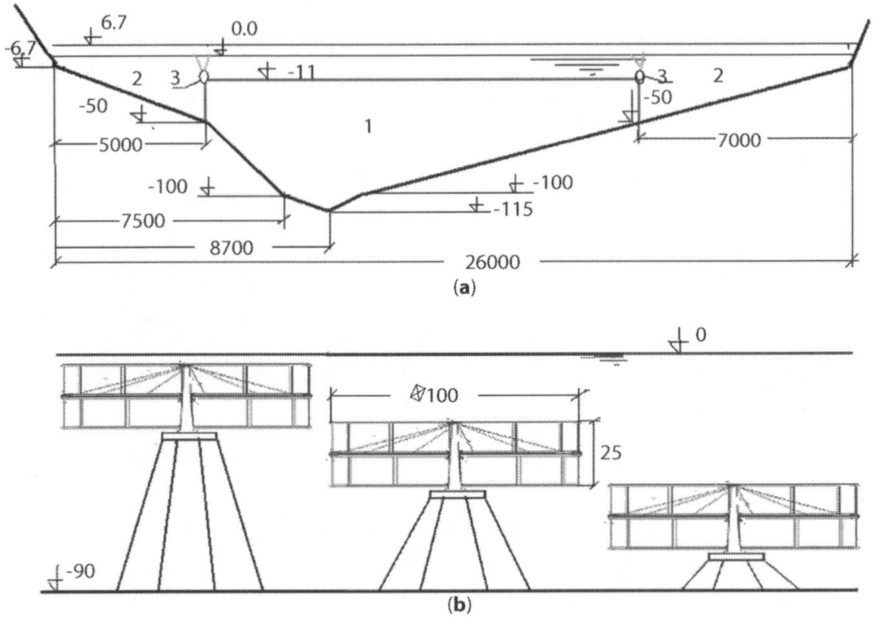

Fig. 7.12 The conditional cross section (a) and the unit of the turbines (b), located in the deep section of TPP site. The power of unit is 225 MW. 1) zone of TPP units, 2) navigation area, 3) pneumatic enclosure of TPP.

area is 4.7 m and capacity of TPP may be 126 MW. On the north, near Stykkisholmur the bay Hvammsfjordur has the square of 400 km^2 with tidal range 5.2 m and phase difference ½ hour. The possible power of TPP at this point may be 1763 MW. The TPP of 130-200 MW will not change the tidal regime here. The complex of these two units will have a base capacity about

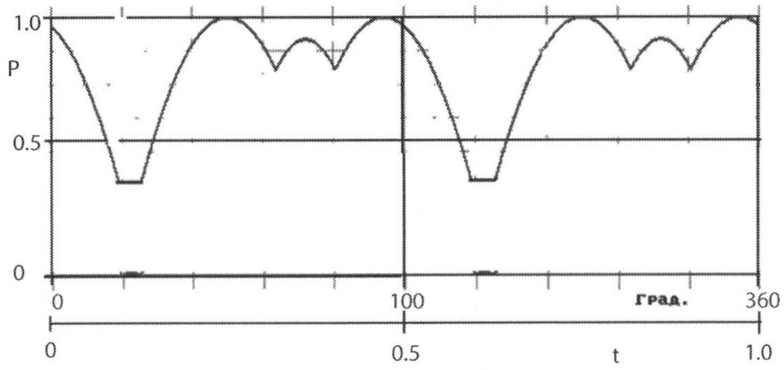

Fig. 7.13 Total power from two TPP. In the inside straight α = 0.2.

Fig. 7.14 The bays of Iceland to build TPP with basis power.

100 MW or more. Many of the same complexes may be built in Iceland and other countries.

For the filling of power dips of off-system individual tidal power stations the integration of tidal power stations of the proposed type with the tidal power station of the traditional type, whose power is proportional to rise of tide at the given moment, can also be used [3]. The universal method of compensating for the changeability of issued power is the organization of production of gaseous or liquid hydrogen, accumulated at the site of production of electric power and transported to the place of consumption by means of pipe-lines, special sea or aviation transport facilities. TPP capacity can be increased gradually and in principle may not be brought to the maximum value, remaining within the limits of economic optimum with simultaneous development of power-consuming production.

For example, such production can be the production of liquid hydrogen for the energy purposes. In view of orientation on the achieved energy intensity indices of this production – 53 kWhr/kg of gaseous or 65 kWhr/kg of liquid hydrogen – Penzhinskaya TPP will allow it to produce almost 4.8 million tons of liquid hydrogen per year or on average 2.14 m^3/s. In the composition of the prime cost of hydrogen the energy costs amount up to 60÷80% [4]. Therefore the price of hydrogen, produced at Penzhinskaya TPS, can be less than 0.5 dollars per kg. There exist experimentally substantiated proposals on abrupt (several times) reduction of energy intensity of the process of obtaining hydrogen from the water (Mogilevsky I.N., Surikov A.K., Ovsiannikov E.M., the patent of Russia 2224051 with the priority from 17.02.03).

Thereafter the volume of hydrogen production can be increased and the costs can be reduced. The obtained hydrogen is either accumulated during the winter period and removed by tankers in the navigation season or removed during the whole year with the aid of specialized aerostatic apparatuses (elaboration of Prof. O.A. Chembrovsky). With the expansion of the scales of the use of ecologically clean renewed energy sources in this region, for example, owing to the building of other tidal or large wind power stations the creation of cryogenic pipeline transport system with the use of high-test and flexible piping from the quartz glass, laid at the bottom of the sea, can be economically expedient. The same cryogenic pipes can also be used as cable channels with superconducting conductors from pure aluminum.

Conclusion

It is proposed to use the energy of tidal flows without the creation of a solid head front and cutting off the area of water of the power station basin from the sea. This approach allows to repeatedly decrease the initial capital investments, drastically reduce the terms of setting in operation the first units, and repeatedly decrease the volumes of the works carried out in TPP sites. In this case the new type of the orthogonal multi-blade turbines of large capacity, proposed by the author, can be very effective. The optimum flow speed for the orthogonal turbines amounts from 2 to 4 m/s. Such speeds have a place in the straits of large bays in Canada and the east seas of Russia, which permits to raise the question about the effective use of these flows with the aid of floating power stations of new type without carrying out the construction work in TPP site.

It is expedient to use the energy of flows in the regime of the maximum turbines effectiveness. In a region with the developed power system (Canada, Australia) this energy can be issued into the electrical network, displacing the capacities of usual hydraulic or fuel power stations. In the uninhabited regions (east of Russia) it is expedient to direct the energy of TPP on the production and accumulation of liquid or gaseous hydrogen. This hydrogen can qualitatively change the energy situation in the Far East of Russia and can also serve as the object of export into Japan, Korea and China, substituting traditional hydrocarbon fuel.

References

1. Lyatkher V.M. Tidal Power Plants without the Dams, *Hydrotechnical Construction* 4, 2006, Moscow.
2. Lyatkher V.M. Power Potential of the Tidal Power Plants, *Izvestia RAS* 4, 2006, Moscow.
3. Lyatkher V.M. Complex of tidal power plants, ensuring the established schedule of issuing of power, *Hydrotechnical Construction,* 12, 1998, Moscow.
4. Walter Peschka, *Liquid Hydrogen: Fuel of* the Future. Springer-Verlag, 1992.

8

High Jet Power Plant

In the North region of Bangladesh, it may be efficient to use large power plants using the power of high jet streams. Our purpose is to propose to create an operating demonstration wind power plant with capacity **50-100 MW**, having high economic efficiency at full ecological safety [1]. It is important that the proposed system is robust even in Autonomous conditions at the breakage of the mechanical linkage with the earth.

The proposal refers to wind-power engineering, namely to wind power plants that use the energy of altitude jet streams. Technical result consists in efficiency coefficient increase, material-output ratio decrease, plant safety improvement, provision of secure communication with the land. Altitude wind power plant includes aerodynamic components in a form of a framework, the units of which are located on the elliptical surface; the framework is connected with ground support by cables and on the framework along its span there are located orthogonal well-balanced bladed rotors arranged at an obtuse angle in relation to each other. The cables are performed in a form of sectional hoses, the sections of which are filled with helium and are connected by gates. Rotors are located between framework units, their blades are equipped with jet devices of circulation control. The jet management in the boundary layer of the blades may be formed from an air tank with high pressure or formed during the combustion of the fuel mixture fed into the internal cavity of the blades.

There is a known altitude aerostatic plant that holds aircraft performed in a form of cargo-carrying kite, the load-bearing surface of which has an annular wing consisting of aerodynamic components with different in curvation and chord by span. Around the periphery of the ring there are wind-engines behind the rear edge of the aerodynamic components and wind-engines in horizontal plan before aerodynamic elements, note that the engines have different rotation direction. Each wind-engine has a compressor of centrifugal type, speed-increasing gear and blade propeller mechanically connected with each other, flow parts are interconnected by connection of the outlet of the following part with the inlet of the previous one. Note that the outlet of the last compressor is connected with the input

Victor M. Lyatkher and Ziaur Rahman. Wave, Wind and Current Power Generation, (257–282) © 2022 Scrivener Publishing LLC

of a power machine located on land by pressure hose [2]. The disadvantage of this decision is the need to use aerostatic structure as well as insufficient security of the system in case of land fixture rupture.

There is a known wind plant for high altitudes [3]. It consists of a load-carrying surface along the span of which is located is a sectional wind-catching device with generators attached with the help of attaching elements on the support. The wind plant has a fixed balloon in the form of a load-carrying surface with an empennage. Along the span of the load-carrying surface there is a sectional wind-catching device formed by several multi-blade rotors with horizontal rotation axis located close to each other. Rotor annular rim is installed with the possibility of rotation at roller supports attached in cantilever fashion on a load-carrying surface, where there is a two-shaft generator. The disadvantage of this decision is the need to use aerostatic structure, rotors with horizontal rotation axis, lack of secure connection with the land.

There is known "the best invention of 2008y" SWP technology (www.skywindpower.com) using 4 HAWT located vertically in the angles of the frame, connected with the ground by the cable-tross. The disadvantage of this decision is the low efficiency of the wind power system in cross wind flow.

Technical results of our proposed unit are: efficiency coefficient increase, material-output ratio decrease, plant safety improvement, provision of secure communication with the land. Technical results are achieved by the fact that in altitude wind power plant consisting of load-carrying aerodynamic component connected with ground support by means of cables and orthogonal well-balanced bladed rotors arranged at an obtuse angle in relation to each other along its span the aerodynamic component is performed in a form of a framework, the units of which are located at elliptical surface, cables are in a form of sectional hoses, the sections of which are filled with helium and are connected by gates, rotors are located between framework units with their blades equipped with jet devices of circulation control. Furthermore, in wind power plants rotor blades are made with channels for air supply control through the holes on blades surface into jets that control the circulation around the blades. This provides the possibility to control the value and direction of total average lifting force that is applied to rotors and plants in general.

Orthogonal rotors rotate in opposite directions providing the balancing of torques and lateral forces. Jet devices in rotor blades designed for local circulation control increase plant power generation and, if necessary, in conditions of no wind at plant lifting and lowering create lifting force, direction and value that can be changed allowing plant control as an aircraft. Rotor axes form an obtuse angle that provides a system turning down the wind and its stable balance. In case of cable rupture or one of rotor destruction the plant overturns and smoothly lands to the ground maintaining automatic control by radio beacon station. The sections of the cable, performed in a form of hoses, and some elements of the frame are filled with helium under excessive pressure and create lifting force of cable allowing optimization of wind plant position by height. The main constructive ideas of the balanced turbines are protected by patents of Russi [4], patents of the USA - Lyatkher V.M. [5] and Gorlov A.M. [6]. New innovative technologies and materials can improve the yield and power of wind turbines and can improve the ratio of energy yield investing costs. Currently parallel developments are the R&D activities in the field of high altitude wind turbines. The power and maximal energy will increase for these turbines, because of the higher average wind speed at higher altitudes. This will increase the economic feasibility of wind energy. In Fig. 8.1 distribution of streams of wind power over Moscow at different times of the year is shown.

In Fig. 8.2 distribution on height of density of a stream of wind power over the different cities of the world is shown [7].

The U.S. jet stream covered the entire Central part. Over Northern California, the energy flux density in the jet reaches a height of 6 km on average 7-10 kW/m2, over new York – up to 16 kW/m2. One of the most favorable places for projects on the use of jet streams is almost the entire East coast from Florida to New England. Total reserves of wind energy in the atmosphere is estimated at $4x10^{12}$ kW or $3.5x10^{16}$ kWh/year, of which less than 7% refers to the bottom layer thickness of 100 m, mostly above the water surface. Energy "wind river" - the jet stream is 100 times more

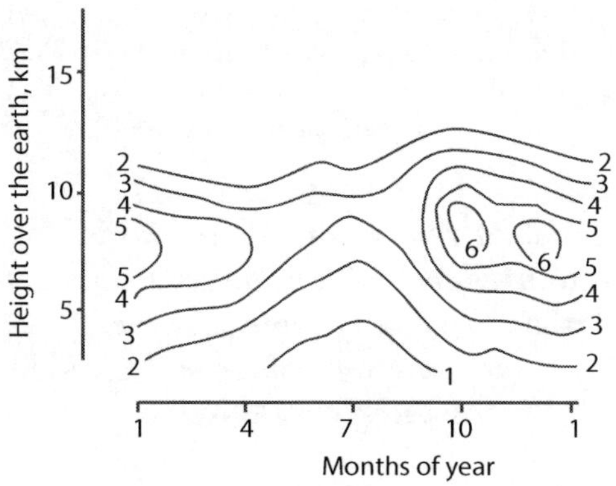

Fig. 8.1 Isolines of mean streams of wind power (kW/sq.m) over Moscow.

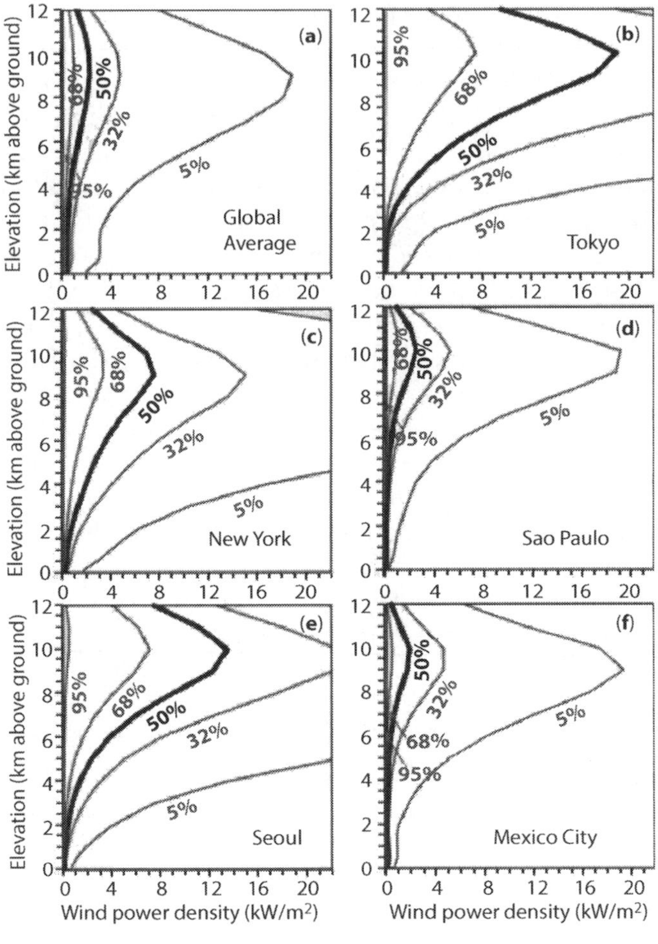

Fig. 8.2 Wind power density (kW/m²) that was exceeded 5%, 32%, 50%, 68%, and 95% of the time during 1979-2006 as a function of altitude from the NCEP/DOE reanalyses by Cristina L. Archer, and Ken Caldeira [7]. The profiles at the five largest cities in the world are shown in (b-f). The global average profile (a) is the area weighted mean of values like those represented in panels (b) through (f) at all grid points.

hydropower potential of rivers around the world. Half percent of this energy is enough to provide 8 billion future inhabitants of the Earth by modern standards the most comfortable countries.

The wind power plant operation principle is the following. Each power plant consists of three structures, connecting in one space building. Each of structures consists of three or more assemblies of orthogonal rotors united by one building unit (Fig. 8.3 shows the structure model with three

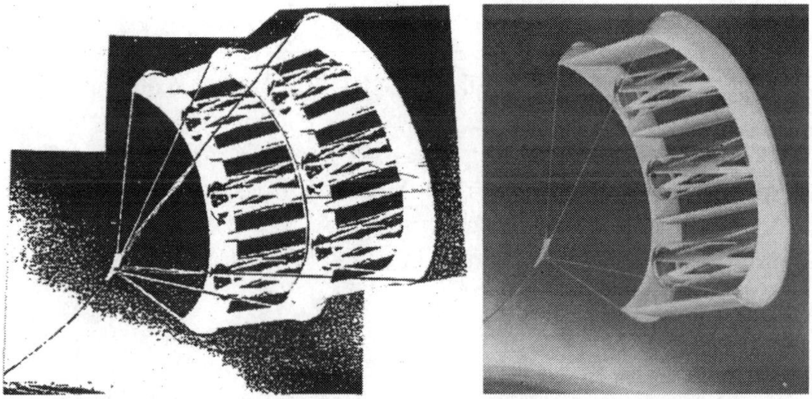

Fig. 8.3 The blocks of High-Altitude Wind Power Plant.

rotors, which is the minimal plant model), the components of which are located at an imaginary elliptical plane (Fig. 8.4). The orthogonal rotors create circulation which assures the necessary lifting capacity in the normal conditions.

The power extracted P(W) of a orthogonal turbine or complex turbines may be defined by the follow relation

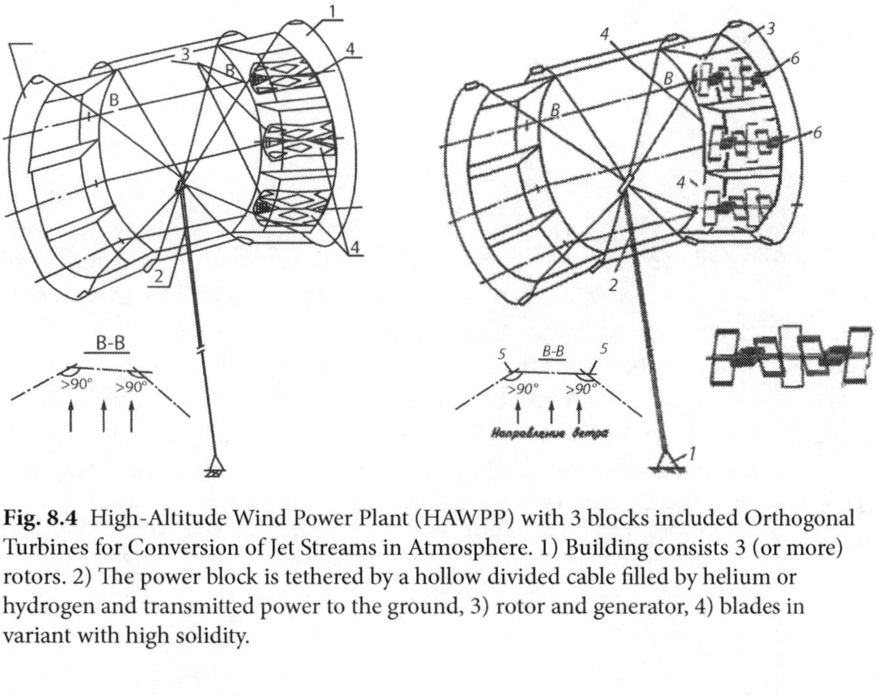

Fig. 8.4 High-Altitude Wind Power Plant (HAWPP) with 3 blocks included Orthogonal Turbines for Conversion of Jet Streams in Atmosphere. 1) Building consists 3 (or more) rotors. 2) The power block is tethered by a hollow divided cable filled by helium or hydrogen and transmitted power to the ground, 3) rotor and generator, 4) blades in variant with high solidity.

$$P = C_N \rho V^3 \Omega_b /2, \text{где } \Omega_b = A\sigma, \sigma = ib/D \quad (8.1)$$

Here V (m/s) = πDn/60 -blades speed, Ω_b (m^2)- area of blades, A = D H– swept area, σ -solidity, i- numeral of blades, b – chord of blade, D – blade's way diameter, ρ (kg/m^3)– density of air. Power factor C_N is a function of the turbine shape, advance speed ratio V/U, roughness of the blades surface, etc. U (m/s) – steady free-stream wind velocity.

Power factor C_N is approximated by linear function

$$C_N = B (U_0/V - B_0) \quad \text{if } 0 < V < U/B_0 \quad (8.2)$$

Parameter $1/B_0$ is the maximum of blade speed without a brake torque moment and constant wind speed.

$$n_{max} = 60U/\pi DB_0 \text{ (rpm)} \quad (8.3)$$

Turbine capacity as a function of frequency of rotation and wind speed is

$$P = B\rho\pi^2 D^3 H\sigma n^2 (U_0 - B_0\pi Dn/60)/2\ 60^2 \quad (8.4)$$

Maximum of turbine capacity P_{max} will be with frequency of rotation n_{opt}

$$n_{opt} = 2/3\ n_{max} = 40U/\pi DB_0 \quad (8.5)$$

$$P_{max} = C_{Pmax}\ \rho U^3 A/2,\ C_{Pmax} = 4B\sigma/27B_0^2 \quad (8.6)$$

For the 3-blades turbine (σ = 0.45), optimized by prof. A.Gorlov was fined: B_0 = 0.336, B = 0.60. Turbines with 3 and 6 curved blades (σ = 0.45 and 0.90), proposed and tested by the author in 1981y were shown: B_0 = 0.28 and 0.34, B = 0.44 and 0.14 for σ = 0.45 и 0.90. These results may be used to estimate the sizes of turbines, blocks and plants. Many other results are published in the books [8, 9]. Real-optimized design of orthogonal wind turbines according to the experimental data are efficiency, reaching 43%. Optimization is achieved by selecting shading, frame rotor design, and reversal of the blades (Fig. 8.5).

Longitudinal F_x and lateral force F_y acting on the orthogonal machines are characterized by the coefficients C_x, C_y (Fig. 8.6). $F_x = C_x \rho U^2 DL/2$, $F_y = C_y \rho U^2 DL/2$, L is the length of the turbine rotor, D is the rotor diameter.

The maximum efficiency of the orthogonal turbine increases with decreasing number of blades while maintaining the optimum shading. However, with the decrease in the number of blades increases the torque pulsations and all the forces acting on the rotor.

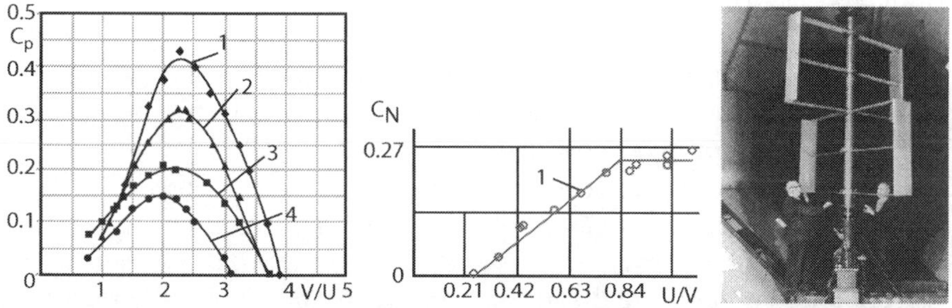

Fig. 8.5 Maximum efficiency of a frame two-bladed rotor with a solidity 0.3 reaches C_p = 0.43 at V/U = 2.2. *Re* = 4,0·10^5. 1: φ = -5^0; 2: φ = 0; 3: φ = -10^0; 4: φ = +3^0. Tests ZAGI 10, 25.

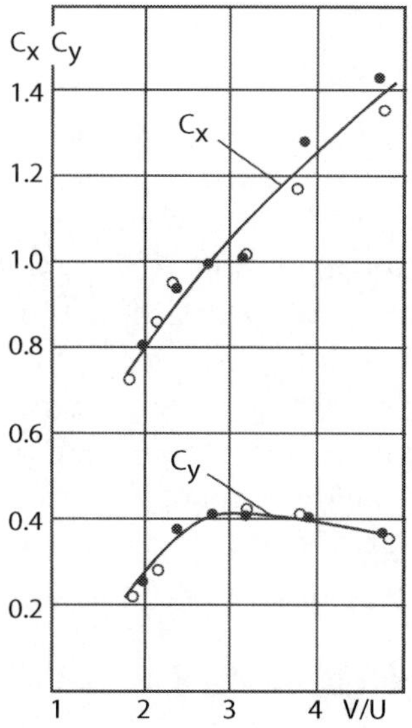

Fig. 8.6 The relative loads acting on one tier of two-bladed turbine. Re_V= 8.5x10^5. Solidity 0.306. Turbine on Fig. 8.5.

If design wind speed according to nominal capacity 50 MW will get 35 m/s, power efficiency blades speed has to be 50 m/s about. The total surface of the blades must be 6,410 m^2, solidity $\sigma = 0.45$ and sizes of the plant will be 120x120 m^2, one block – 40x120 m^2.

Besides, the rotor blades are equipped with jet local circulation control devices. These devices are to boost the energy generation by the plant, and when it is necessary (there is no wind, or the plant is being lifted or lowered), they can create the lifting capacity the direction and value of which can be modified, which permits to guide the plant like an aircraft. The axes of the rotors form an obtuse angle, which makes the plant turn downwind automatically and keep its stable equilibrium. In case of emergency (cable breakdown or destruction of one of the rotors) the plant turns over and smoothly goes down to the ground or to the lake. It can still be automatically controlled through radio beacons.

The operation of the circulation control system is based on the idea that letting out an air stream to the blades at the right moment can change considerably the lifting capacity of the blade in the wind flow, change its direction and even create the lifting capacity on the rotor revolving in the still air. Jet devices in rotor blades designed for local circulation control increase plant power generation and, if necessary, in condition of no wind at plant lifting and lowering create lifting force, direction and value that can be changed allowing plant control as an aircraft. Rotor axes form an obtuse angle that provides a system turning down the wind and its stable balance. In case of cable rupture or one of rotor destruction the plant overturns and smoothly lands to the ground maintaining automatic control by radio beacon station. The sections of the cable, performed in a form of hoses, are filled with helium under excessive pressure and create lifting force of cable allowing optimization of wind plant position by height.

Rotor blades are equipped with jet devices of local circulation control. These devices should increase power generation and, if necessary, in conditions of no wind at plant lifting and lowering create lifting force, direction and value that can be changed allowing plant control as an aircraft. Rotor axes (in plan) form an obtuse angle that provides system automatic turning down the wind and its stable balance. In case of an accident (for example, cable rupture or one of rotors destruction) the plant overturns and smoothly lands to the ground maintaining automatic control by radio beacon station.

The idea of a circulation control system on the blades consists in the fact that at the right moment of jet supply on the blade it is possible significantly to change blade lifting force in the wind stream, to change its direction and even to create lifting force at rotor that is rotated in stationary air. In real altitude wind station jet control schemes should be more complicated. To simple elements there added are elements that in case of emergency cable rupture allow the system not only smooth descending, but movement in specified direction so that the landing is done in specified place, for example, on the surface of the specified water body. Such complicated movement is performed due to the use of independent circulation control systems at each rotor (Fig. 8.7). This allows provision of necessary lifting force vectors and moments that deliver the station to specified landing point during falling (or lifting). Altitude wind power plant consists of carrying aerodynamic element 3 connected with ground support 1 with the help of hollow cable 2 with orthogonal blade rotors 4 arranged along the element at an obtuse angle in relation to each other, carrying inductors and rotor of line (curved) electric generators. Aerodynamic element 3 is performed in a form of framework, the units 5 of which are located at elliptical surfaces, the cable 2 is performed in a form of sectional hoses, the sections of which are filled with gas and connected between themselves by gates. Rotors 4 are located between framework 3 units 5, their blades 6 are equipped with jet devices 7 of circulation control. Rotor 4 blades are supported at the shaft

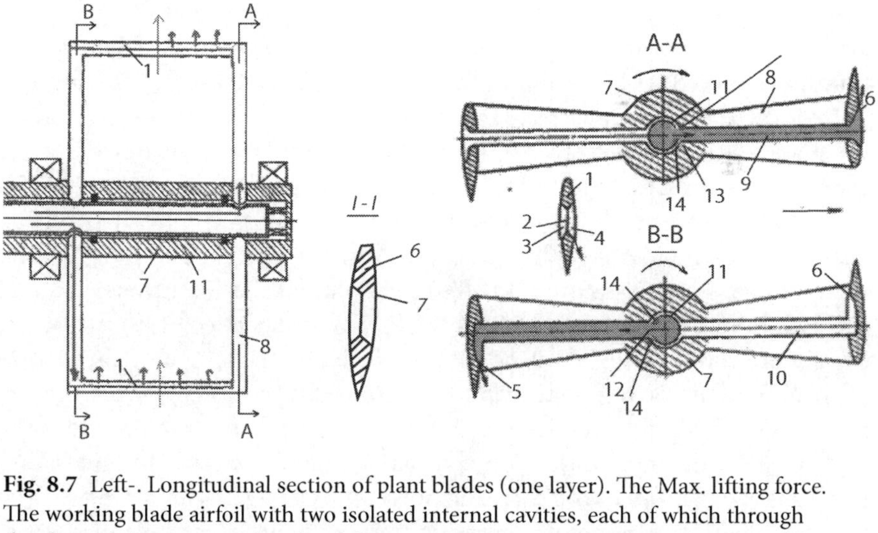

Fig. 8.7 Left-. Longitudinal section of plant blades (one layer). The Max. lifting force. The working blade airfoil with two isolated internal cavities, each of which through the channels in traverse 8 has the ability to communicate with the inside cavity of the main shaft 11, placed inside the rotating turbine shaft 7. When lifting forces on the blades is determined by the position of the holes in the control shaft 11, in front of the inlet channels in 8 traverses. Right - Sections of channels supplying air to blade surface for circulation control. The numbering of the items in Fig. 8.20 and patent [13]. 14 - inductance coil, ignition spark external or internal contour in the variant with the feed gas mixture.

8 by cross bars 9, 10. Hole 11, 12 position adjustment for controlling the maximum lifting force point is performed by distributing shaft 13 or special system of automatically controlled valves.

The plant is operated as follows.

Original lifting force is provided by coactions of lifting force of gas-filled hoses 2 and lifting force created by the work of generators of the rotors 4 in engine mode at adequate action of jet devices 7 of circulation control at rotor 4 blades 6. With the help of jet devices, for example, changing the position of holes 11, 12 of distribution shaft 13 in relation to the land, it is possible to change "support" direction and, thus, control the position of the wind plant in relation to ground.

Torque moment developed by rotors 4 at wind stream approaching is transmitted to generators, and from generators the generated power is supplied to the ground through current-conducting elements located in rod 2.

The arrangement of each pair of rotors 4 axes at an obtuse angle in relation to each other provides automatic turning of the whole system down the wind and its stable balance. In case of an accident (cable 2 rupture) the plant turns and smoothly lands maintaining automatic control by radio beacon stations from the land. Rotor 4 rotation and plant smooth descending is performed due to power batteries arranged at the plant and generators operation in engine mode. Each blade (1) has two separate volumes, connected with the shaft of the turbine (2) by traverses (3) and (4). Tube located in the traverse (3) is connected with the outside volume of the blades. Tube in the traverse (4) is connected with the inside volume of the blades. The distribution shaft (8) gets inside the shaft of the turbine (2). The distribution shaft has two opposite holes (6) and (7). The distribution shaft is connected with high pressure volume or pump by line (5). When the high pressure air is going through the hole 6 at the right traverse 3 and through hole 7 at the left traverse 4 the lift force acts at right. In the opposite situation the lift force will act at the left. Jet management of circulation on blades of the orthogonal turbine is an essential element of the offered project of the highly effective wind power station transforming energy of high-rise jet currents. In this project management of circulation on blades is used for formation of the carrying power providing maneuvering (lifting lowering) with the block of power plant irrespective of a mode of a wind. Stationary modes of a flow of profiles with jet control circulation are studied quite in detail (Fig. 8.8) [11].

Influence of a jet of management in non-stationary conditions, characteristic for modes of a flow of blades in orthogonal turbines, on the instructions of the author was studied on oscillatory installation in a

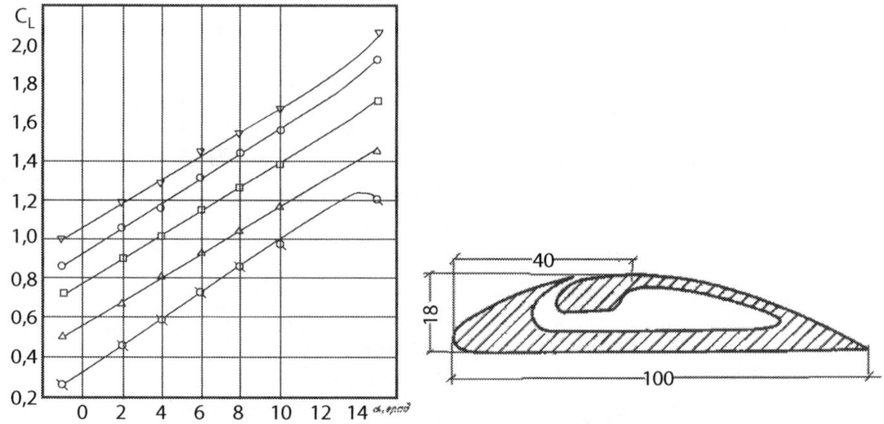

Fig. 8.8 Coefficient of a lifting force of a wing on tests in water as an angle of attack at impulses of a jet 0, 0.05, 0.10, 0.15 and 0.20.

flat hydrodynamic pipe at Institute of mechanics of the Moscow State University in 1989-1991 (supervisors – V.P. Karlikov, A.N. Khomyakov, G.I. Sholomovich). Tests are carried out in the conditions of a flat task with a considerable excessive pressure of water (2.5 bar), excluding aeration and cavitation on the blade, with the careful accounting of possible mistakes. Non-stationary and stationary characteristics were received for the NACA-0010, 0015, 0017.5, 0020 profiles with a chord of b=0.2m at a stream speed to m U=10.2 m/s.

Change of an angle of attack was carried out or with sign preservation - under the law

$$\alpha = 15^0 \mathrm{Sin}\ (2\pi ft) + 15^0 \tag{8.7}$$

or according to the scheme closer to real operating conditions of turbines, - with sign change:

$$\alpha = 14^0 \mathrm{Sin}\ (2\pi ft) - 1^0 \tag{8.8}$$

Not stationarity is characterized by number of Strukhal

$$Sh = bf/U \tag{8.9}$$

In the same hydrodynamic pipe with considerable excessive pressure of water (2.5 ати), on the same oscillatory system on which were tested

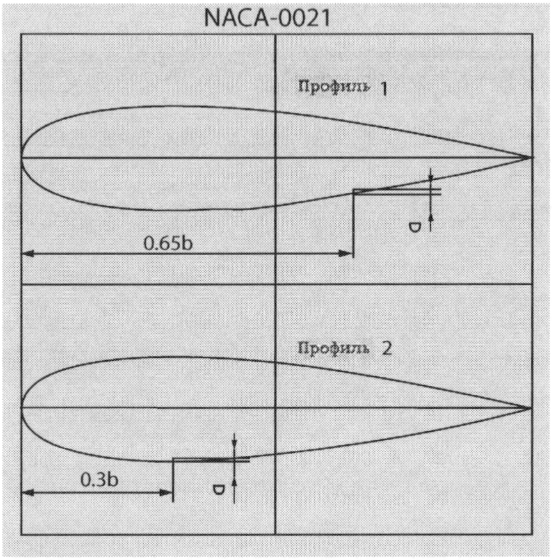

Fig. 8.9 Options of an arrangement of a crack for giving of a jet, b = 200 mm, a = 0.83 mm.

a profile without jet giving, characteristics of two options of blades of the NACA 0021 profile with a chord of b=200 mm with release of a flat jet by thickness of a=0.83mm in tail part of a profile (at distance 0.65b from a profile sock) – option 1 in Fig. 8.8 and in head part of a profile at distance 0.30b from a profile sock – option 2 were studied. Influence of a jet is characterized by the relative size of an impulse

$$C_s = (U_1/U)^2(a/b), \tag{8.10}$$

U_1 – expiration speed of jet, U speed of a running stream.

Change of an angle of attack was set by blade turn concerning an axis located in the middle of the blade (0.50b from a sock):

$$\alpha = 14.5^0\text{Sin}(2\pi ft) \tag{8.11}$$

The general conclusion of the carried-out tests consists that at both options of an arrangement of a jet its action on shady side very effectively increases lifting and pulling forces in stationary and in non-stationary modes (Figs. 8.10–8.14).

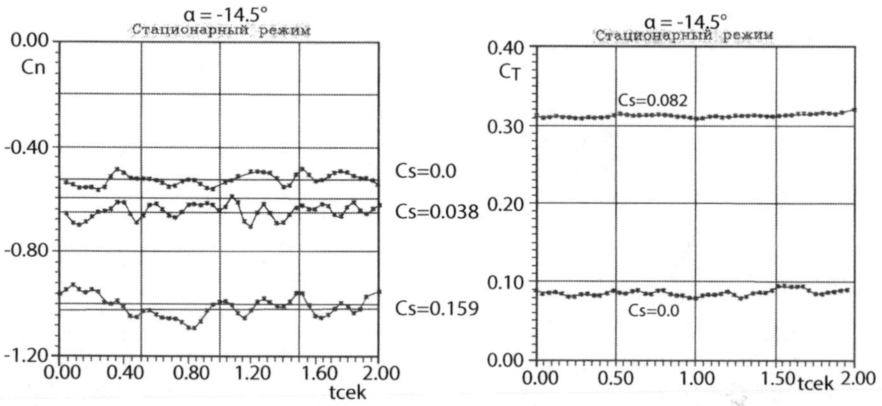

Fig. 8.10 Option 1. Jet on shady side at the end of a wing.

In the studied range of change of number Strukhal $0<Sh<0.113$ and a relative impulse of a jet $0<C_s<0.155$ influence of a jet is considerably shown only when it is on the shady side of a profile. In this case in both constructive options – at a stream arrangement at the beginning or at the end of a profile giving of a stream increases lifting and pulling forces of a profile, especially on big (subcritical) angles of attack in direct ratio to a jet impulse. At the maximum tested angle of attack 14.5^0 at a relative impulse of a jet 0.085-0.088 increase in carrying power made 60%, and increase in the pulling force – about 200% from values of these sizes without jet. Criterion influence Strukhal on this result was insignificant. Thus, nearly,

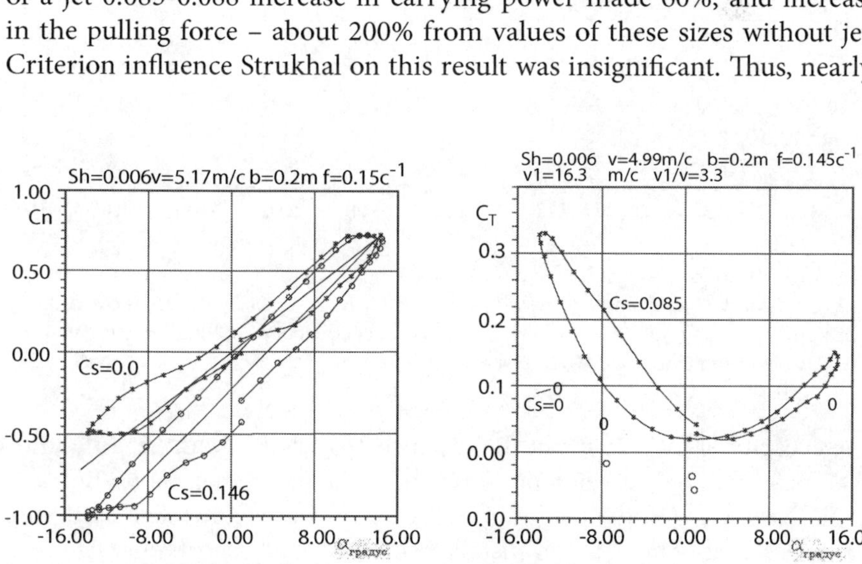

Fig. 8.11 At negative corners a jet on the shaded party. Jet influence at small numbers Strukhal.

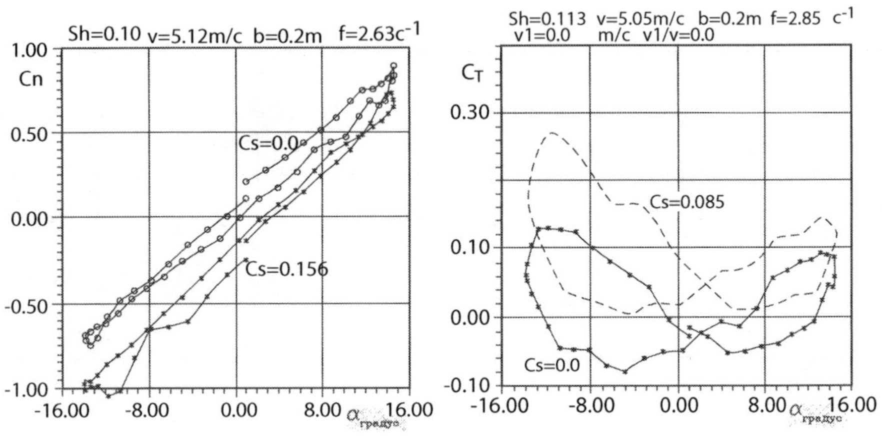

Fig. 8.12 Jet influence at large numbers Strukhal.

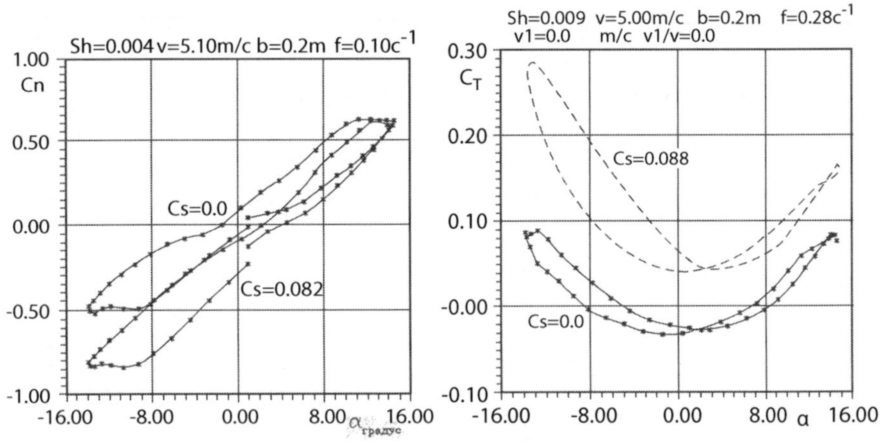

Fig. 8.13 Option 2 – a jet at the beginning of a wing. Small numbers Strukhalya. At a stationary mode and an angle of attack-14.5° coefficient of normal pressure is equal-0.5, -0.86 and -1.08 at an impulse 0, 0.082 and 0.15, the coefficient of pulling force is equal 0.09 and 0.29 at an impulse 0 and 0.086 respectively.

the capacity of the turbine at the expense of a jet can be increased almost by 3 times and the carrying power of the blade more, than by 1.5 times.

Inkjet devices in the blades of the rotor to control the local circulation increases energy production installation. Published reports that jet circulation control on the blades of the orthogonal model of wind turbine with shading 0.1 can increase the maximum efficiency of the unit CP from 0.32

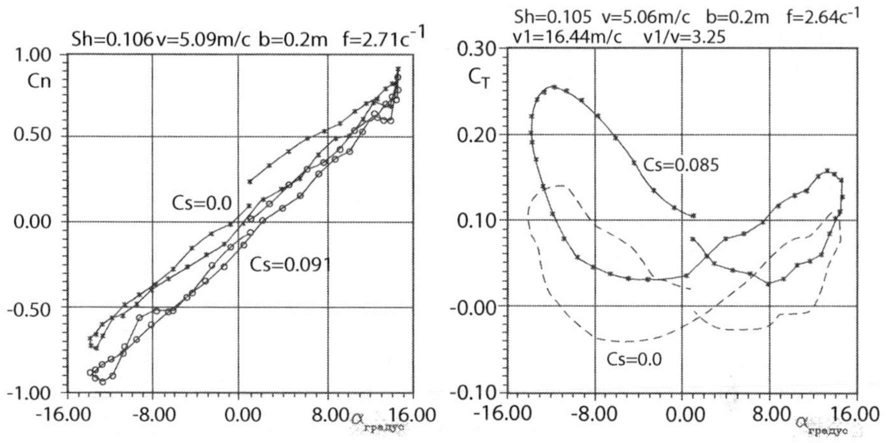

Fig. 8.14 Option 2. Large numbers Strukhal. The jet on shady side of a profile considerably increases the lifting and pulling force of a wing.

to 0.42 with a decrease in the optimal rotor speed from V/U= 5 to 4 [26] [12]. In our design control jets on the blades in the absence of wind, when the raising and lowering of the installation, create a lifting force, the direction and magnitude of which can vary, allowing you to control the installation as aircraft.

The main ideas of the turbine with operated circulation on the blades, providing any choice of the direction of the resultant force operating on the turbine as a whole, were checked on a large model of a fragment of the two-bladed unit (Fig. 8.15).

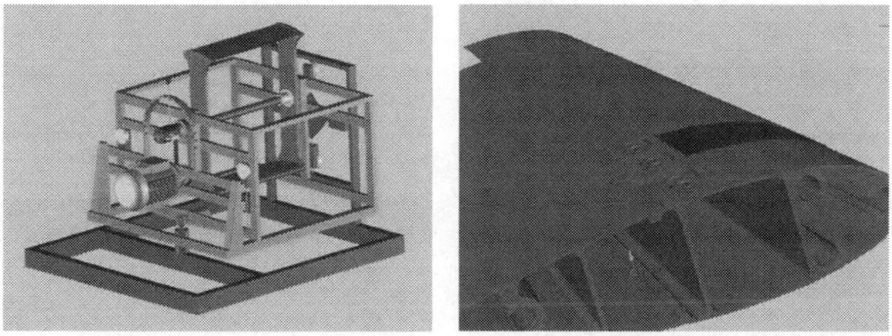

Fig. 8.15 General view of model and Scheme of giving of a stream. On the opposite side of the blade the stream moves from other channel in the blade.

The turbine model with hollow blades in which air through cuts in distributed to a shaft is brought, placed in a bearing shaft of the turbine, was made in the dimensions shown in Fig. 8.16.

The system of a suspension bracket of model was made in two options – with the external electric drive in the form of a high-speed electric motor and without it. In the latter case dispersal of the turbine is carried out at the expense of reaction of the operating stream which is let out on the external or internal parties of blades through cracks about 1 mm high (Fig. 8.17).

In any option the model was carefully balanced and hung out on electronic and mechanical scales. The consider results of the experiments made without the external electric drive – the rotor rotated under the influence of reaction of the currents of air coming to a surfaces of blades during the periods when openings in walls of an air bringing pipe in an axis of a rotor coincided with entrance openings in channels inside a traverse, giving air in internal cavities of blades. The compressed air moved from a high-capacity receiver through the system of air ducts. The moment (phase) of air supply in a cavity of blades and further – in a stream of management was established by means of the rotary lever, rigidly connected with by the hollow managing director of shaft (airgiving pipe) in a turbine axis. The results of the experiments conducted without external actuator - rotor is rotated under the action of reaction jets air-facing surface of the blades during periods when the holes in the walls air pipe inside the rotor axis coincident with the inlet holes in the channels inside the traverse, feed air in the inner cavity of the blades, is described in detail in the book of author. The main conclusion from the experiments and calculations is that the proposed scheme is operational and can be effective.

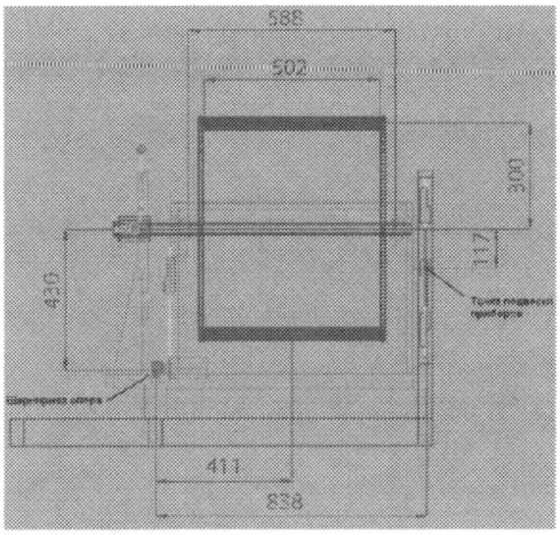

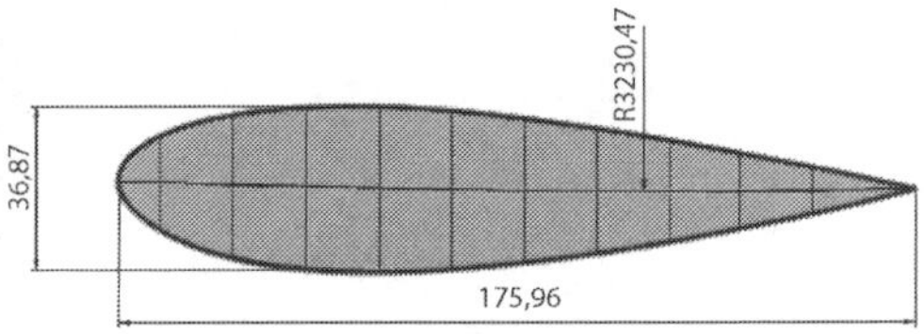

Fig. 8.16 The turbine model with hollow blades and shaft.

Fig. 8.17 1) sock of the blade, 2) an alignment of the greatest thickness o the blade, 3) cracks for release of an operating current of air.

The pressure source to create a jet in this case was external (compressor and pressure tank). However, in real high-pressure chambers of the blade can be generated due to the combustion of the fuel mixture or hydrogen supplied to the blade and ignited there at the right time due to an electric pulse (Fig. 8.18).

Complex movement when lifting or emergency lowering station is carried out through the use of two control system circulation on each rotor. One use of the system v modes fall (or rise) provides the necessary lift force, and the other forms a horizontal component of force, delivering station v set the wheel landing. Accordingly, within one of the hollow shaft

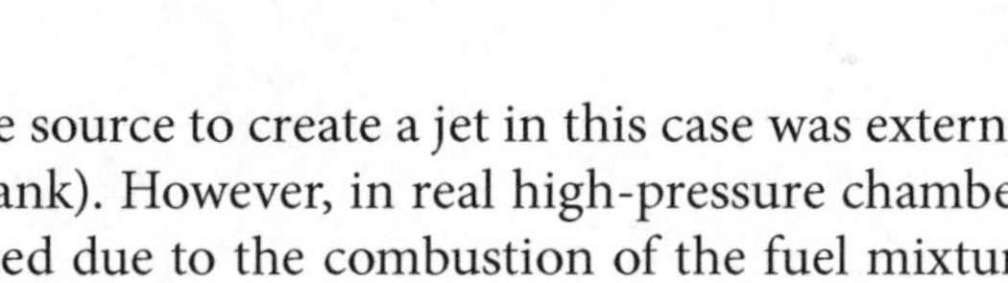

Fig. 8.18 Example of the blade's profile (top)1,2 - nozzles, forming a wall jets on the blade surfaces, 3,4 - spark ignition fuel mixture in the cavities of the blades [14].

shown in Fig. 8.7 add one or two more internal hollow shaft, and v the blades have additional partitions (Fig. 8.19).

Each blade (1) has two separate cavities (2 and 3) separated (4) and (5) are connected with the cavity in the Central turbine shaft (6) through channels (pipes) in the traverse, for example, (7). Pipe in one traverse is capable of feeding air to the outer side of the blade, and a trumpet in the other traverse - on the inner side of the blade. Distribution system located inside the shaft of the turbine, has two oppositely located openings (10) and (11). Inside the shaft is constantly maintained high pressure air (for air duct). When this pressure is fed through a hole 10 in the upper yoke and through the hole (11) in the lower yoke, the lifting force is initiated and jets on the blades will act up. Changing the position of the holes of the camshaft (8) relative to the earth or relative to the wind), you can change the direction

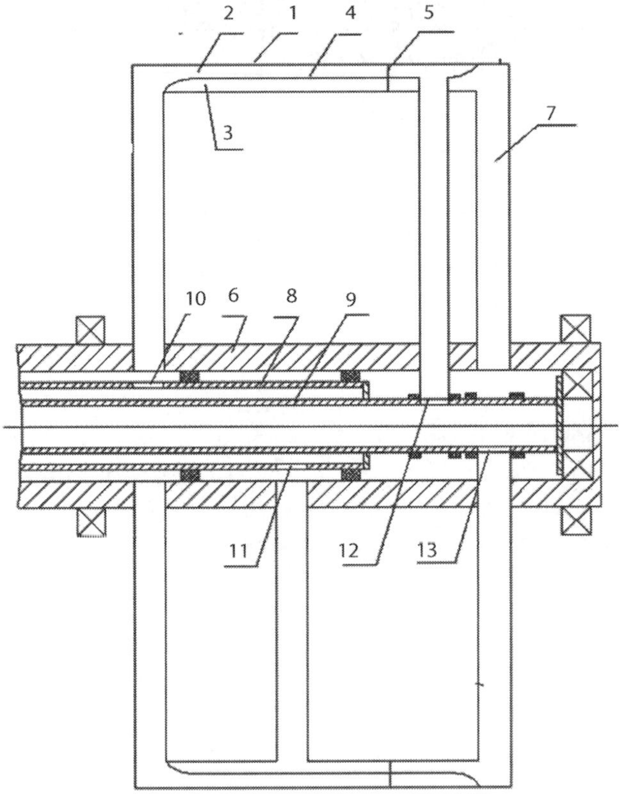

Fig. 8.19 Sections of channels supplying air to blade surface for circulation control. The numbering of the items in Fig. 8.7 and patent [13]. 14 - inductance coil, ignition spark external or internal contour in the variant with the feed gas mixture.

of "lock". Additional flexibility of the system is achieved by the presence of cavities in the blade, the selected partition (5), and the second inner shaft (9). The pressure in these cavities is passed through the holes (12) and (13) in the inner shaft, the position of which does not depend on the position of the main control shaft (8).

A scheme of this type – with a turning wing and a flap controlled by the traction from the center offset from the rotor axis – has been successfully applied in practice. The jet control system of the circulation on the wing in this case may be easier and more long-lived than the ordinary high-lift device system. The averaged value of the lifting force F acting upon the rotor per one revolution is a little greater

$$F= 1/6\, C_{Lmax}\rho\{(V+U)^2 + (V-U)^2\}\Omega_b \tag{8.12}$$

in which V is the blade speed, U is the wind speed inside the rotor and Ω_b is total surface of the rotor blades. For the designed wind speed $U_0 = 35$ m/s wind speed inside the rotor will be

$$U = 0.8U_0 = 28 \text{ m/s}$$

And averaged lifting force may be

$$\begin{aligned} F &= 1/6\, C_{Lmax} 0.75\, \{(50+28)^2 + (50-28)^2\}\, 6{,}410 \text{ N} \\ &= 5{,}262{,}610 C_{Lmax} = C_{Lmax} 526 \text{ (ton)} \end{aligned} \tag{8.13}$$

At the ground when a wind is absent the averaged lifting force may be

$$F_0 = 1/6\, C_{Lmax} 1.23\, \{50^2 + 50^2\}\, 6{,}410 \text{ N} = C_{Lmax} 657 \text{ (ton)} \tag{8.14}$$

The orthogonal turbine is connected with an induction generator. It may be two variants – 3 or 6 turbines in the block (9 or 18 units in the plant). In variant 1 (3 turbines in block – Fig. 8.3) the turbine diameter will be D= 40 m and frequency of rotation n = 23.9 rpm. The weight of generators with so slow rotation will be too much. In the second variant (6 turbines in block) frequency of rotation will be 47.8 rpm and weight of generators will be 18 kg/KW about. Total weight is 900 ton. To choose the kind and numeral of rotors and generators will be compared to the efficiency of each variant. In the first variant with 3 turbines in block the overload on the blades by the inertia force ($2V^2/gD$) reachs 12.75, in other variants (6 turbines) it may be twice more –25.5. In modern sea fighters (as Sea Hawk or Sca Fury) the real overload reached 10.3 and more. For the all aircrafts safety factor has to be 2 or more. It means that in turbine blade construction the overload even 25.5 may be admitted. In the modern large aircrafts with cantilever wing length up to 24 m the mass of the wing reaches 29 to 35 kg per sq.m of the wing surface. The design load on the wing is 350 to 580 kg per sq.m. The real experience of the constructing and testing the plastic blades for the orthogonal turbines with cantilever blade length 2.3 m is shown that the mass of the blades may be up to 12 kg per sq.m of blade with overload 26.5 and static loads up to 656 kg per sq.m. In the proposed system the blades surface is 6410 sq.m in total. The mass of blades will be less than 225 metric ton, but really up to 77 ton. If the mass of the connected elements will be 20% of the rotor mass the total mass of the browing power system will be

$$1.2\,(125 + 225) + 900 = 1320\text{t}$$

Using the plastic in the power construction we will be able to decrease the mass to 155+900=1055т.

In variant with smaller turbines the total mass will decrease at least on 300t reaches to 1020 (755)t.

To lift the system from the ground in accordance to (14) needs

$$C_{Lmax} = 1.6 - 2.0, \tag{8.15}$$

It reaches by any way with control of the blade circulation that provides $C_{Lmax} = 2.4$ and more.

In the nominal operation position the averaged lift force will be 1262t. It is more than the weight of system made by high strength plastic on the surplus vertical force

$$F_v = 207 - 507\text{t}. \tag{8.16}$$

The longitudinal force acted on the system in the nominal position is

$$F_x = C_x \rho U_0^2 A/2 \tag{8.17}$$

Here: A - swept area of system, blowing by wind, C_x – drag factor of system. For the design wind velocity $U_0 = 35$ м/сек, blades speed V = 50 м/сек and solidity $\sigma = 0.45$:

$$C_x < 0.5 \text{ and } F_x = 331\text{t}. \tag{8.18}$$

The power blocks are tethered by a hollow divided cable filled by hydrogen (or helium). This cable is made by the kevlar. Kevlar is an Aramid fiber produced by the E.l. Dupont Company to offer a light-weight replacement for steel wire in products such as tires, conveyor belts and cables. KEVLAR's high strength and modulus, combined with ease of manufacturing. Results in high efficiency at competitive costs. Ropes and cables using KEVLAR have excellent fatigue resistance and are non-corrosive, non-magnetic and non-conductive, with low elongation, and have the highest combined specific strength and modulus. Specifically KEVLAR are: stronger than steel (3X stronger than Nylon and Polyester) – allowable cyclic stress 5250 kg/sm^2 at least, at 1/5th the weight (density 1.6 g/sm^3).

The cable consists of connected separate parts filled by hydrogen with small surplus pressure. Diameter of cable is 6 m or more. The lift force acted on the one meter of the length of the filled cable is 18.7 kg on the elevation of 6 km and much more near the ground. Chart 8.1 shows the total force acted on the cable from the power units with different surplus vertical forces. In accordance with this total force was calculated the cross section of the kevlar cable, the mass of cable per one meter of length and total length of cable. The power blocks are located on elevation 6 km in all variants.

Chart 8.1.

Surplus lift force F_v, t	200	300	400	500
Total force acted on cable, t	387	447	519	600
Square section of Kevlar, sm^2	74	85	99	114
Mass of Kevlar per 1 m. of cable, kg	11.8	13.6	15.8	18.2
Cable length, km	11.6	8.94	7.79	7.2

On the surface of the cable will be fixed the aluminium wires to transmit the power from generators with generators voltage 10 kV. If the density of current will be 4A/mm^2 the weight of these wires has to be 3.4 kg per one meter of the cable length.

Thus, if the cable diameter is 6 m, surplus lift force has not been more 300 t and cable length has to be 8940 m. or more. When the wind speed will increase more design (35 mps) the power system will go down to the area with slower winds itself. Opposite, when the wind speed will decrease the power system will go up itself. Additional control would be made by changing the lift force acting on the power system. The needed length of the cable may be decreased by choosing the fixed point on the high elevation of ground (in the mountain area of California).

The surplus pressure of hydrogen in the parts of cable is made by the pumping to these parts through the individual pipes. If the losses of hydrogen will be about 0.8% per day, to keep the normal pressure it will be needed to use one unit of IMET-30 discharged 30 cub.m. of hydrogen per hour with installed capacity 200 kW.

On the outside of the cable surface may be moved the cabins to serve the power units and to go up the tourist groups.

The lifting capacity would be sufficient to operate the plant in emergency situations and while lifting it from the ground. In practice, circulation control is to be fulfilled automatically with the plant oriented by land-based radio beacons and the HAWPP gyroscope systems. In case of the cable breakdown, the revolution of rotors and smooth lowering of the plant to an assigned area is fulfilled with the aid of the HAWPP electric accumulators and operation of the HAWPP generators as engines. A similar routine is used while lifting the plant; the difference is that in this case line supply is employed.

The plant component being discussed is quite a complex structure and is the least developed one. Although its elaboration presents many difficulties, there is a practical solution to the problem. The comparison of a possible mass of the HAWPP and that of the cable shows that the HAWPP optimal unit power is to be about 70-100 or more MW, as the increase in the power plant capacity does not exert any significant influence on the cable mass, which makes the cost per unit of capacity drop. Preliminary calculations have shown that being production-run the plants of the given capacity would require specific capital investment of not more that 600 US dollars/kW.

It is advisable to accept the first experimental plant capacity of not less than 50 MW, as a considerable share of the plant development and manufacture self-cost would fall on the gas-filled cable. Further on, as the plant capacity doubles, the cable self-cost increases by not more than 20-30%, as most of the cable operation systems do not depend on the plant capacity. **The first experimental model of 50 MW capacity would cost approximately 2.4 thousand US dollars/kW, including all the design and research expenses and costs of the equipment manufacture and plant construction, installation and commissioning. Taking into account the economic and ecological prospects of the proposed energy source, the above-named cost of the first model should not be considered excessive.**

The whole complex of work including the plant turnkey and simultaneous preparation of further line production of the plant can be fulfilled within 3 years. The first step, including the check all ideas and numerals, in accordance with this applicant can be fulfilled within one to two years.

Further on, high-altitude wind power plants can be developed, which would have the capacity of 100-500 MW and annual using time up to 7,000 hours/year and cost about $600 /kW.

The combination of lower investment costs and extra yield bring the payback time back with a factor 4, compared to traditional wind technologies. Due to this positive influence on investment costs, offshore wind energy can compete with traditional energy production, such as gas or coal fired power plants. Therefore wind can play an important role in the renewable energy mix. It can even provide a solution for substitution of more conventional power plants.

The created demonstration windmill (model) has the same external form, as well as future is skilled electric power plant (Fig. 8.4 right).

We accept following preliminary parameters of model:

Chord-0.160 m

Diameter of a line of blades - 1.2 m

Number of blades in one circle – 2

Number of circles – 7

Length of the blade-0.6 m

Length of the turbine-4.2 m

Weight of the complete set of blades of one turbine-14 kg

Nominal frequency of rotation - 750 rpm

Settlement speed of blades (sliding 1.08) 50 m/s

The speed of a wind answering to the maximum capacity of 40 m/s

The speed of a wind answering to the beginning of delivery of capacity of 10 m/s

Power factor C_N=0.25

Efficiency of the turbine at an optimum mode (a wind of 23 m/s) $C_p = 0.43$ (see p.4).

Forces, the moments, the capacity, turbines operating on one store by calculation (length 0.6 m), the Moment, Nm, Power, kW. Forces, N along, across Forces on the blade, N Radial Tang

Averaged	4.8	0.4	142	-45	-1.6	2.4
Max.	28.5	2.4	558	187	188	28.2
Minimum	-9	-0.8	-91	-276	-339	-6

The force of inertia operating on one blade of 375 kg
The maximum averaged capacity of the turbine according to experiences

$$7x\ 0.67x0.25x503x2x0.16x0.6/2=14kW$$

Parameters of a wind (Moscow area, datum for different regions are in "Energies", 2009, 2, pp. 307-319)

Height over the earth, km		Season of year			
		Winter	Spring	Summer	Autumn
6	Average speeds of a wind, m/s	19.4	17.9	11.3	20.3
	Energy stream, kW/m^2	3.8	2.7	1.8	4.3

Average temperature of air a minus 24^{0}C, average density of air of 0.67 kg/m^3
Capacity of power station (maximum) 126 kW
Capacity at an optimum mode (a wind of 23 m/s) 79 kW
Considering high reloading ability of asynchronous generators, we accept for model the engine-generator the AIR 132M2 in weight 54 kg with an admissible overload 3.5.

Management control jet.
Thickness of a jet 2mm
Air density in jet 1.25 kg/m^3
Speed of air in jet 200 m/s
Relative impulse of a stream/forces C_L
On 6 km 0.75/2.0
At the earth 0.40/1.6
Average carrying power on one turbine at the earth at speed of a stream of 200 m/s of 84 kg
The expense of air 16.8 litre/sec
At speed of a stream about 120 kg

The expense of air 25 litre/sec

The capacity is necessary for braking of the station at the earth during 20 min. 2-3 м^3 with pressure of air of 10 bar. Such capacity is located in dividing constructive elements in width 3m and height 0.15-0.25m.

The wind station is fixed on the earth through the hollow cable filled with helium. At superfluous pressure of helium 0.5 bar superfluous carrying power makes 1 kg/m^3 at the earth and 0.4 kg/m^3 at height of 6 km. The current strength from generators at voltage 660 v is 114 A. The section of copper wires is nearly 28 mm^2 with a weight 249 g/m of length. For maintenance of deduction of such weight it is necessary to increase diameter of a cable with helium to 0.57 m at the earth and to 1m at height of 6 km.

The length of a cable will be defined by superfluous carrying power which is formed both at the expense of turbines and at the expense of gliders between turbines. The minimum length of a cable is 6.5-7 km. Superfluous "buoyancy" of station is reached at the expense of an aerodynamic flow of turbines and dividing profiles, and also (if necessary) at the expense of filling by helium of vertical elements of a design.

Each turbine has the open one-stage animator with transfer number 4 and the individual asynchronous generator with shortly closed rotor and nominal frequency of rotation of 3000 rpm.

All generators are united also by a uniform cable connected to a powerful electric network on the earth. In case of breakage of a cable generators are instantly switched to the battery of condensers through which power supply of elements of the management located directly on wind power model proceeds, and opening of valves for giving of operating streams, electric deduction of turbines from excessive dispersal, orientation model in space and its direction on an airfield is carried out. Optimization of a design of the orthogonal unit is reached in a series of experiences with the orthogonal units having the form of a framework.

The turbine with operated jet circulation is as a project basis of "the flying car" [13].

References

1. V. M. Lyatkher., High-rise wind power installation, News of the Russian Academy of Sciences. Power, 2006, No. 4, p. 47-57., V. M. Lyatkher, High-rise wind power installation, Patent RF 2240444, May 05.2003. V.M.Lyatkher and

V.L.Smirnov, High-rise wind power station, C.C. USSR 1765495, October 03. 1989.

2. G. V. Demidov, E. S. Osipov. High tall aerostaticaldynamical power plant, Patent RF 2064085, 1996.
3. D.H. Shepard, Power generation from high altitude winds, Patent USA 4659940, 1987.
4. V.M. Lyatkher, Altitude wind turbine, Patents RF 1150395, 2240444, 2004.
5. Viktor Lyatkher, Non-vibrating units for conversion of fluid stream energy. Patent USA 7741729, 2008.
6. A.M. Gorlov, Unidirectional helical reaction turbine operable under reversible fluid flow for power systems, Patent of the USA No.5451137, 1995.
7. Cristina L. Archer, and K. Caldeira, Global Assessment of High-Altitude Wind Power, *Energies* 2009, *2*, 307-319.
8. V.M. Lyatkher, *Renewable Power. Effective Design*, 2011 (M-Izevsk, Russia), p. 172.
9. V.M. Lyatkher, *Wind Power. Turbine Design, Selection and Optimization*, Wiley, 2013.
10. K.P. Vashkevich, V.V. Samsonov, The calculation of aerodynamic characteristics of windwheels vertical-axis type using the method of discrete vortices, Industrial aerodynamics, 1988, N 3(35). P.G.Baklushin, K.P.Vashkevich and V.V.Samsonov. *J. of Wind Engineering and Industrial Aerodynamics*. Vol. 39, N° 1-3, May 1992.
11. V.B Amfilokhiyev. hp Artyushkov, B.A. Barbanel, A.I. Korotkin. K.M. Mazayev. L.I. Maltsev, B. N. Semenov Current state of the theory of management of an interface. – St. Petersburg: Prod. SPMBM "Malachite", 2000. –415 pages 5. P.A.Bogdanov, P.G.Kozhukharov, L.I.Maltsev, V.I.Mikuta, V.Kh. Khadzhimikhalev, Underwater wing with jet management of its hydrodynamic characteristics, Sb. Liquid currents with free surfaces and polymeric additives, Novosibirsk, 1986, pp. 36-73.
12. Angle II G.M., Pertl F.A., Mary Ann Clarke, Smith J.E., Lift Augmentation for Vertical Axis Wind Turbines, International Journal of Engineering, vol.4, Issue5, pp. 430-442, D.McGrain, G.M.Angle II, J.P.Wilhelm, Circulation Control Applied to Wind Turbines, *ASME 2009 3rd International Conference on Energy Sustainability*, Volume 2, Paper no. ES2009-90076 pp. 905-910
13. Lyatkher V.M. Power installation for the vehicle drive, Patent RF 2327059, Dec. 14.2006r.
14. Lyatkher V.M. Power plant with thrust vectoring, Patent RF 2558716, 2015.
15. Lyatkher V.M. Wind energy unit, Patent RF 2558491, 2015.

9

Power Unit with a Controlled Thrust Vector – The Base for a Vehicle of Absolute Cross-Country Capability

On the basis of the general results of this book may be considered cylindrical and spiral rotors with bearing blades of an aerodynamic profile, equipped with jet circulation control systems that generate forces normal to the axes of rotation of the rotors. These rotors, placed above the vehicle body, allow it to move and maneuver in any direction with or without contact with the support surface. The proposal applies to energy and transport, in particular, to universal power plants with a controlled thrust vector, which can be used in transport equipment-applicable to land and water transport.

There are known designs of aircraft systems that include wings that form devices that resemble degraded aircraft circuits, with the placement of one or more fans above the cabin, similar to the known wind turbines with a horizontal axis. An example of such a solution is the "Airbus" and "Sure Fly" systems.

A common disadvantage of these systems is the large size, high complexity and low reliability of the mechanical drive, and poor maneuverability of the system.

It is a well-known aerial vehicle in which the lifting force of the desired direction is created by two symmetrically arranged orthogonal wind turbines with adjustable rotation of the blades [1]. The disadvantage of this solution is the mechanical blade control system, high material consumption due to the presence of a shaft and numerous turbine traverses.

There are proposals for the use of jet circulation control on the blades of orthogonal rotors, which creates a lift of the desired direction [2]. Variants of such a system with increased efficiency are implemented in patents [3]. The disadvantage of these solutions is the high material consumption of the structure, which includes the bearing shaft and traverse, and the slow response of the system to control commands associated with changing the

Victor M. Lyatkher and Ziaur Rahman. Wave, Wind and Current Power Generation, (283–292) © 2022 Scrivener Publishing LLC

mode of fuel supply to the combustion chambers. It is proposed to eliminate these shortcomings by increasing the speed and reliability of the control system and reducing the material consumption of the structure [4]. As a result, you can get an economical and maneuverable vehicle that combines the advantages of a car and an aircraft.

Vehicle of absolute cross-country capability (option 1-Fig. 9.1) equipped with a rotor of two symmetrical balanced spirals of the aerodynamic profile, each of which has a relative rotation of the end sections by 1800. The spirals are fixed above the vehicle body so that the rotor axis is positioned horizontally a short distance in front of the vehicle's center of gravity. Since the position of the center of gravity can change, the spiral attachment system is equipped with movable supports and horizontal hydraulic cylinders (not shown in the figure), which automatically move the supports of the working spiral to a position that ensures the horizontal position of the machine during movement (flight).

Channels 1 through valves 2 are connected to isolated chambers 3 located symmetrically in six identical sections of the cavity of each of the blades. The tanks 3 are provided with smoothly outlined channels for the output of control jets, which are blocked by valves 4.

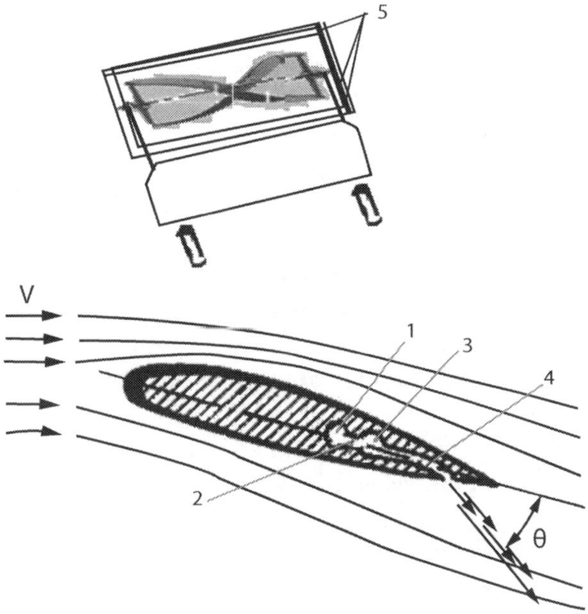

Fig. 9.1 General view of the car with spirals (up) and a section along the spiral blade (down).

Chambers 3 are equipped with spark plugs connected to the car's battery and a computer control system that selects the moment of ignition of the candles at the driver's command, causing a sharp increase in pressure in the desired chamber 3, the formation of a jet and a sharp increase in lift on the selected sections of the blades, which determine the direction and amount of forces acting on the car from the spirals.

Spirals rotate under the action of jet pulses so that the parts of the blades located at the top point move in the direction opposite to the direction of movement of the car at a linear speed significantly higher than the speed of the car.

The valves 2 are closed, and the valves 4 are opened simultaneously with the ignition of the candle in the corresponding chambers 3 of both spirals. The computer system selects 3 cameras in accordance with the driver's commands and synchronizes the operations in the spirals, providing the desired direction and magnitude of the force vectors developing in the chambers of 3 different spirals, ensuring that the required maneuver of the car is performed. The resulting pulse jets ensure continuous rotation of the spirals and a sharp increase in lift at the right time in the right direction.

The engine can consist of a single balanced spiral, the end sections of which are rotated by 3600 (option 2-Fig. 9.2).

Spirals consisting of any variant of 6 compartments are fixed on hollow end traverses, through which the fuel mixture is continuously fed into the channels of the blades (Fig. 9.1). To prevent air flow along the forming part, the internal separate walls perpendicular to the axis of the spiral are brought out to a small distance exceeding the thickness of the boundary layer (Fig. 9.2).

To avoid excessive acoustic radiation, the working spirals are surrounded on top and sides by sound-absorbing smoothly streamlined walls 5 (Fig. 9.1). In front of the spirals, these walls form a smoothly outlined confusor, and

Fig. 9.2 Fragment of a spiral turbine in a single-blade rotor variant. The blade shows thin protrusions perpendicular to the axis of the turbine, preventing flow in the boundary layer. Elements that provide system balancing are installed in the support nodes.

behind the spirals – an effective, continuous diffuser. The closest to the described solution is the Russian patent 2558716 C1 dated 13.05.2014.

The difference of this proposal is that

- blade (blade) separated into isolated compartments, the expiration of the jet of which is determined by a computer program forming the direction and magnitude of force – direction of force is determined by the time of the opening gap (position of the compartment of the blade on the track), and the magnitude of the force – number of compartments from which expires jet at the right time;
- the working gas supply is carried out in the cavity of the blade at all times, except for the moments of ignition in a particular compartment, causing the closure of the check valve in the compartment;
- the separation walls of the compartments protrude beyond the surface of the blade, preventing the longitudinal flow of air along the blade;

Fig. 9.3 shows a simplified possible scheme of a vehicle (TS) with a simplified engine having a single helix. It is important that in the described scheme (in any version), the maneuverability of the vehicle does not depend on contact with the ground (wheels may be missing) and does not require large dimensions like a helicopter.

A quantitative assessment of the action of a jet released at high speed at the surface of the blade in its tail section was carried out for one version of the scheme [5]. A straight blade of length L=1m with a chord b=0.6 m and a naca0021 profile moves along a circular track of radius R=1m with a constant linear velocity V=12.5 m/s in a uniform air flow at a speed U=5 m/s (U/V=0.4) (Fig. 9.4). The angle of the blade to the tangent is 6 degrees.

Fig. 9.3 High maneuverability vehicle that does not require expensive equipment.

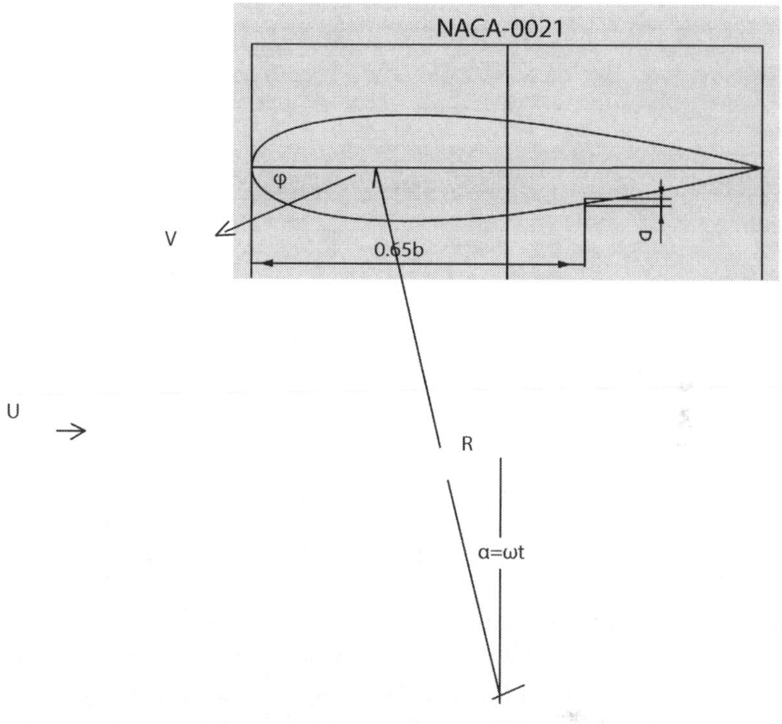

Fig. 9.4 Geometric parameters and flow diagram of the wing profile.

Ambient air density r=1.225 kg/m3, kinematic viscosity μ=1.7894·10-5. The Reynolds number calculated with respect to the chord of the blade and the speed of the incoming flow is equal in this case.

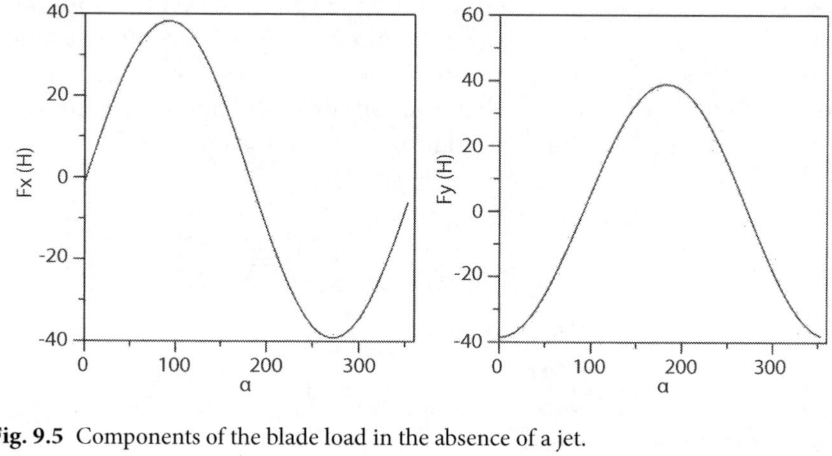

Fig. 9.5 Components of the blade load in the absence of a jet.

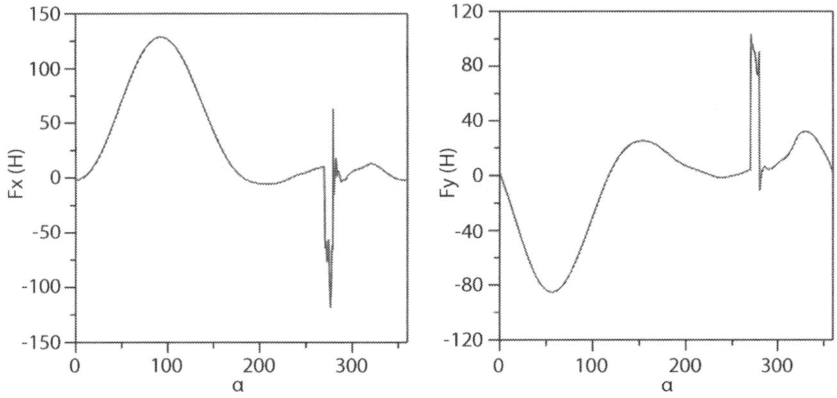

Fig. 9.6 Components of the load on the blade under the action of the jet on the rear section of the blade route.

Fig. 9.5 shows the loads on the blade along the route of its movement in the absence of a jet.

In Fig. 9.6, the load on the blade under the action of the jet at a speed of 123 m/s for 0.015 seconds at the moment when it passes the back side of the rotor. The duration of the jet action corresponds to approximately 100 rotations of the rotor. The mass of the gas entering through the nozzle is kg, and the relative value of the pulse.

The thrusts shown in Fig. 9.6 (forward and upward) in the turbine diagram in Figs. 9.2 and 9.3 should be repeated on the side of different sections (sections) of the blade, as soon as this section is in the desired position. The formation of jets (lighting candles in the combustion chambers) is controlled by an electronic system that monitors the steering commands of the car. At the same time, the total (in time) vertical component of the force for different parts of the rotor can remain almost constant or change smoothly, providing a smooth vertical maneuver of the machine. On the contrary, the total horizontal component of different sections of the rotor can change at the driver's command, forming a torque that provides

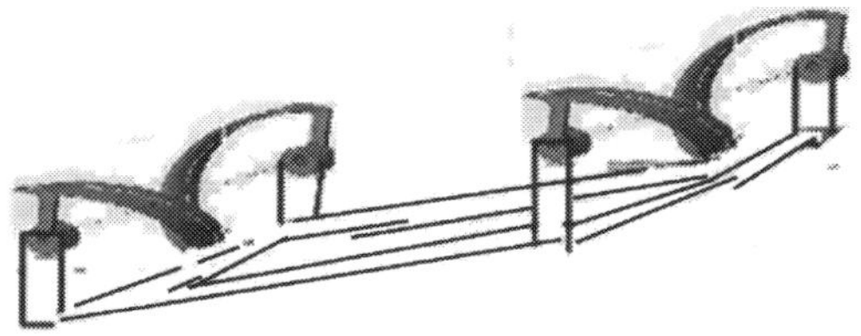

Fig. 9.7 Two rotors with parallel axes above the roof of the vehicle.

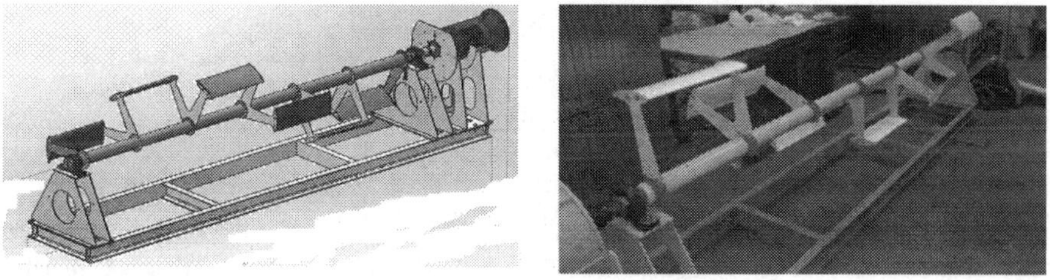

Fig. 9.8 Turbines with straight blades approximating the spiral.

sharper maneuvers of the car in the plan. To ensure the stability of the machine in "flight" over its roof can be located not one (Fig. 9.3), but two C rotors with parallel axes (Fig. 9.7).

At the first stages of design and development of the system, it is possible to replace the spiral blade with different versions of straight blade systems, for example, according to the scheme implemented on an orthogonal hydraulic turbine (Fig. 9.8).

A turbine with straight blades does not have to repeat the shape of the spiral. A variant with pairwise balancing of the centrifugal forces acting on the blade is quite possible (Fig. 9.9).

In any case, the ignition of the jet, the opening (closing) of the valves is controlled by a special program associated with the movement of the steering wheel of the device. In versions with a shaft and straight blades, fuel enters the blades from the channel in the Central shaft through channels in the traverse. Maneuvering the vehicle is performed through an electronic circuit that controls and controls the lighting of candles in different chambers of the system. According to Fig. 9.6, the ejection of a jet with a mass of 11 g in a certain position will cause a pushing force of 12-15 kg. one meter wide of the car. These pulses, acting one after the other as the turbine turns, form the total thrust of the car with a width of 2m about 30kg.

For a preliminary demonstration of the effectiveness of the proposed scheme, it is possible to use a simple system of blades according to Figs. 9.8, 9.9

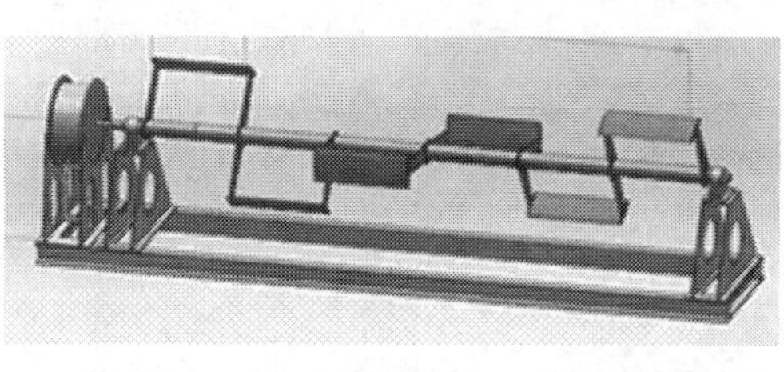

Fig. 9.9 Arrangement of turbine blades with compensation of centrifugal forces.

with fast-acting individual flaps equipped with an electromagnetic or hydraulic drive.

Conclusion

It is proposed to consider and develop a new type of vehicle that combines the advantages of a conventional car and an airplane.

References

1. Xisto, C., Pascoa, J., Masarati, P., Dumas, A., Parametric Analysis of a Large-Scale Cycloidal Rotor in Hovering Conditions. *J. Aerospace Eng.*, 04016066_1–14, July 2016.
2. Kweder, J., Panther, C.C., Smith, J.E., Application of Circulation Control, Yesterday and Today. *Int. J. Eng.*, 4, 5, 411–429, 2016.
3. В.М. Лятхер, Энергоустановка с управляемым вектором тяги, Патент РФ №2558716, приоритет от 13.05.2014, В.М. Лятхер, Крыльчатый ортогональный движитель, АС СССР №1634567, приоритет от 20.02.1989, В.М. Лятхер, Энергетическая установка для привода транмпортного средства, Патент РФ №2327059, приоритет от 14.12.2006, В.М. Лятхер, Энергоустановка с управляемой реактивной тягой, Патент РФ№2687190, приоритет от 26.01.2017.
4. В.М. Лятхер and А.В. Лятхер, Энергоустановка с управляемым вектором тяги для привода в движение транспортного средства. ПатентРФ2019112356. Дата начала отсчета срока действия патента 23.04.2019.
5. В.А. Дегтярев, Разработка и адаптация математической модели для двухмерного расчета обтекания крылового профиля NACA0021 совершающего движение по кольцевой трассе. ООО «ТОРИНС» Отчет по договору №3 от 31 марта 2017 г. Красноярск. (see Application).

Conclusion General

It is estimated that total power may be transformed from the wind power to the other without climate changes. The main part of this power content is in the high wind areas in the boundary layer of atmosphere and much more in the high layers of it.

The new technologies of the wind power near the earth and jet stream of atmosphere and hydrogen manufacture with low cost and low initial investment, are described. To converse the power from the wind the new flying turbine (Russian Patent of author 2240444, priority of 05.05.03) and new multi-blades turbine (Russia Patent of author 2245456, priority of 05.22.02) may be used.

To converse jet stream power the horizontal orthogonal turbine is proposed. In these units the axis of the blades follows the spiral line ("twist" - turbine). The torque moment of these turbines is constant (without pulsation) and vibration of turbines is small. The units can be installed inside the block of the power station floating in the free jet stream. The turbines have a system with the perforated surfaces of blades to control the lift forces by the variable jets in the boundary layers on the blades. This system may be used for high efficiency conversion of flow energy in the jet stream and generate the additional lift force to support the power units in the stream. The power block is tethered by the cable that is fixed on the earth.

The averaged density of the maximum energy flow in the jet stream above California may be 15-30 kW/m^2 or more. The same situation is above Russia and Moscow where the average wind speed in winter is 19.4 m/s at an elevation of 6 km.

We propose to project and to construct the jet stream power plants with 100–500 MW capacity including 3 or more blocks with 3 horizontal orthogonal "twist" turbines with control circulation on the blades.

The other construction proposes for the wind near the earth or water surfaces. The new solutions, which are free from the deficiencies of traditional schemes and allow to create effective wind power unit of large capacity, is described. The application of very large installations is expedient by itself, since these installations smooth out the pulsations of wind speed, but it is still more effective in the powerful electrical systems, where the energy flows from the wind power units, located in different regions, can be considered as statistically independent ones and their summary relative fluctuation is the less, the greater is the number of the sources. In the submitted report wind power units with unit capacity up to 10 MW are examined as an example, although the larger installations can be more effective counting on the unit of capacity and production.

The hydrogen, produced by power from these windmills, will be of the lowest cost – not more than $0.5/kg.

10

High Altitude Turbine (HAT): The Future of Wind Energy

The jet power plant, although very efficient, cannot be implemented all over the world because of the sheer nature of the jet stream. A natural phenomenon, the jet stream is a result of solar radiation heating the earth and the axial rotation of the earth. When warm air masses meet the cold air masses due to uneven heating of the solar radiation it creates narrow bands of strong wind in the upper troposphere. Near the pole the current is strongest at a relative lower height of 9-12 km above sea level. This is labeled as polar jet stream. A relatively weaker jet stream, called subtropical jet stream, is found above 10-16 km above sea level.

A unique characteristic of the jet stream is that the wind speed increases almost by double, and scientists are trying to utilize this source of energy. This energy has already been proven to be very beneficial for the aviation industry and goes back to World War II, in which the strong flow of wind current was used for bombing. However, as a source of renewable energy it is still in its infancy. To explore the untapped opportunity of wind power along with high jet power plants the high altitude turbines (HAT) pose great potential.

Harnessing the Wind Energy

The potential of HAT remains high because of the limitations of the jet stream and diminishing opportunity for ground-based wind turbine systems. The drawbacks associated with Conventional Wind Turbine (CWT) ground-based turbine systems include turbine hub heights, blade lengths and location restrictions linked to the environment and permitting issues including special areas of conservation and social acceptance due to the visual and noise impacts and also adverse impact on avian wildlife [1]. Thus the high altitude wind power potential is significant.

Victor M. Lyatkher and Ziaur Rahman. Wave, Wind and Current Power Generation, (293–318) © 2022 Scrivener Publishing LLC

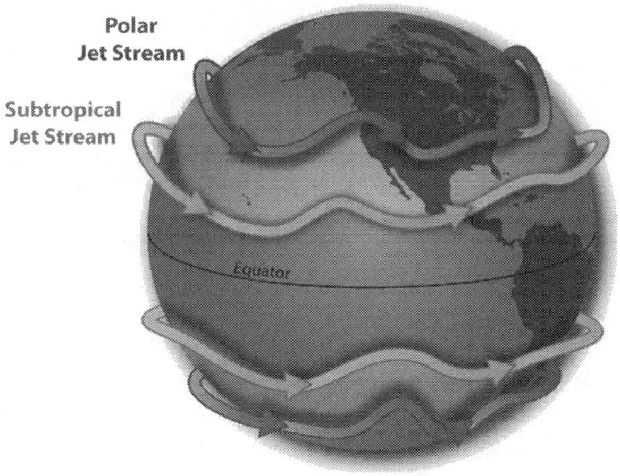

Fig. 10.1 Jet stream.

The total power that is available to a wind is given as:

Total wind power in Watts, Pw = $(mwV^2)/2 = (\rho\ A\ V^3)/2$ [1] where mw= ρ A V, where ρ is the density of the air in kg/m^3, A is the exposed area in m^2, and V is the velocity in m/s. The density is a function of pressure, temperature, and relative humidity [2].

The wind energy potential is often estimated based mainly on wind speed and the power curve of the turbine. It only marginally depends on the temporal and spatial variations in air density. Vertical changes in air density are usually taken into consideration. The annual average air density of a site is estimated as a function of height when applying well-known models such as the standard atmospheric profile to obtain the mean pressure and temperature at that height and, as a result, the constant air density for that site [3].

At a height of 1300m wind velocity is steadier, higher, and more persistent, thus resulting in a higher consistency of power generation [4]. The profile of wind power densities with respect to altitudes between 500 m and 12,000 m have been assessed globally [5]. It has been found that the highest wind power densities are at altitudes between 8,000 and 10,000 m above ground level (agl). This is primarily due to the intercepting fast winds of both the polar jet and sub-tropical jet streams. These high power densities are hugely important for wind energy development as it is estimated that wind power in the jet stream is roughly 100 times the power expended on the planet by human activity [6].

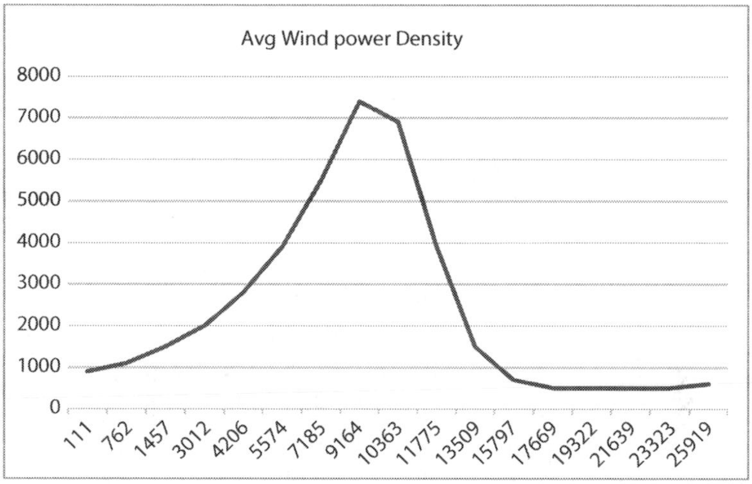

Fig. 10.2 Average wind power density [1].

Basic Architecture of HAT

It has been found that several HAT designs are currently being researched, but to date no fully operational models have emerged into the market. The roots of these high altitude wind power (HAWP) devices originate largely from the work carried out by Loyd [7]. In the 1980s he proposed the concept of 'crosswind kite power' using tethered airplanes to harvest energy. To produce electricity from such a system two main techniques were proposed; the first technique known as the 'lift mode' uses kites to create tension within a tether to drive a ground-based generator to produce electricity. The second technique called the 'drag mode' employs a generator on-board the Air Borne Module (ABM) driven by flying blades and transmits electricity to the ground via a tether [8].

Despite the difference in design and technology a HAT consists of five parts. One ABM that floats in the air. One tether which is used to tie the ABM to the ground. This tether can be a means of generating power or transferring electricity in an on-board electricity-generating HAT. There is a conversion unit made of rotor and a generator to convert wind power to electricity. This is the same as a conventional wind turbine. There are also launching and landing systems and lastly, to support the overall operation, a base station.

Air Borne Module (ABM)

The three most notable approaches to couple high altitude winds using an ABM are tethered kites, rotorcraft and balloons [1]. The kite approach has generated a growing interest as a result of the urgency in the need to generate renewable energy. Kites are considered a favourable concept as it is believed they would survive crashes better. Kites can be designed to be extremely lightweight for a given surface area but rely on the aerodynamic load distributed airflow to maintain their shape [11]. Different types of commercially available kites include ram air kites, with air inlets at the leading edge, and leading edge inflatable kites. Delft University of Technology has developed a concept called the 'kite plane' for high altitude winds. The design is primarily based on 'tube kites' that are used for surfing [9]. The structure of the kite plane consists of 'inflatable beams and canopy panels.' In this structure the inflatable beams provide rigidity and can be sealed airtight to contain a gas. Woven fibres are commonly used for kite applications, such as 'rip-stop nylon' for parafoil type kites and 'rip-stop polyester' as the canopy material for tube kites. Dacron is a heavier polyester cloth used to reinforce the tubular frame of tube-kites containing Kevlar [10].

A company called the Makani Power has utilised carbon fibre for tethered rotor craft for its rigid wing spar due. This provides high tensile strength-to-weight ratio and for the wing skin, the company used e-glass

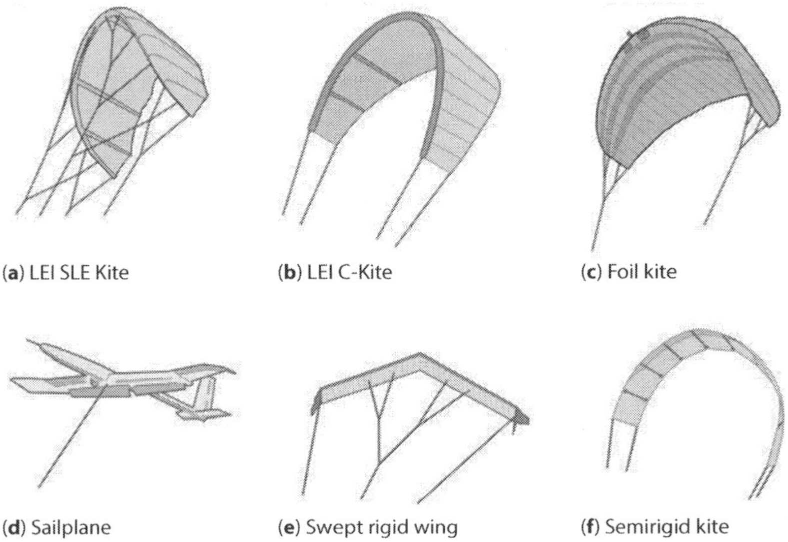

Fig. 10.3 Different types of kites.

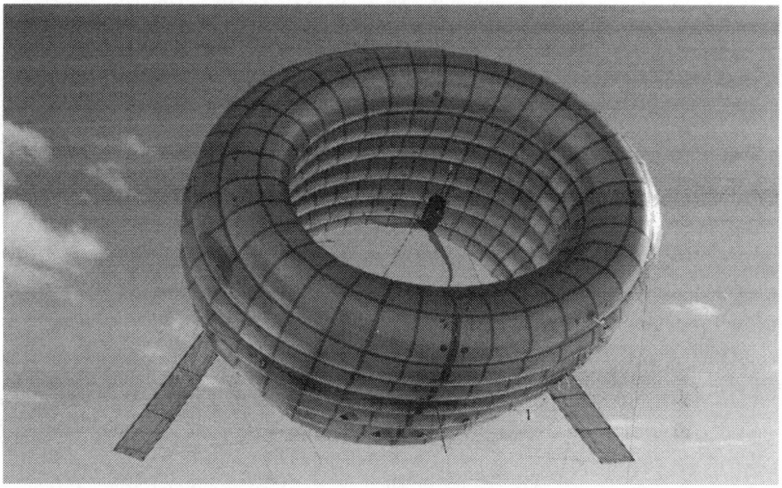

Fig. 10.4 Buoyant Airborne Turbine.

as it reflects heat by restricting the amount of ultraviolet and infrared light that can pass through [12].

A rigid wing ABM has numerous benefits. The shape is maintained independent of the wind conditions and because of its higher lift-to-drag ratio it can reach high velocities resulting in a significantly higher power output per wing area. However, in the event of a crash a rigid wing structure has the potential to have major consequences, thus causing considerable damage. Laminates can be tailored by layering materials with varying

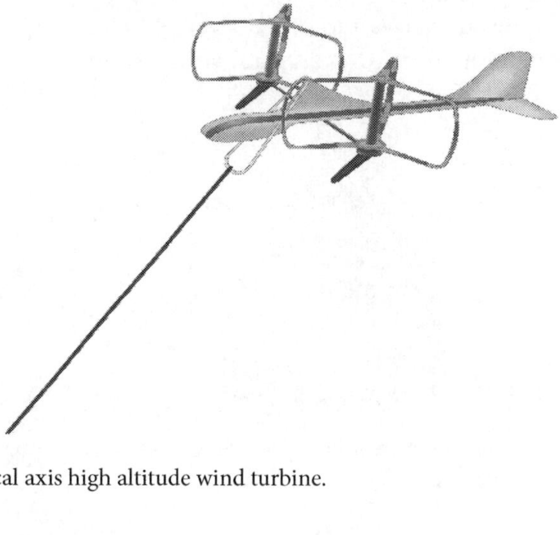

Fig. 10.5 Vertical axis high altitude wind turbine.

properties into a material with particular properties. Laminates are commonly used in lighter-than-air applications, for example in DBR devices that combine high modulus fibres such as Vectran or Kevlar to carry load [9].

Tether

There are two primary objectives for tethers used for harvesting HAWP: 1) to hold the ABM at a given altitude; 2) transmit the power efficiently to the ground-based station [12].

The properties of the tether material are important as a tether needs to withstand high stress from the powerful, high-speed winds. Furthermore a tether is subject to strain due to ascending and descending the ABM load, stretching the tether [14]. In order to transmit electricity to ground level a high-voltage tether is required although the weight of the tether is an important constraint due to the effects of tether drag. Makani Power has chosen a copper conductor within a dyneema fibre for its current prototype [15]. This was chosen because of the tensile strength of 11,000 MPa, which is '15 times stronger than steel' and its load distribution properties. The Sky Wind Power team have instead utilized a 10 mm tether with lightweight aluminum conductors in a Vectran fibre composite. Vectran is a multi-filament yarn spun from liquid crystal polymer similar to that of Kevlar with 'exceptional strength, rigidity and modulus' [16]. Nonetheless, both of these concepts are still in their prototyping stage and it is not known how strong these materials will be in the tough jet-stream conditions. Another potential developing solution is carbon nanotube technology [17] which is an extremely strong (up to 60 GPa) yet lightweight and highly conductive tube-shaped material that would allow HAWP technologies to operate at greater altitudes with lower drag and transmission losses.

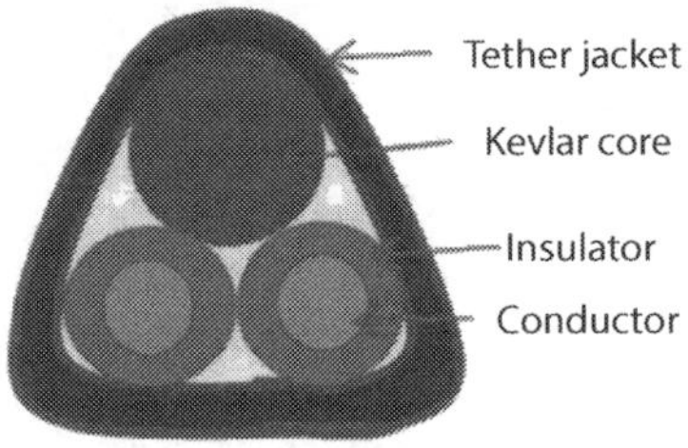

Fig. 10.6 Tether designed for high altitude wind power generation with two power cables and one tensile strength core [23].

A simplistic mathematical representation of tether flow and velocity has been articulated by Dunker (2018) [36]. The tether length is generally not constant but varies as a result of the reeling motion and to a minor degree also the strain of the tether. It is important to note that the radial coordinate r is a geometric measure which does not describe material points on the tether. Because of the reeling motion, the radial velocity $\mathbf{v}_{t,r}$ of material points is constant along the tether when neglecting strain, and equal to the radial velocity $\mathbf{v}_{k,r}$ of the kite **K**. A point on the tether moves with the material velocity $\mathbf{v}_t$, consisting of radial and tangential components $\mathbf{v}_{t,r}$ and $\mathbf{v}_{t,\tau}$, respectively. The tip of the tether at $r = l_t$ moves with the kite velocity, i.e. $\mathbf{v}_t = \mathbf{v}_k$. The radial unit vector $\mathbf{e}_r = \mathbf{v}_{k,r}/v_{k,r}$ is aligned with the tether, while the tangential unit vector $\mathbf{e}_\tau = \mathbf{v}_{k,\tau}/v_{k,\tau}$ is perpendicular and pointing in the flight direction of the kite

$$\mathbf{v}_{t,r} = \mathbf{v}_{k,r}, \text{ for } 0 \leq R \leq 1. \tag{10.1}$$

The tangential velocity $\mathbf{v}_{t,\tau}$ is constrained to zero at the ground attachment point **O** and is identical to the tangential velocity $\mathbf{v}_{k,\tau}$ of the kite at the kite attachment point **K**. It can be formulated as a linear function of the radial coordinate

$$\mathbf{v}_{t,\tau} = \mathbf{v}_{k,\tau} R, \text{ for } 0 \leq R \leq 1. \tag{10.2}$$

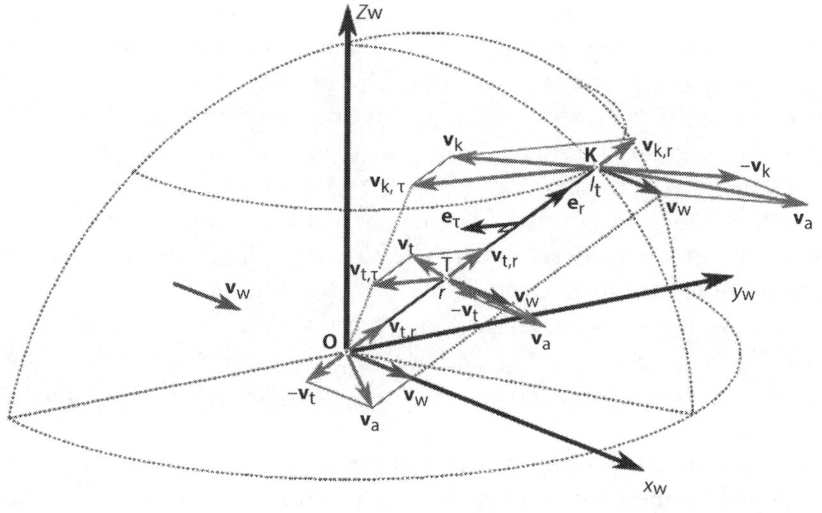

Fig. 10.7 Relative flow conditions at the origin **O**, an arbitrary point **T** along the tether and the kite.

The material velocity $\mathbf{v}_t = \mathbf{v}_{t,r} + \mathbf{v}_{t,\tau}$ of a point on the tether can thus be related to the radial and tangential velocity components of the kite by

$$\mathbf{v}_t = \mathbf{v}_{k,r} + \mathbf{v}_{k,\tau} R, \text{ for } 0 \leq R \leq 1. \tag{10.3}$$

The apparent wind velocity of a material point on the tether is defined as

$$\mathbf{v}_a = \mathbf{v}_w - \mathbf{v}_t, \tag{10.4}$$

$$= \mathbf{v}_w - \mathbf{v}_{k,r} - \mathbf{v}_{k,\tau} R, \text{ for } 0 \leq R \leq 1, \tag{10.5}$$

The Conversion Unit

There are different types of conversion unit. A short introduction to each is given below.

Ground-Based Power Generation

HAT devices with ground-based power generation use wind energy from kites. This device drives a ground-based generator using a tethered wing that flies in a lying-eight orbit taking advantage of high crosswind speeds [17]. At the ground station, the lower portion of the tether is tied around a drum that is connected to an electric generator. When the kite rises into high altitudes, the high pull causes the tether to reel out, rotating the drum and subsequently generating power. The tether is reeled back onto the drum using the generator as a motor in order to control the kite [1]. The larger the tension difference during the reeling-in and -out phases of the tether, the greater the net power that can be generated [18].

Light flexible wings often designed by surf kite manufacturers are used for pumping kite generators (PKG), although there are marked differences in how the kite is steered and how many tether lines are used [11]. Opposed to flexible wings rigid wings can also avail of the pumping mode available to ground-based generation. These concepts include autopilot glider planes that create tension on a tether by flying through high crosswind speeds [19].

An alternative ground-based concept is the carousel configuration developed for medium-to-large-scale HAWP generation. This HAWP proposal consists of several kites placed on the arms of a vertical-axis rotor positioned on vehicles moving along a circular rail path. A control system

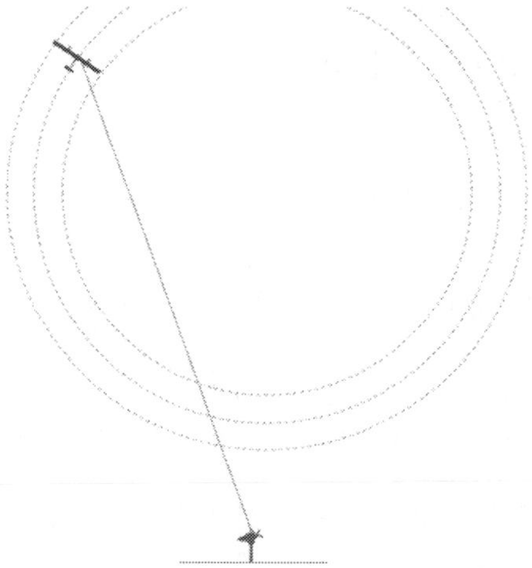

Fig. 10.8 Ground-based power generation.

is designed to drive the flight of the kites so that the maximum torque can be exerted on each of the rotors to furthermore drive an electric generator. The torque opposed to the motion by the electric generator is controlled to have a constant rotation speed. The cycle is comprised of two phases, the traction phase and the drag phase. For a specified wind direction, each kite can generate electricity for 300° of the carousel rotation (traction phase) and only a small fraction of the generated electricity is consumed during the drag phase to pull the airfoil against the wind through the remaining 60° [20]. The major obstacle for this HAWP concept is the

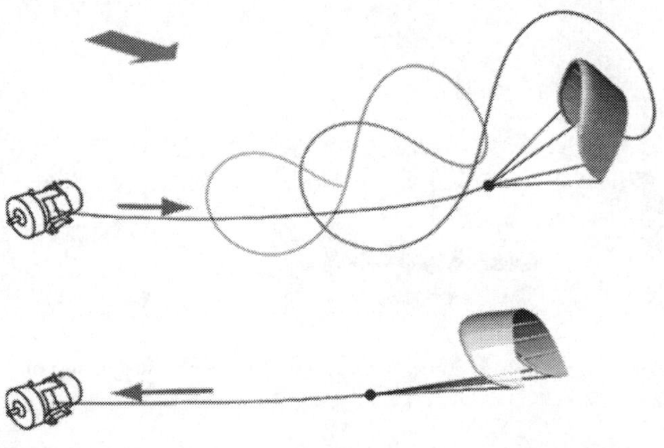

Fig. 10.9 Pumping cycle with reel-out stage (top) and reel-in stage (bottom) [29].

design of suitable robust kites and optimal flight control systems to obtain maximum generated power [1].

On-Board Power Generation

This type of HAWP device is associated with Loyd's 'drag mode' technique as it requires the ABM to carry a turbine on-board. It generates crosswind power through exposure to high relative airspeeds forcing the rotors to rotate and consequently produce electricity. This electricity is then transmitted to the ground by means of a flexible tether. This idea of using 'propellers on-board a self-erecting structure' was first introduced by [21]. A 'variant of a gyroplane' with quadruple rotor arrangements mounted on an airframe to simultaneously generate lift and electricity was developed by Roberts *et al.* [6]. This concept is that this HAWP device would be located in the upper atmosphere, 4.6 km and above, harnessing wind energy from the vigorous jet streams. The rotors are controlled and optimally adjusted into the oncoming wind. This ABM can launch and land similar to a helicopter by supplying a small proportion of power from the ground with the generators operating as motors.

Permanent magnet synchronous generator (PMSG) gives better power to weight ratio for a HAWP generation [28]. The generated AC voltage needs to be converted into optimal medium voltage DC in the airborne unit for efficient transmission. So, the airborne PEC is designed to convert low voltage AC to medium voltage DC efficiently. Flying electric generator supported by an ABM has an electric machine

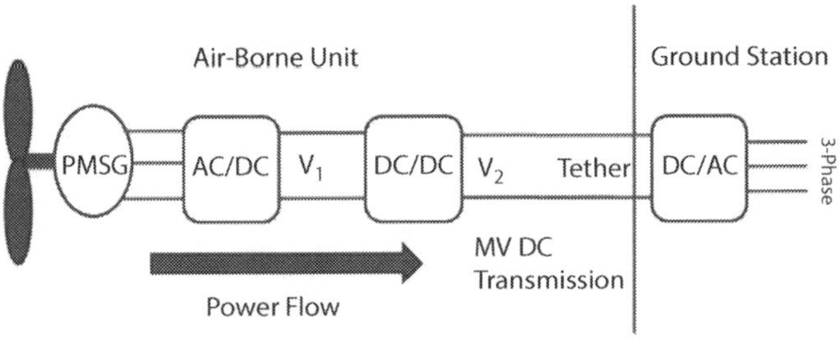

Fig. 10.10 Converter topology in the airborne unit for ABM supported on board power generation [23].

that works in generation mode only, so the bidirectional power flow is not a requirement [23].

Dirigible-Based Rotors (DBR)

These HAWP devices are advantageous in low intermittent winds, as the aerostat can remain airborne without consuming power. This on-board generated power approach is based on aerostat technology, which relies on a lighter-than-air gas-filled balloon to keep the heavy payloads aloft instead of relying on lift generated via rotors [18]. Researchers at Massachusetts Institute of Technology have developed a functional prototype, comparable to an observation aircraft concept developed by NASA, which is powered by a traditional horizontal axis turbine [22]. This device is designed for 'fast, passive alignment into the wind' that operates at high altitudes of 600 m. It does not avail of crosswind power but functions by means of a stationary aerostat [23]. This HAWP device is an ideal application for emergency power relief in disaster stricken areas as it has been designed

Fig. 10.11 HAT in low intermittent wind.

to be fully autonomous, from take-off, through operation, to landing onto the transportable ground station [24]. Other approaches have combined the 'pumping mode' with the aerodynamic phenomenon known as the Magnus effect. This occurs when a DBR device is rotated while ascending to higher altitudes, which in turn lowers the drag effect of the wind, thus reverting it to lift mode [25].

The Launcher and Landing System

The landing system allows safe retrieval of the wing system from any direction. It consists of a steel structure with wooden deck and a foundation. Additional equipment for deceleration the wing system is mounted on the deck.

The launcher launches the system independently of the wind speed and direction.

Balance-of-Station

The balance-of-station will contain various supporting and auxiliary components like:

- Internal cabling
- Energy storage (buffer)
- Hangar
- Control tower
- Transformer station
- External cabling (connection of the wind farm to grid).

Comparison of HAT with Conventional Wind Turbine (CWT)

Cost of Power: Over time conventional wind turbines have grown taller, heavier and expensive in order to generate more power. Conventional wind turbines are expensive and complex to construct and maintain. Because of its simple construction a HAT is inexpensive as well as portable.

Scalability: With height density of power production increase it also widens the opportunity for establishing more sites for power production.

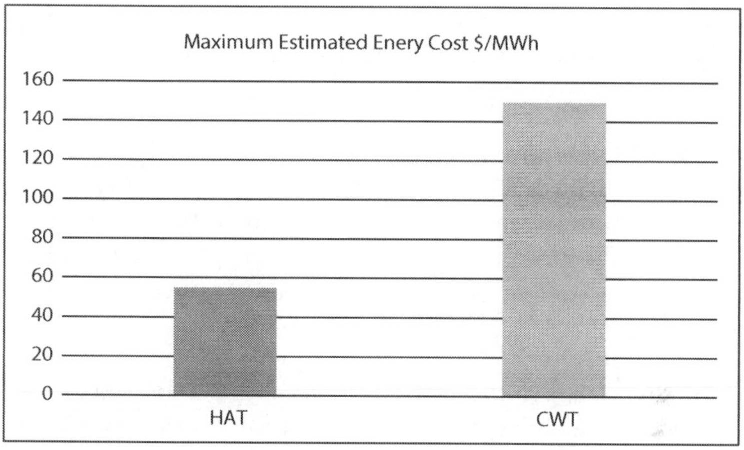

Fig. 10.12 Production cost [33].

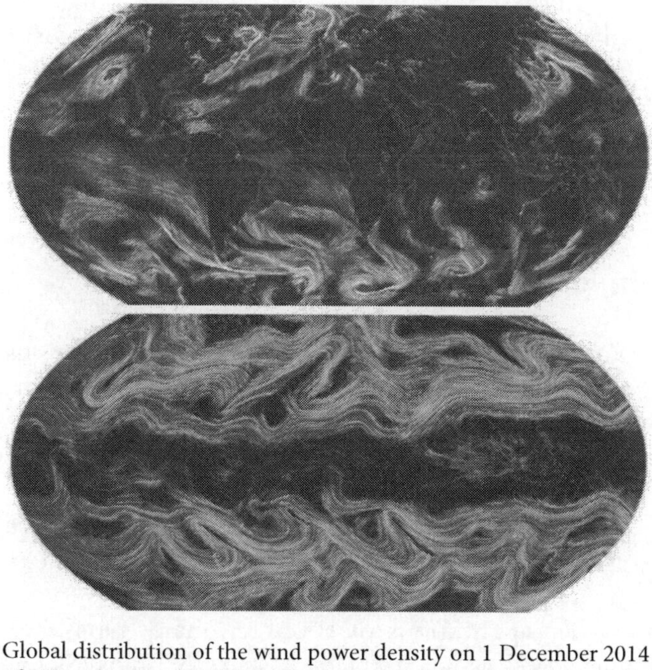

Fig. 10.13 Global distribution of the wind power density on 1 December 2014 9:00 UTC at altitudes of 100 m (top) and 10,500 m (bottom). Dark blue, no stripes: negligible, dark blue = below 0.5 kW/m2, brown = 0.5 to 2 kW/m2, medium blue = 2 to 10 kW/m2, turquoise = 10 to 40 kW/m2, red = 40 to 70 kW/m2, pink = over 70 kW/m2 [34].

A site deemed to be inefficient for CWT might be very effective for HAT. Zillmann, U., & Bechtle, P. (2018) produced a heat map comparing CWT and HAT.

Environmental Footprint: HAT with a simple design consumes less material to construct. In comparison with other renewable energy sources it uses fewer nonrenewable materials. It means it has lower footprint on production and disposal as well. It can also withstand adverse weather conditions, which helps it to be a smaller contributor to environmental footprint.

Environmental Disturbance: Conventional wind turbine takes a substantial toll on avian life. It has also been responsible for noise pollution. High altitude turbines overcome these issues though they have their own limitations regarding the impact on precipitation and heat increase.

Land Required for Power Generation: Only 5-8 MW of wind power generation is possible per square kilometer of land [35]. Total area required is 20,000 m^2 (approximately twice the area of football field) for a small-scale wind power of 100 kW. High altitude wind power has a small power station at the ground with blimp in the air. Airborne wind turbine for 100 kW, has ABM of 517m^3 which is tethered to the ground station. The ground station along with small power station also has a blimp landing station. So, around 2000 m^2 area (10 times less than CWP) is sufficient for extracting power using blimp-supported high altitude flying generator [12].

Multidimensional Scope of HAT

The power generated by HAT can be used in diverse situations due to its low cost and simplicity in infrastructure development. It can be a great source of energy during disaster recovery. Also, in events where temporary electric supply is needed, this can be an easy alternative to a non-renewable energy source.

It can also be an alternate source of mobile 5G and broadcasting. Simultaneously surveillances and remote sensing will reach to a next generation through HAT.

The innovative applications of HATs are many. One alternative use for HAWP devices is as a propulsion system for ships. Sky Sails Power, a German company [26] is the leader in this domain and has a patented 'towing kite propulsion system' for cargo ships. The company uses a fixed length tether line to connect a para foil to a ship that has no relative translational motion other than to pull the ship. Sky Sails Power claims that in good wind conditions this technology can replace 2 MW of power from the engine, thus daily saving '10 t of oil'. Sky Sails Power has also developed an offshore power system operating at altitudes up to 800 m. It can be installed on both CWT offshore platforms or floating platforms anchored to the seabed. This technology uses the pumping mode as a means of electricity generation.

Another interesting proposed use for electrical power produced from HAWP devices is in naval propulsion systems applications. It is used to electrolyse seawater on-board to produce hydrogen or methanol or 'convert carbon dioxide into storable forms of liquid' [27]. These by-products could be stored within tanks in a ship and be unloaded to either a courier tanker or at port.

These innovative technology proposals are a prime example of clean renewable energy obtained from the wind. They do not release greenhouse gases (GHG) into the atmosphere. However, many challenges must be overcome in order for this technology to become feasible. It must rely primarily on advancements in research and technology and policy measures to protect mankind and the earth. Hence the promise of inexpensive zero-carbon and abundant wind energy will continue to drive the development of HAWP devices.

Probable Drawbacks

Energy extraction from the kinetic speed of the wind can have unforeseen climatic consequences [30]. None of the HATs are efficient enough to convert even 60% of the wind energy. In the case of wind turbines, the kinetic energy would be converted to mechanical or electrical energy, most of which would ultimately be dissipated as heat [31].

A study by the same group of researchers also found that civilization-scale reliance on wind power, if uniformly distributed, might change zonal mean temperature by ~0.1 K and zonal mean precipitation by ~1%.

During a thunderstorm, the conductivity of the tether may increase [32]. Hence if a lightning discharge occurs, there will be a possibility that the tether attracts the lightning leader to develop via it. Or the strong impulsive electric field generated by the lightning may cause a surface flashover along the tether. The strong static electric field generated by the accumulated charge in the thundercloud may also cause a breakdown on the tether. If the tether reaches the charge center of the cloud, new lightning may be triggered by the breakdown on the tether. The breakdown on the tether will also attract the lightning channel to approach. Whether a new lightning is triggered by the tether or the formed lightning channel attracted by the breakdown on tether, the breakdown could develop continually along the tether, finally throughout the tether and the current is conducted downward. If the strong lightning current is conducted via the tethers to the ground station, the current and the lightning high voltage accompanied will bring destruction to the upper system and the ground-based generator.

Commercial Endeavors

A lot has happened in the last decade in the area of HAT [37] though commercial production has not been successful. Small-scale wind drone prototypes were built by many research teams and companies and have been flying for many years. Makani, which had been acquired in 2013 by Google and is currently one of the "moonshot" projects of the Alphabet subsidiary X, is the first company to take the next step towards the approximate scale of present commercial wind turbines. It has already built a fully functional demonstrator with 600 kW output, 26 m wingspan and a mass of 2 tons. Illustrated in Fig. 10.14 are some of the commercial names in HAT [34].

Fig. 10.14 Overview of major companies [34].

Case Study

1. IRAN [38]

With growing wind turbine sizes and increasing demands for electricity, it is important to determine accurately wind shear profile, up to high altitudes, for large-scale wind turbine installation. The installed wind turbine towers in Iran have hardly exceeded the height of 40 m. Wind data was collected from the center of Iran which consists of 35 wind stations in nine provinces. For precise evaluation, the wind speed data were recorded based on 10-min average at various heights ranging from 10m to 100m and the wind power potential was obtained on both a seasonally and daily basis. The widely used Weibull distribution function is selected for which the shape factor and scale parameter are determined using the standard deviation method, the empirical method of Lysen and the power density method at different heights. From the extrapolated wind data, the mean wind speed, wind power density, optimal wind speed, probability and cumulative density distributions are obtained and the geographic information systems maps are produced and discussed. The results indicate that the power density method is the most accurate method for extrapolating wind characteristics. From the results, it is also observed that variation of power density with time is significant. In period I (spring and summer), most parts of the center of Iran have significant potential for installing large-scale wind turbines. However, in period II (fall and winter), the amount of wind power density decreases considerably. In overall, the northwestern and northeastern parts of the center of Iran are recommended for large-scale wind turbine installation.

The land area of Iran is 1,648,195 km^2 and is located between latitude 25.00–39.47 N and longitude 44.02–63.20E. Iran is divided into 31 provinces, and the aim of this paper is to evaluate wind potential at nine central provinces: Tehran, Isfahan, Yazd, Ghom, Ghazvin, Markazi, Alborz, Semnan and Chaharmahal and Bakhtiari. There are 35 wind stations in this region from which wind data were collected. In most areas of the center of Iran, the climate is warm and dry and differences in temperature throughout a day are considerable. The center of Iran consists of several closed basins that are referred to as Central Plateau. The average elevation of this plateau is about 900 meters, but some of the mountains that tower over the plateau exceed 3,000 meters. The eastern part of the plateau is covered by two salt deserts, Dasht-e Kavir and Dasht-e Lut. These deserts are uninhabited except for some scattered oases.

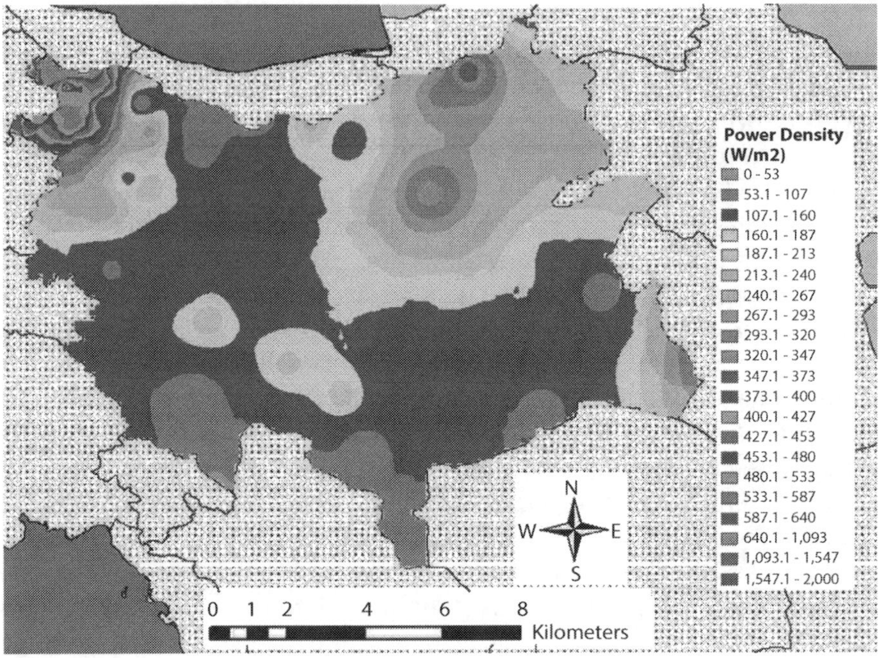

Fig. 10.15 GIS map of wind power in period I, 30 m.

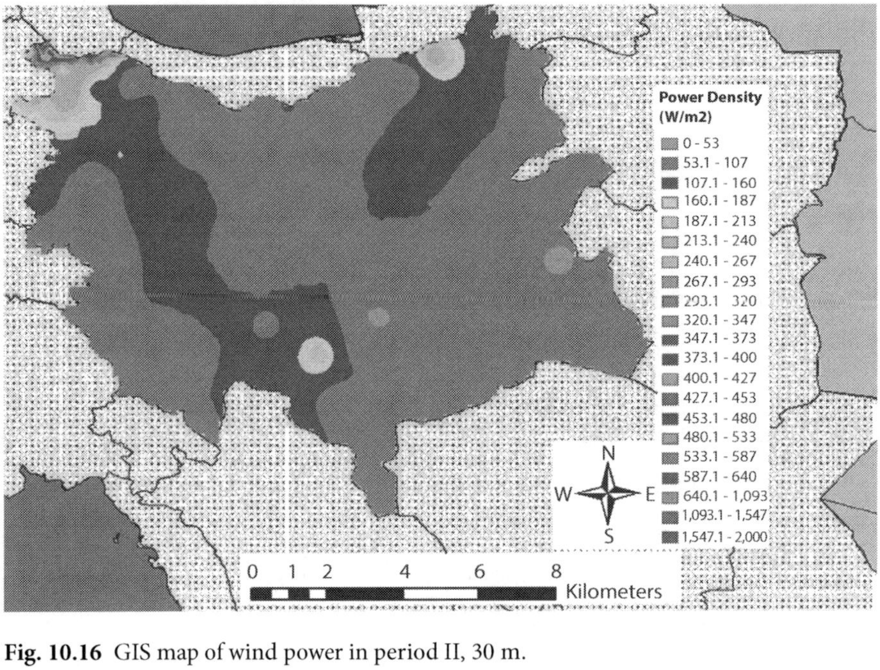

Fig. 10.16 GIS map of wind power in period II, 30 m.

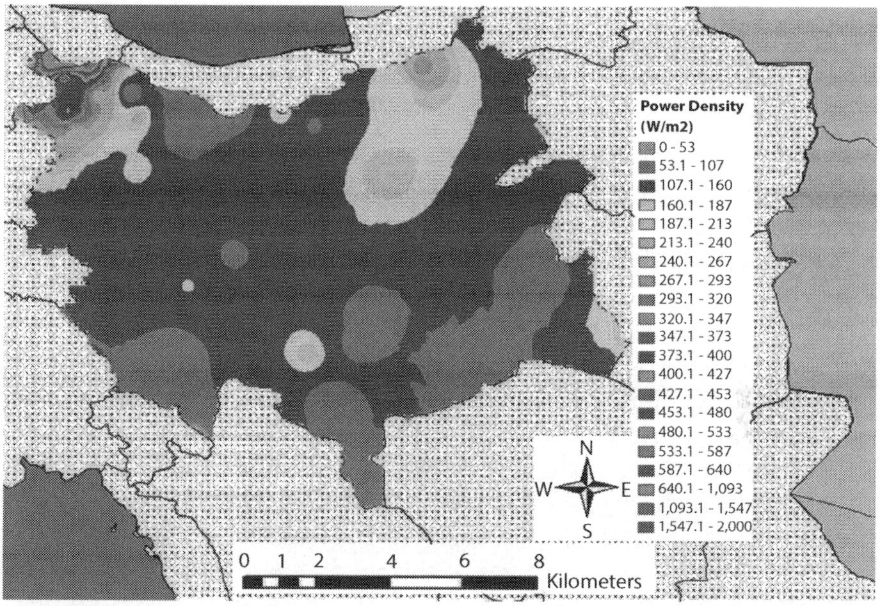

Fig. 10.17 The overall GIS map of wind power for the center of Iran, 30 m.

The main results of this study can be summarized as follows:

- In most wind stations the amount of wind power density is low and 22 wind sites stand in class 1 (poor condition) at height of 80 m. However, there are eight wind sites with high wind power densities and for better evaluation a detailed analysis for them has been done.
- GIS maps show that northeastern and northwestern parts of the center of Iran have good potential for installing large-scale wind turbines. Moreover, in period I (spring and summer) the amount of wind power density is very high and this amount is reduce strikingly in period II (fall and winter).
- In most wind sites, the power density method can estimate Weibull parameters with the most accuracy and a detailed study showed that this method can be used in all wind stations. By extrapolating shape and scale parameters, surface roughness exponent, feature of terrain and wind power density at heights of 50–100m are determined.
- Histograms of wind data are used in order to specify the probability distributions of wind speeds at eight selected

wind stations and from cumulative density diagrams the time that wind speed exceeds cut-in wind speed (3 m/s) is computed.

- Short time interval is very important for calculating wind power density and the results show that by increasing time interval the growth of error is significant.
- The findings of this study can be extended and implemented for any wind turbine heights and any optional wind regions.

2. Northern Ireland [1]

In many countries wind energy has become an indispensable part of the electricity generation mix. The opportunity for ground-based wind turbine systems is becoming more and more constrained due to limitations on turbine hub heights, blade lengths and location restrictions linked to environmental and permitting issues including special areas of conservation and social acceptance due to the visual and noise impacts.

As one of the main objectives of this work was to assess the potential of HAWP in Northern Ireland code was developed in MATLAB to manage, filter and pre-process the wind data from the NetCDF file format and then to undertake the subsequent data analysis and visualization of high-altitude wind speeds and power for the study area. Once the wind power density across Northern Ireland was determined the data was averaged over height, available area and for a time span of 20 years. Furthermore the results were plotted on a coloured contour spatial chart using the Mapping Toolbox within MATLAB. The maximum height for averaging was set to 3000 m as this covers the majority of proposed operating altitudes by established HAWP companies in the market presently.

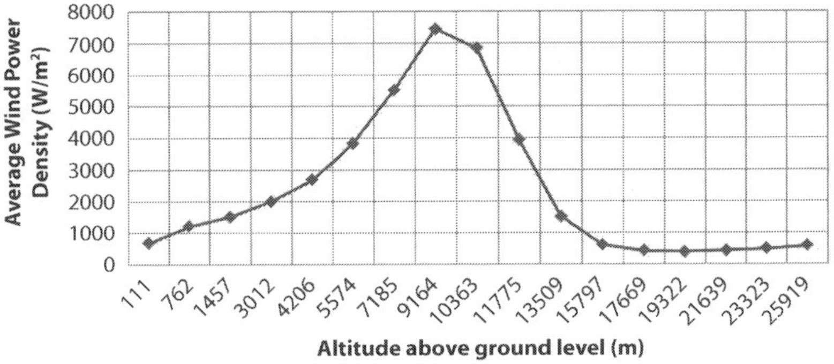

Fig. 10.18 Wind power density averaged over a 20-year time span across various pressure level heights.

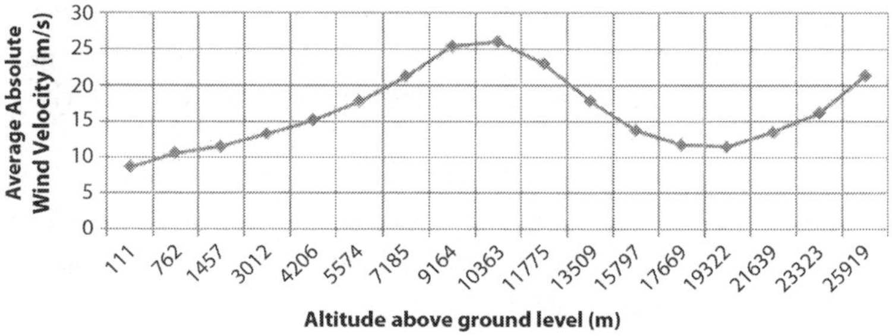

Fig. 10.19 Absolute wind velocities averaged over a 20-year time span across various pressure levels.

Several 'geographical limitations' have been identified as layers within the test system analysis to ensure the safety of society and infrastructure surrounding a potential HAWP farm. These geographical limitations are:

- Motorways – These high-speed lanes have been selected in precedence to regional roads, as the traffic is more concentrated and has a shorter time to react if a HAWP device was to undergo system failure and crash. This data was sourced from a 1:50,000 full vector dataset of Northern Ireland provided by Ordnance Survey of Northern Ireland (OSNI).
- Railways – This data was sourced from 1:50,000 full vector dataset of Northern Ireland provided by OSNI.
- Settlements – 'Large towns' and 'towns' were filtered from the settlement data contained within the OSNI 1:50,000 full vector dataset of Northern Ireland. These areas take precedence over 'villages' and 'small villages', not only due to greater population concentrations but also due to their transport infrastructure.
- Airports and Military Training Areas – The data for civil aerodrome locations were sourced through the Google Earth application and plotted onto an OSNI outline map. Military training areas were sourced from a chart published by the National Air Traffic Services (NATS) that displays Airspace Restrictions and Hazardous Areas in the UK.
- Controlled Airspace – Special Use Airspace (SUA) and Digital Aeronautical Flight Information Files (DAFIF) data was sourced for the Northern Ireland region identifying restricted air corridors in the vicinity of airports but also segregated areas restricted for alternative aviation activities such as hand gliding, paragliding, and sky diving.

- Protected Areas – To ensure both wildlife and natural landscapes are preserved, special areas are designated for protection, including Areas of Outstanding Beauty, World Heritage Sites, Areas of Special Scientific Interest and Special Areas of Conservation. These areas have been included as limiting layers due to the stringent planning permission restrictions to any major developments. This digital data was sourced from the Department of the Environment (DOE).

The specific technical findings of the study are summarized as follows:

- There is huge potential for HAWP in Northern Ireland with a wind power density ranging from 1850 W/m^2 to 2100 W/m^2, with an average of 1998 W/m^2, at a set height of 3000 m, averaged over a 20-year span.
- The highest wind power densities occur off the west coast of Northern Ireland attaining wind power densities in excess of 2050 W/m^2.
- The onshore western areas of Northern Ireland are subjected to the highest wind power densities of 2050 W/m^2 and above with the most suitable site for a demonstrator high altitude wind project located in the west of County Tyrone.
- The total viable area across Northern Ireland for siting a HAWP device was determined to be 5109.6 km^2.
- At 3000 m, the maximum height chosen for the study, the average absolute wind velocity was 13.25 m/s.
- These technical findings support the initial considered locality for this innovative technology to be above the earth's neutral ABL, at 1,300 m above ground level.

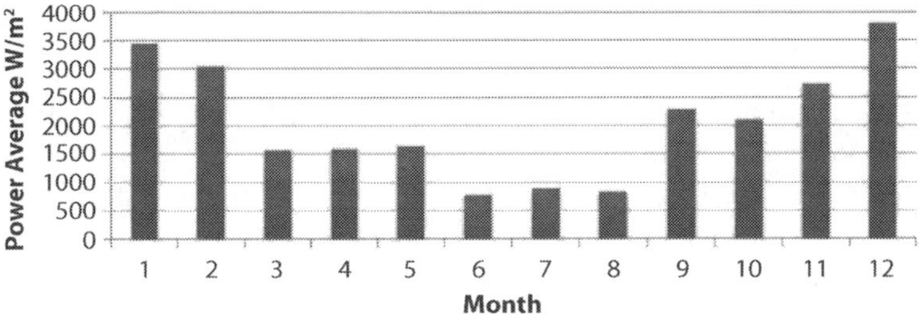

Fig. 10.20 Wind power density averaged by month from 2010 to 2013 to a height of 3000 m over Northern Ireland.

- The maximum optimal altitude height worth exploring for HAWP technologies in Northern Ireland is 10,000 m.
- The average power density variations were found to be below.

References

1. Lunney A, Ban M, Duic N, Foley A (2017), A State-of-the-art review and feasibility analysis of high altitude wind power in Northern Ireland. *Renewable and Sustainable Energy Reviews* 68.
2. Bansal R.C., Zobaa A. F., and Saket R.K. (2005), Some issue related to power generation using wind energy conversation system: An overview, *International Journal of Emerging Electric Power Generation* Vol 3, No 2.
3. Ulazia A., Saenz J., Ibarra-Berastegi G., Gonzalez-Roji S. J. (2019), Global estimations of wind energy potential considering seasonal air density changes, *Energy* 187.
4. Arya P. *Introduction to micrometeorology*. 2nd ed. Academic Press; 1988.
5. Archer CL, Caldeira K. Global assessment of high-altitude wind power. ISSN 1996–1073. *Energies* 2009;2:307–19.
6. Roberts BW, Shepard DH, Caldeira K, Cannon ME, Eccles DG, Grenier AJ, Freidin JF. Harnessing high-altitude wind power. *IEEE Trans Energy Convers* 2007;22(1).
7. Loyd ML. Crosswind kite power. *J Energy* 1980;IV(3):106–11.
8. Gillis J, Goos J, Geebelen K, Swevers J, Diehl M. Optimal periodic control of power harvesting tethered airplanes: how to fly fast without wind and without propellor? In: *Proceedings of the American Control Conference; 2012.* p. 2527–32.
9. Verheul RF, Breukels J, Ockels WJ. Material selection and joining methods for the purpose of a high-altitude inflatable kite. In: *Proceedings of the Structures, Structural Dynamics, and Materials Conference. Palm Springs, California*; 2009
10. Doyle Sailmakers, Dacron. Available at: ⟨http://www.doylesails.com/design/fabric.html⟩ [accessed 24/04/2021).
11. Ahrens U, Diehl M, Schmehl R. *Airburne wind energy*.Dordrecht: Springer; 2013.
12. Adhikari J. & Panda S. K. (2013), Overview of high altitude wind energy harvesting system, *5th International Conference on Power Electronics Systems and Applications (PESA)*.
13. P.P.G. Ideascapes, Glass Education Center, Glass Topics, Available at: ⟨http:// educationcenter.ppg.com/glasstopics/how_lowe_works.aspx⟩ [accessed 21/04/2021].
14. Lansdorp B, Ockels WJ. Comparison of concepts for high-altitude wind energy generation with ground based generator. In: *Proceedings of the 2nd*

China international renewable energy equipment & technology exhibition and conference. Beijing; 2005.
15. Dyneema, Dyneema Fiber. Available at: ⟨http://www.dyneema.com/americas/ product-technologies/fiber.aspx⟩ [accessed 22/04/2021].
16. Vectran Liquid Crystal Polymer Fibre, About Vectran, Penvision Internet Marketing [Online]. Available at: ⟨http://www.vectranfiber.com/⟩ [accessed 23/04/2021].
17. Argatov I, Rautakorpi P, Silvennoinen R. Estimation of the mechanical energy output of the kite wind generator. *Renew Energy* 2009;34:1525–32.
18. Williams P, Lansdorp B, Ockels W. Optimal cross-wind towing and power generation wtih tethered kites. American Institute of Aeronautics and Astronautics Inc.; 2007.
19. Power Ampyx, Ampyx Power, Airborne wind energy, Available at: ⟨http://www.ampyxpower.com/OurTechnology.html⟩ [accessed 22/04/2021].
20. Ockels WJ. Laddermill, a novel concept to exploit the energy in the airspace. *Aircr Des* 2001;4:81–97.
21. Payne PR, McCutcheon C. Self-erecting windmill. USA Patent 3,987,987; 26 October 1976.
22. Colozza A, Dolce JL. High-altitude, long-endurance airships for coastal surveillance. NASA, Glenn Research Center; 2005.
23. Adhikari J, Panda SK, Rathore AK. Harnessing high altitude wind power using light gas filled blimp. National University of Singapore: Department of Electrical and Computer Engineering; 2.
24. Vermillion C, Grunnagle T, Kolmanovsky I. Modeling and control design for a prototype lighter-than-air wind energy system. In: *Proceedings of the American Control Conference*. Montreal, Canada; 2012.
25. Perkovic L, Silva P, Ban M, Kranjcevic N, Duic N. Harvesting high altitude wind energy for power production: the concept based on Magnus' effect. *Appl Energy* 2012;101:151–60.
26. Sky Sails. Power, wind power – next level. Available at: ⟨http://www.skysails.info/fileadmin/user_upload/Power/Presse/EN_SkySails_Power_Wind_Power_ Next_Level.pdf⟩ [accessed 05.11.14].
27. Kim J, Park C. Wind power generation with a parawing on ships, a proposal. *Energy* 2009:1425–32.
28. Kolar, J.W.; Friedli, T.; Krismer, F.; Looser, A.; Schweizer, M.; Steimer, P.; Bevirt, J., "Conceptualization and multi-objective optimization of the electric system of an Airborne Wind Turbine," Industrial Electronics (ISIE), 2011 IEEE International Symposium on, vol., no., pp. 32,55, 27-30 June 2011.
29. Mondal, A. K., Mondal S., Devalla V., Sharma P. and Gupta M. K. (2016); Advances in floating aerogenerators: present status and future. *International Journal of Precision Engineering and Manufacturing* Vol 17 No 11.
30. Marvel, K., Kravitz, B., & Caldeira, K. (2012). Geophysical limits to global wind power. *Nature Climate Change*, 3(2), 118–121. https://doi.org/10.1038/nclimate1683.

31. Boville, B. & Bretherton, C. Heating and kinetic energy dissipation in the NCAR community atmosphere model. *J. Clim.* 16, 3877_3887 (2003).
32. Ji, Y., He, J., 2013. Analysis on lightning triggering possibility along transmission tethers of high altitude wind energy exploitation system. *Electric Power Systems Research* 94, 16–23. https://doi.org/10.1016/j.epsr.2012.05.021.
33. M. Canale, L. Fagiano and M. Milanese, (2009), High Altitude Wind Energy Generation Using Controlled Power Kites, *IEEE Transactions on Control Systems Technology* 18, 279.
34. Zillmann, U., & Bechtle, P. (2018). Emergence and Economic Dimension of Airborne Wind Energy. In *Airborne Wind Energy* (pp. 1–25). Springer Singapore. https://doi.org/10.1007/978-981-10-1947-0_1].
35. Paul Denholm, Maureen Hand, Maddalena Jackson, and Sean Ong, "Land-Use Requirements of Modern Wind Power Plants in the United States".
36. Dunker, S. (2018). Tether and Bridle Line Drag in Airborne Wind Energy Applications. In *Airborne Wind Energy* (pp. 29–56). Springer Singapore. https://doi.org/10.1007/978-981-10-1947-0_2.
37. Cherubini, A., Papini, A., Vertechy, R., Fontana, M.: Airborne Wind Energy Systems: A review of the technologies. *Renewable and Sustainable Energy Reviews* 51, 1461–1476 (2015). doi: 10.1016/j.rser.2015.07.053.
38. Faghani, GH.R., Ashrafi, Z.N., Sedaghat, A., 2018. Extrapolating wind data at high altitudes with high precision methods for accurate evaluation of wind power density, case study: Center of Iran. *Energy Conversion and Management* 157, 317–338. https://doi.org/10.1016/j.enconman.2017.12.029.

Application 1

Development and Adaptation of a Mathematical Model for a Two-Dimensional Calculation of the Flow Around an NACA0021 Airfoil Moving Along a Circular Track*

Mathematical Model

Gas flow regimes near the blade system of wind turbines occur in the subsonic region; therefore, the equations of an incompressible fluid can be used to model the flow. These are equations of conservation of mass, momentum and transfer that describe the turbulent characteristics of flows.

Mass conservation equation

$$\frac{\partial \rho}{\partial t} = \nabla \cdot (\rho \vec{u}) = S_m \tag{Ap1.1}$$

Equation (Ap1.1) represents the general form of the mass conservation equation and is applicable to both incompressible and compressible flows. S_m is the source term. For our case, it can be written as:

$$\nabla \cdot (\rho \vec{u}) = 0$$

*Work was made by "Torins" LTD, Krasnoyarsk, RF (Poushkarev V.A., Dekterev A.A.) in accordance with the technical assignment of the BAT.

Victor M. Lyatkher and Ziaur Rahman. Wave, Wind and Current Power Generation, (319–354)

The momentum conservation equation. Conservation of momentum in an inertial (non-accelerating) frame of reference in general form is described by the equation:

$$\frac{\partial}{\partial t}(\rho\vec{u})+\nabla\cdot(\rho\vec{u}\vec{u})=-\nabla p+\nabla\cdot(\bar{\bar{\tau}})+\rho\vec{g}+\vec{F} \qquad \text{(Ap1.2)}$$

Where p is the static pressure, $\bar{\bar{\tau}}$ - viscous stress tensor, ρg and F - gravitational force and external mass force. For our case, in the absence of mass forces and constant density:

$$\frac{\partial}{\partial t}(\vec{u})+\nabla\cdot(\vec{u}\vec{u})=-\frac{1}{\rho}(\nabla p+\nabla\cdot(\bar{\bar{\tau}}))$$

The viscous stress tensor is given by:

$$\bar{\bar{\tau}}=\mu\left[(\nabla\vec{u}+\nabla\vec{u}^{T})-\frac{2}{3}\nabla\cdot\vec{u}I\right] \qquad \text{(Ap1.3)}$$

where μ is the molecular viscosity, I is the unit tensor, and the second term on the right is the volume expansion effect.

For our case: $\bar{\bar{\tau}}=\mu(\nabla\vec{u})$

Turbulence Modeling

From modern concepts of the dynamics of liquid and gas, it follows that both laminar and turbulent flows are described by equations (Ap1.1) and (Ap1.2) - the so-called direct numerical simulation (DNS). However, the need to resolve all spatial scales (from dissipative scales to the scale of the study area), the use of DNS is still limited only by methodological calculations, and to solve practical problems, an averaging of equations is introduced, leading to the need to use the turbulence model. In practice, the Reynolds averaging method is widely used to describe turbulence, which consists in replacing randomly changing flow characteristics (velocity, pressure, density) with the sums of averaged and pulsation components. So for the speed components:

$$u_i=\bar{u}_i+u_i'$$

where $\bar{u}_i$ and u_i' are the average and pulsation velocity components (i = 1,2,3). Substituting expressions in this form into the equations of conservation of mass and momentum and applying averaging over time (and omitting the line in the designation of the average velocity) we obtain the averaged equations:

$$\frac{\partial \rho}{\partial t}+\frac{\partial}{\partial x_i}(\rho u_i)=0$$

$$\frac{\partial}{\partial t}(\rho u_i)+\frac{\partial}{\partial x_j}(\rho u_i u_j)=-\frac{\partial p}{\partial x_j}+\frac{\partial}{\partial x_j}\left[\mu\left(\frac{\partial u_i}{\partial x_j}+\frac{\partial u_j}{\partial x_i}-\frac{2}{3}\delta_{ij}\frac{\partial u_l}{\partial x_l}\right)\right]+\frac{\partial}{\partial x_j}\left(-\rho\overline{u_i'u_j'}\right)$$

These equations are called Reynolds-averaged Navier-Stokes equations (RANS). They have the same basic form as the instantaneous Navier-Stokes equations, but with velocities and other calculated quantities now representing time-averaged quantities. Additional added terms describe the effects of turbulence. These Reynolds stresses $-\rho\overline{u_i'u_j'}$ must be properly modeled to close the equations. A common method uses the Boussinesq hypothesis to relate the Reynolds stresses to mean velocity gradients

$$-\rho\overline{u_i'u_j'}=\mu_\tau\left(\frac{\partial u_i}{\partial x_j}+\frac{\partial u_j}{\partial x_i}\right)-\frac{2}{3}\left(\rho k+\mu_\tau\frac{\partial u_k}{\partial x_k}\right)\delta_{ij}$$

where μ_τ is the turbulent (vortex) viscosity. Boussinesq's hypothesis is used in many turbulence models such as Spalart-Allmaras (SA), *k*-ε, *k*-ω, MSST. The advantage of this method is the relatively low computational cost. The disadvantages are associated with the non-universality of such models. The disadvantage of the Boussinesq hypothesis is that it assumes turbulent viscosity as an isotropic scalar, which is not strictly true. However, the assumption of isotropic turbulent viscosity usually works well for shear flows where only one of the turbulent shear stresses is dominant. This includes many technical streams such as near-wall boundaries, jets, etc.

The Spalart-Allmaras model [1] is a single equation model that solves the model transport equation for kinematic vortex (turbulent) viscosity. The Spalart-Allmaras model was developed specifically for aerospace applications with limited wall flow and has been shown to give good results for boundary layers subject to unfavorable pressure gradients. It is also popular in turbomachine applications.

The carryover value in the Spalart-Allmaras model, $\tilde{\nu}$, is identical to the turbulent kinematic viscosity, excluding the wall region. Transfer equation for modified turbulent viscosity $\tilde{\nu}$:

$$\frac{\partial}{\partial t}(\rho\tilde{\nu})+\frac{\partial}{\partial x_i}(\rho\tilde{\nu}u_i)=G_v+\frac{1}{\sigma_{\tilde{v}}}\left[\frac{\partial}{\partial x_j}\left\{(\mu_\tau+\rho\tilde{\nu})\frac{\partial\tilde{\nu}}{\partial x_j}\right\}+C_{b2}\rho\left(\frac{\partial\tilde{\nu}}{\partial x_j}\right)^2\right]-Y_v+S_{\tilde{v}}$$

where G_v is the generation of turbulent viscosity, Y_v is the destruction of turbulent viscosity that occurs in the near-wall region. $\sigma_{\tilde{v}}$ and C_{b2}-constants and ν -molecular turbulent viscosity. $S_{\tilde{v}}$ - source term.

Turbulent viscosity μ_τ is calculated from

$$\mu_\tau = \rho\tilde{\nu} f_{v1}$$

where the viscosity damping function f_{v1} is written as: $f_{v1}=\frac{\chi^3}{\chi^3+C_{v1}}$, and $\chi=\frac{\tilde{\nu}}{\nu}$.

Generating viscosity G_v to the simulator $G_v = C_{b1}\rho\tilde{S}\tilde{\nu}$, where $\tilde{S}\equiv S+\frac{\nu}{k^2d^2}f_{v2}$ and $f_{v2}=1-\frac{\chi}{1+\chi f_{v1}}$.

C_{b1} и k – constants, d – distance from board, and S – scalyar measure of deformations tensor.

$S\equiv\sqrt{2\Omega_{ij}\Omega_{ij}}$, $\Omega_{ij}=\frac{1}{2}\left(\frac{\partial u_i}{\partial x_j}-\frac{\partial u_j}{\partial x_i}\right)$. In our calculations are:

$$Y_v = C_{w1}\rho f_w\left(\frac{\tilde{\nu}}{d}\right),\quad f_w = g\left[\frac{1+C_{w3}{}^6}{g^6+C_{w3}{}^6}\right]^{1/6},\quad g = rC_{w2}(r^6-r),\quad r\equiv\frac{\tilde{\nu}}{\tilde{S}k^2d^2},$$

C_{w1}, C_{w2} and C_{w3} - constantes.

$C_{b1} = 0.1355$, $C_{b2} = 0.622$, $\sigma_{\tilde{v}} = \frac{2}{3}$, $C_{v1} = 7.1$

$$C_{w1} = \frac{C_{b1}}{k^2}+\frac{(1+C_{b2})}{\sigma_{\tilde{v}}},\quad C_{w2} = 0.3,\ C_{w3} = 2.0,\ k = 0.4187.$$

Calculation Method

To solve the resulting system of partial differential equations, different discretization methods are used - finite-difference, finite-volume, or finite-element methods. When solving problems of industrial hydro-aerodynamics, finite-volume methods are widely used, which have the second order of accuracy in space and time. In our case, we use the finite volume method implemented for the calculation on an arbitrary unstructured computational grid. The SIMPLE procedure [2] is used to jointly solve the equations of continuity and motion. The typical calculation time for one variant was about 20 hours. The calculation process is parallel and was carried out on a personal computer with 6 computing cores. The typical calculation time for one variant was about 20 hours.

Testing the Algorithm

The parameters and experimental data on the cycloidal rotor IAT21 L3 [3] (Fig. Ap1.1-Ap1.3) were used as a test problem for testing the calculation method.

$$\theta = \frac{\pi}{2} - \sin^{-1}\left[\frac{e}{a}\cos(\Psi + \varepsilon)\right] - \cos^{-1}\left[\frac{a^2 + d^2 - L^2}{2ad}\right]$$

$$a^2 = e^2 + R^2 - 2eR\cos\left(\psi + \varepsilon + \frac{\pi}{2}\right)$$

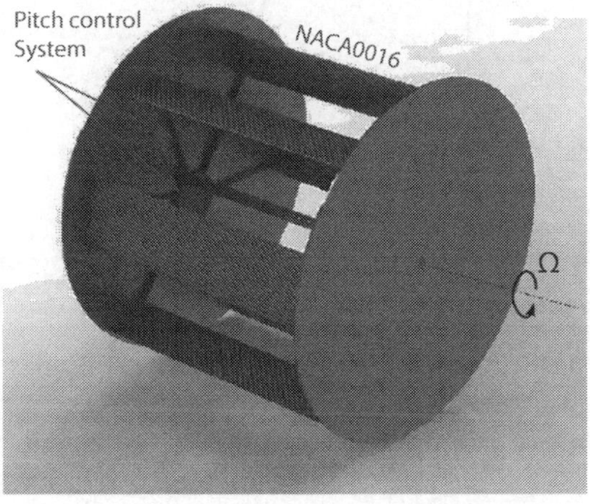

Fig. Ap1.1 Rotor IAT21 L3.

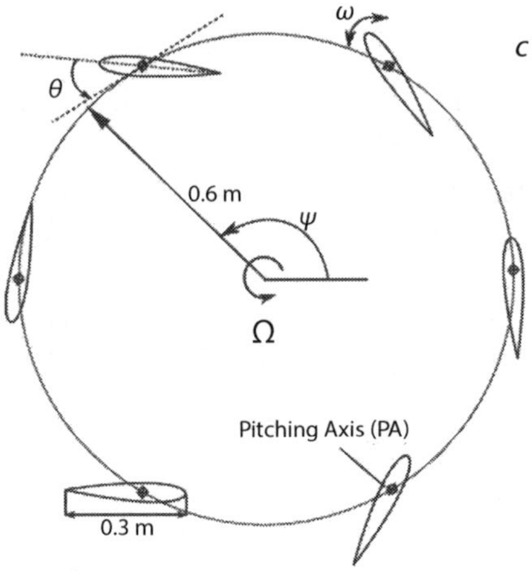

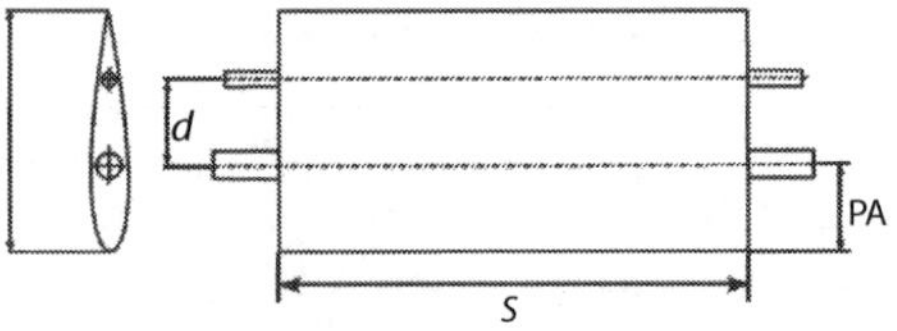

L3: 200 kgf lift / rotor	
Airfoil	NACA0016
Span (*S*)	1.2 m
Chord (*c*)	0.3 m
Rotor Diameter	1.2 m
Pitching axis PA	0.105 m
Control Rod distance (*d*)	0.120 m
θ_{max} (Ψ=90°)	+36°
θ_{min} (Ψ=270°)	-39°
Number of blades	6

Fig. Ap1.2 Parameters of rotor.

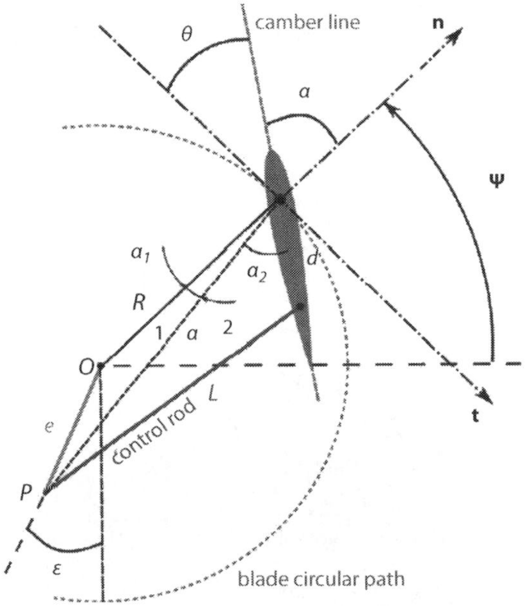

Fig. Ap1.3 Flow around the blade.

The calculation was carried out in a two-dimensional setting. The outer computational domain is circular and communicates with the domain of the rotor rotating at a constant angular velocity through the sliding grid interface. In turn, the areas of each of the blades rotate according to the law determined by the kinematic scheme of the system and also have a grid interface with the area of the rotor. The time step was dt = T/100, where T = 2□/□ is the period of the rotor revolution. Fig. Ap1.4 shows a two-dimensional computational grid consisting of 250,000 tetrahedral cells.

Figs. Ap1.5-Ap1.7 show the instantaneous calculated velocity and pressure fields. Figs. Ap1.8 and Ap1.9 show the graphs of the rotor thrust and power versus speed, respectively. It can be seen from the graphs that the results of our calculation are in good agreement in terms of rotor thrust and power with experimental data and data from other authors.

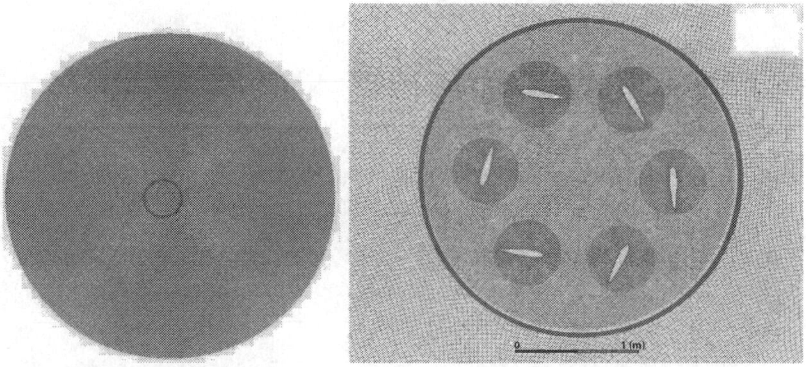

Fig. Ap1.4 2D rotor grid IAT21 L3.

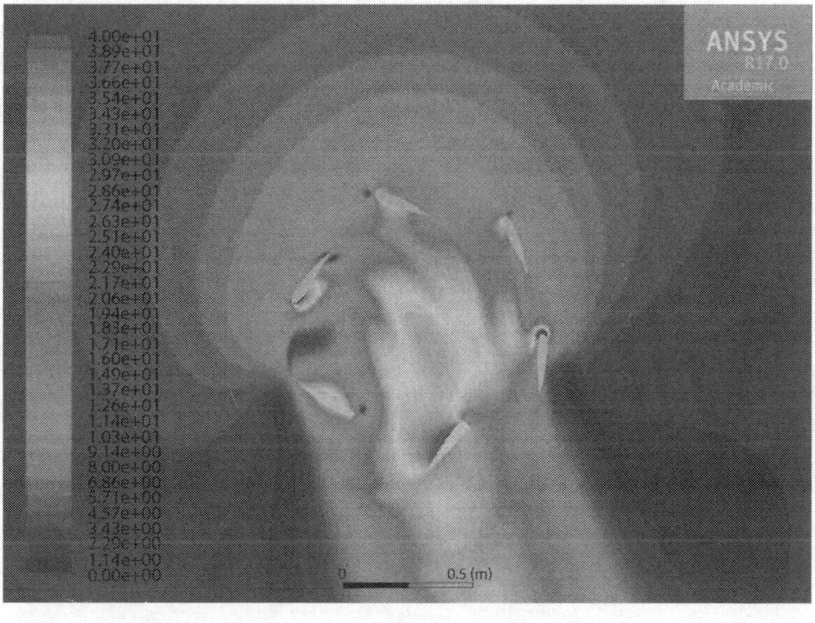

Fig. Ap1.5 Instant speed distribution near rotor (Ω=500 rpm).

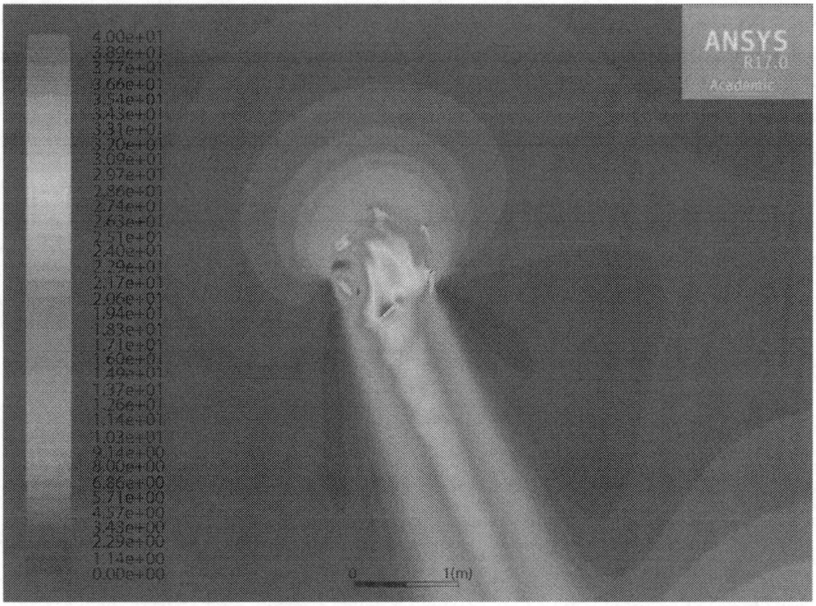

Fig. Ap1.6 Instant speed distribution (Ω=500 rpm).

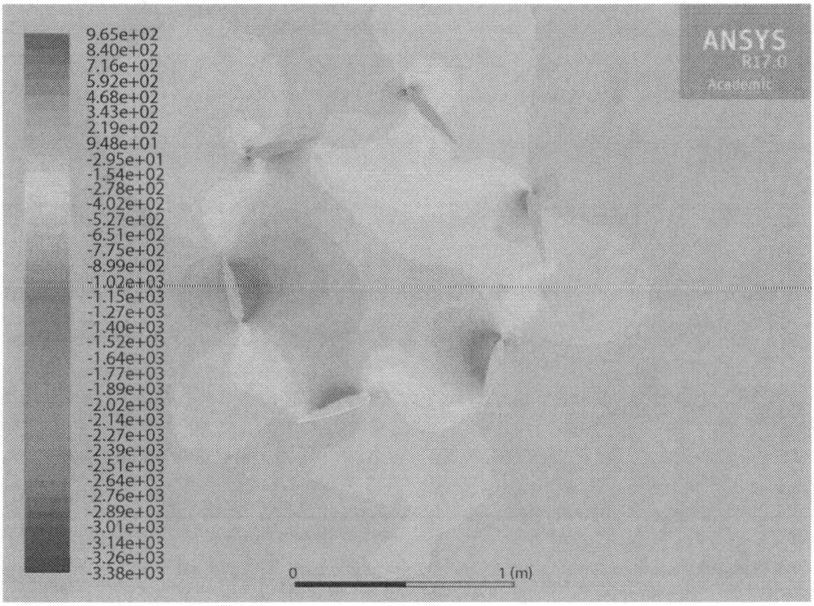

Fig. Ap1.7 Instant pressure distribution (Ω=500 rpm).

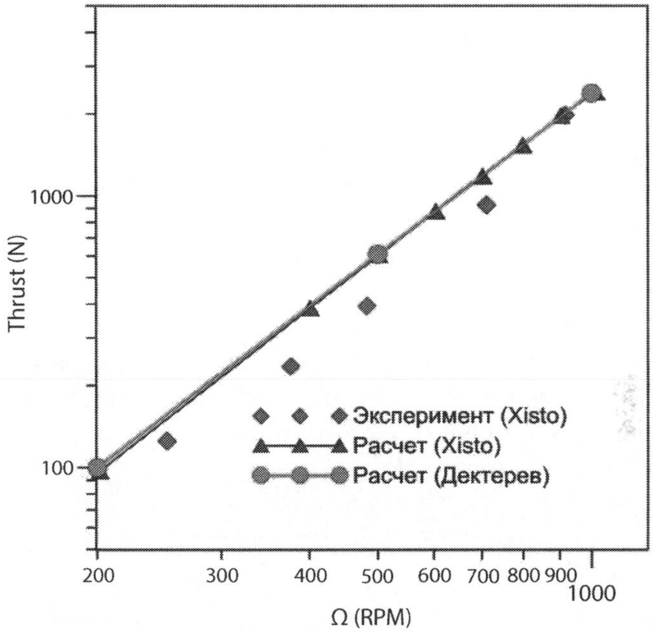

Fig. Ap1.8 Dependence of rotor thrust on rotation speed.

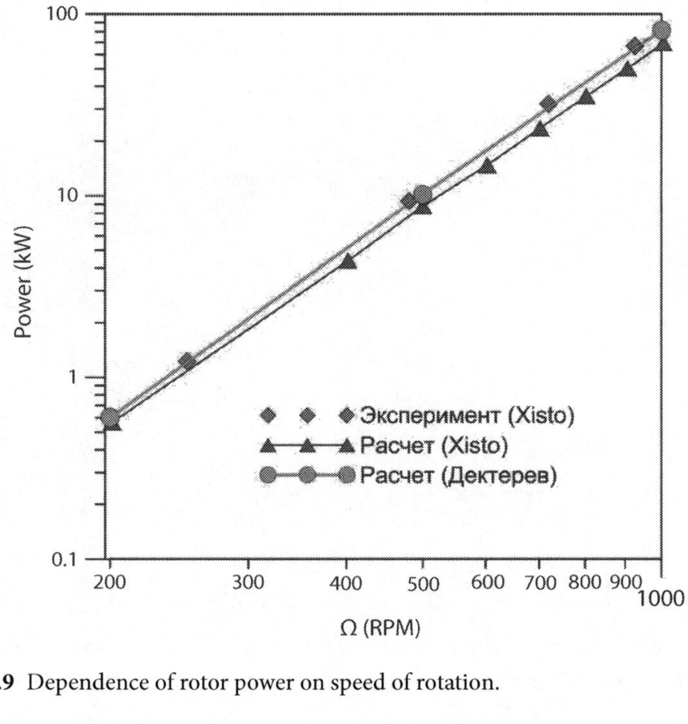

Fig. Ap1.9 Dependence of rotor power on speed of rotation.

To meet the requirements of the TOR for the problem, the following parameters were selected. A straight blade with a length of L = 1 with a chord b = 0.6m, and a profile NACA0021 moves along a circular track with a radius of R = 1m at a constant linear speed V = 12.5 m/s in a uniform air flow with a speed of U = 5 m/s (Fig. Ap1.10) (U/V = 0.4). The angle of the blade to the tangent is 6 degrees. Ambient air density 1.225 kg/m3, kinematic viscosity 1.789 · 10^{-5}. The Reynolds number calculated with respect to the chord of the blade and the speed of the incident flow is in this case

$$Re = \frac{\rho U b}{\propto} = 2.065 \times 10^5.$$

For option no. 2 with impulse delivery, the following parameters were selected: The slot width is a = 0.005 m, the outflow velocity is Vj = 109 m/s, and the pulse duration is 0.015 s, which corresponds to approximately 10 degrees of blade rotation. Jet starts at α = 90°.

The mass of jet is

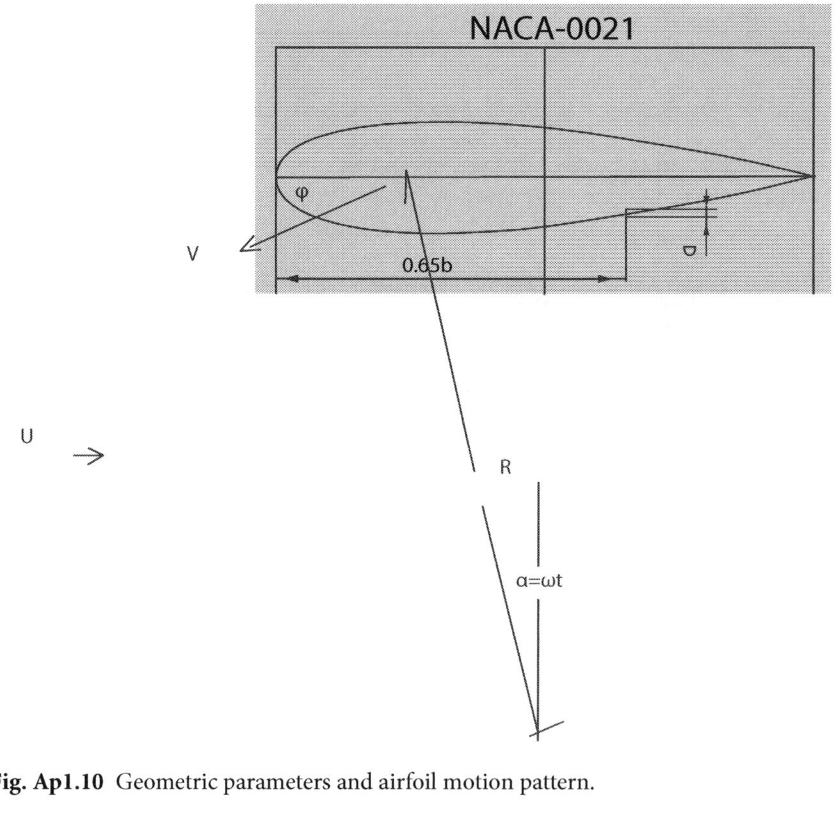

Fig. Ap1.10 Geometric parameters and airfoil motion pattern.

$$m_j = \tau a L V_j \rho = 0.01$$

And relative impuls:

$$J = m_j \cdot a \cdot \frac{V_j^2}{M \cdot b \cdot U^2} = 0.04$$

In variant No. 3, the linear speed of the blade was V = 10 m/s in a uniform air flow with a speed of U = 6 m/s, the outflow velocity of the jet was Vj = 123 m/s, the duration of the pulse was $\tau = 0.015$ s, which corresponds to about 10 degrees of blade rotation, Jet start at $\alpha = 270°$. Thus, the mass of the jet is

$$m_j = \tau a L V_j \rho = 0.011$$

The relative impuls is:

$$J = m_j \cdot a \cdot \frac{V_j^2}{M \cdot b \cdot U^2} = 0.04$$

Option no. 4 is similar in parameters to option no. 3, but the jet feed slot was located on the back side of the blade. To solve the problem, a two-dimensional non-stationary setting was used. The computational grid consisted of 170,000 computational cells (Fig. Ap1.11).

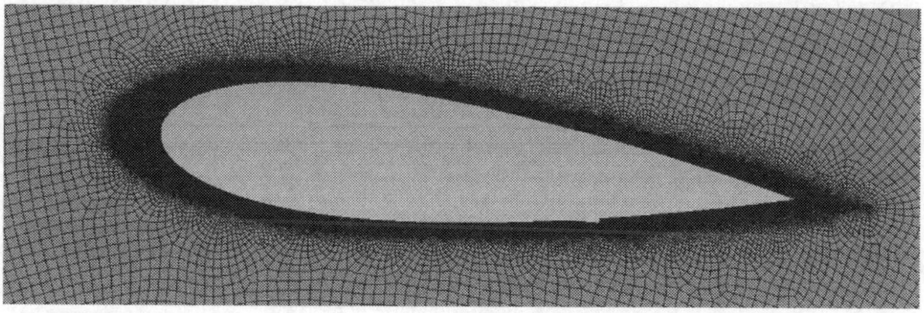

Fig. Ap1.11 Computational grid.

At the front boundary of the computational domain, a flow with a velocity U is set. At the upper, lower and rear boundaries, soft conditions are set -. $\frac{\partial}{\partial n} v = 0.$

As a result of the CFD calculation, the components of the forces acting on the blade surface and the rotational moment due to the aerodynamics of the system are obtained. To take into account the forces due to jet thrust, it is necessary to additionally calculate and add them. In our case, the force will be determined as follows.

$$F_i = V_j \frac{m_j}{\tau}$$

Having decomposed it into components, we added it to Fx and Fy, and took it into account in the torque. Modeling is carried out with the formation of a steady flow, after about N revolutions (N> 10). For the analysis, the results obtained on the last revolution are used.

Results of calculating the motion of the blade without the incident flow
As another test of the algorithm, the problem was calculated, with the conditions corresponding to Option 1, but without the oncoming flow.

$$b/R = 0.6; \quad U=0; \quad V=12.5 \text{ м/с}; \quad J = 0.$$

Below are the average and instantaneous speeds in the area of the blade. Figure Ap1.12 shows that the average velocity field has a symmetrical pattern. Figure Ap1.13 shows the instantaneous velocity field. Figures Ap1.15-Ap1.16 show graphs of the distribution of forces per revolution. Forces change sinusoidally. The moment has a small negative value (friction costs when flowing around the blade) slightly pulsating in time, which is associated with the formation of an unsteady turbulent wake. For comparison, a graph of the moment at the oncoming flow is shown.

Results of modeling. Var. 1

$$b/R = 0.6; \quad U/V= 0.4; \quad J = 0; \quad \alpha = \pi/2.$$

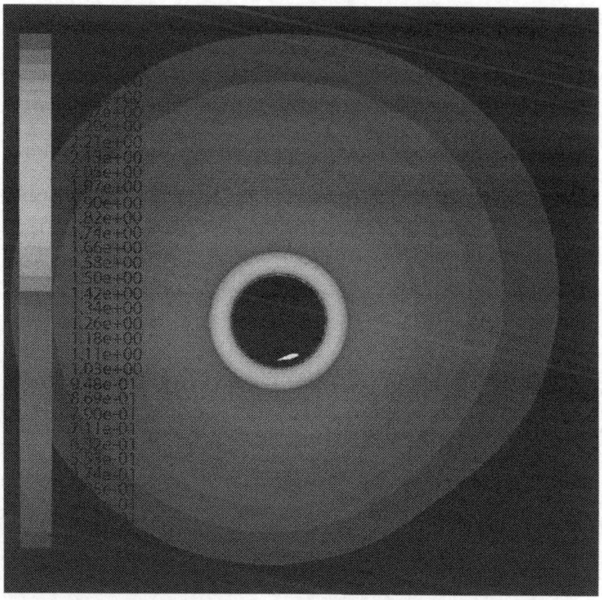

Fig. Ap1.12 Average speed (m/s).

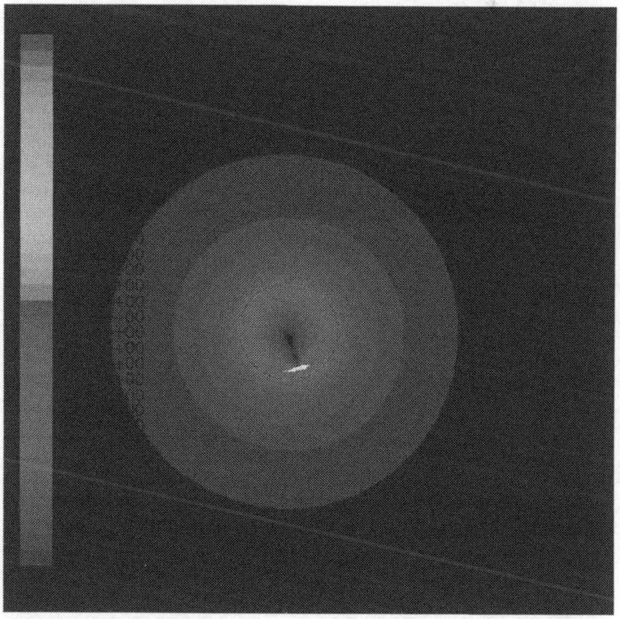

Fig. Ap1.13 Instantaneous speed (m/s).

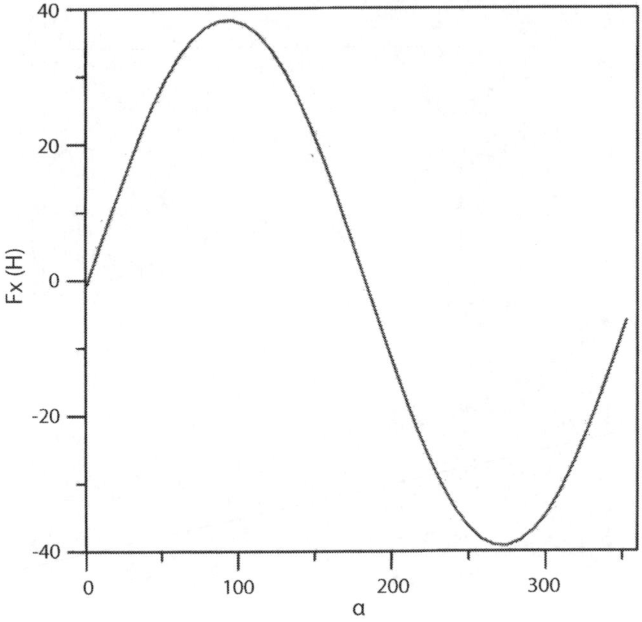

Fig. Ap1.14 Horizontal force (N).

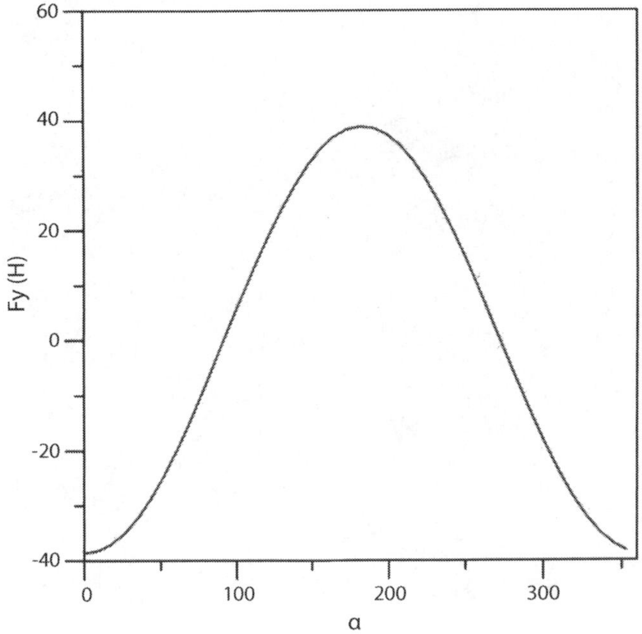

Fig. Ap1.15 Vertical force (N).

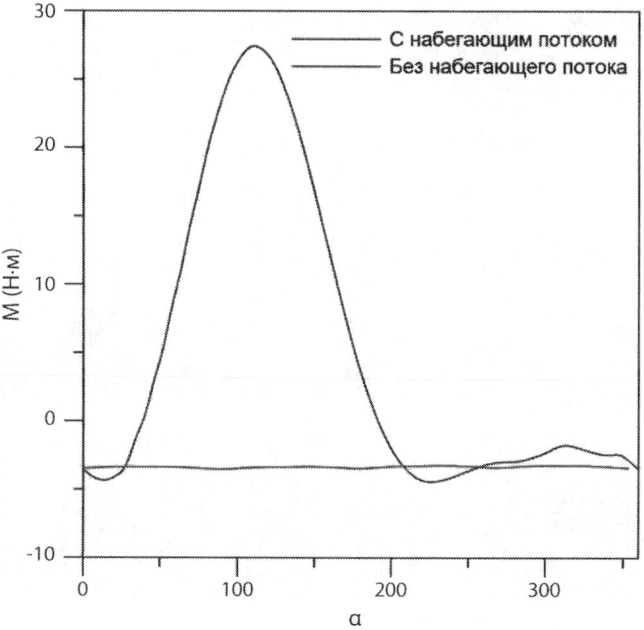

Fig. Ap1.16 Moment (N m).

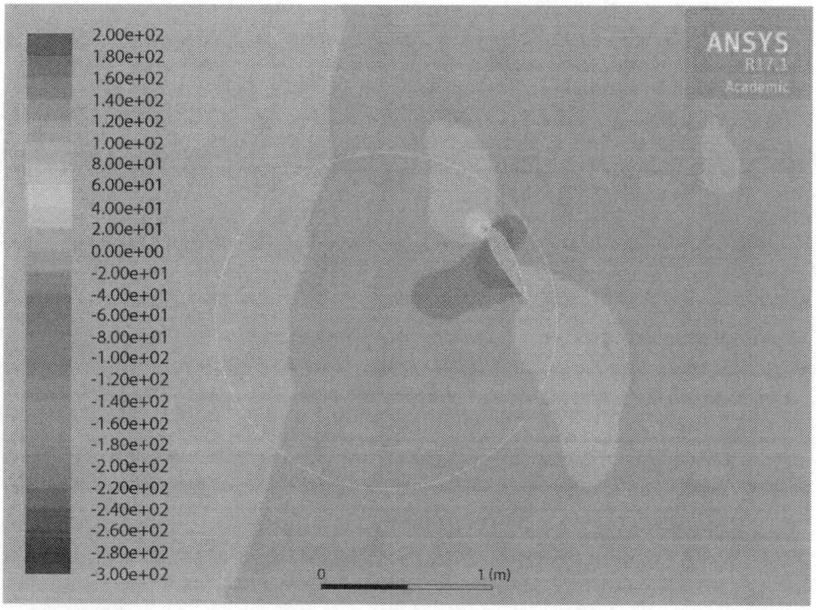

Fig. Ap1.17 Typical pressure field at the moment of jet outflow, Pa.

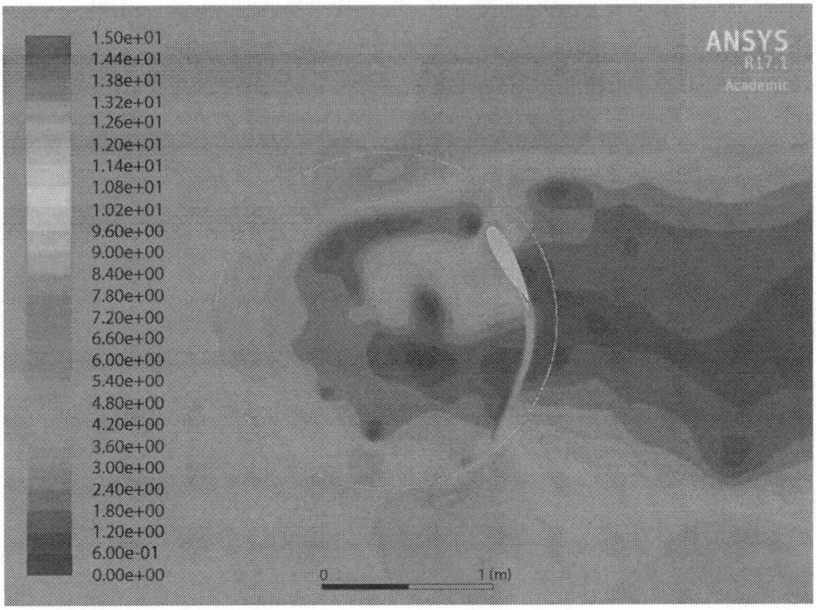

Fig. Ap1.18 The typical velocity field at the moment of the jet outflow, m/s.

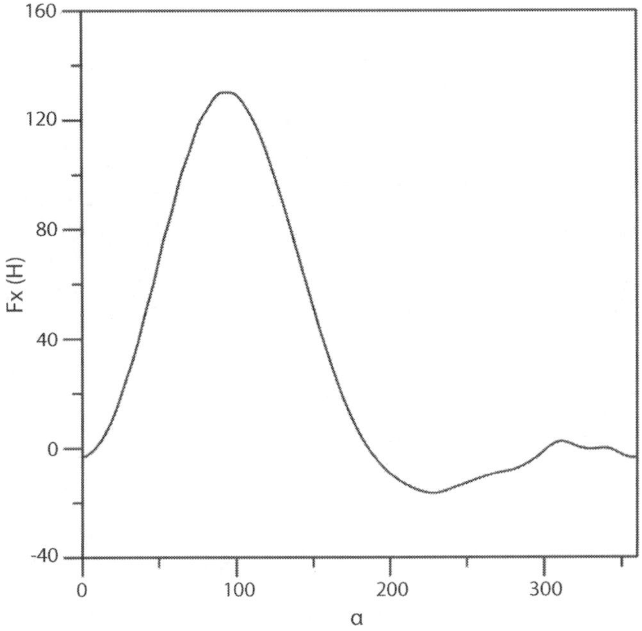

Fig. Ap1.19 Horizontal force.

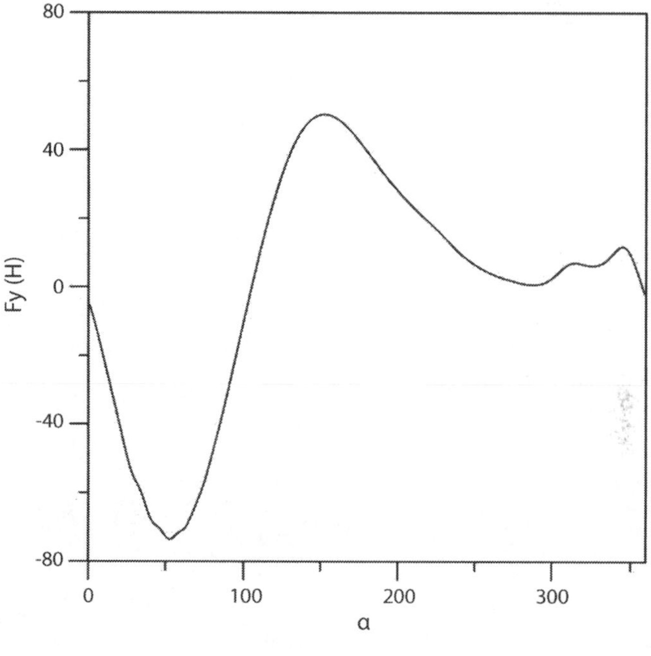

Fig. Ap1.20 Vertical force.

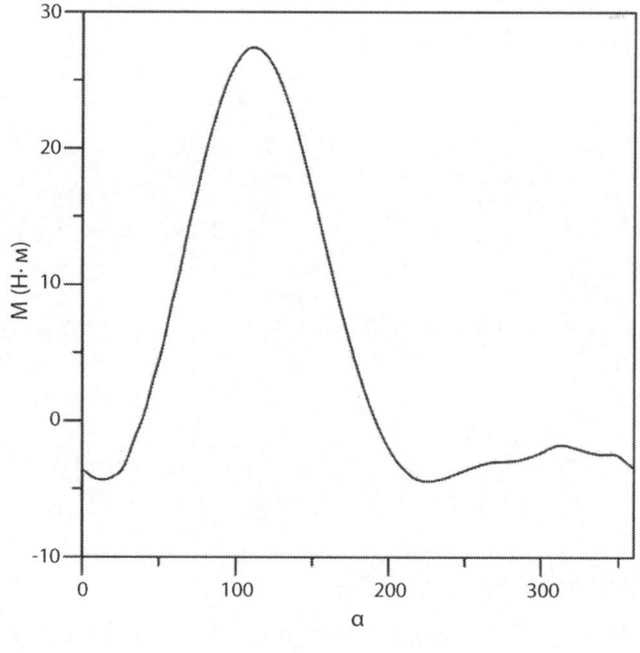

Fig. Ap1.21 Moment.

Results of modelling are agree with the tests (Fig. Ap1.22)

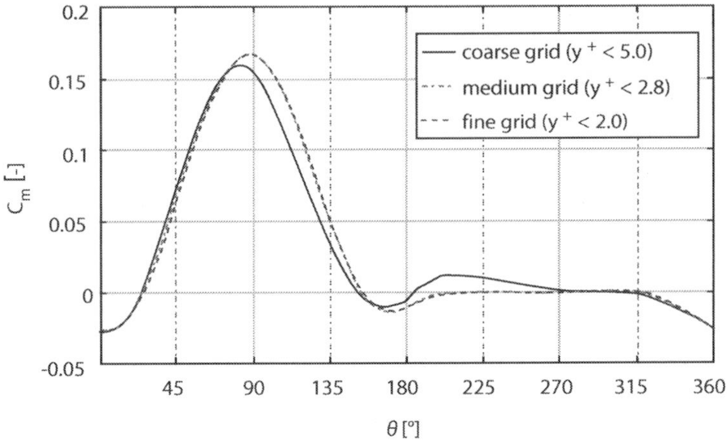

Figl. Ap1.22 Moment [4].

Modeling results for option 2.

The simulation results of the 2nd option are presented in Figs. Ap1.23-Ap1.27.

$$b/R = 0.6; \quad U/V = 0.4; \quad J = 0.04; \quad \alpha = \pi/2.$$

The simulation results of the 3nd option are presented in Figures Ap1.28-Ap1.32

$$b/R = 0.6; \quad U/V = 0.6; \quad J = 0.04; \quad \alpha = 3\pi/2$$

Modeling results for option 4.

The simulation results of the 4nd option are presented in Figs. Ap1.33-Ap1.37

Jet slot on the back of the blade. $b/R = 0.6$; $U/V= 0.6$; $J = 0.04$; $\alpha = 3\pi/2$.

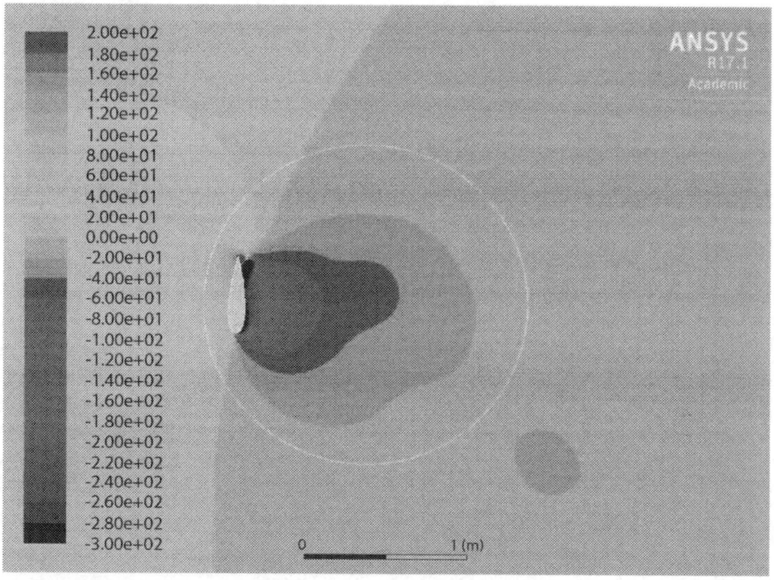

Fig. Ap1.23 Typical pressure field at the moment of jet outflow, Pa.

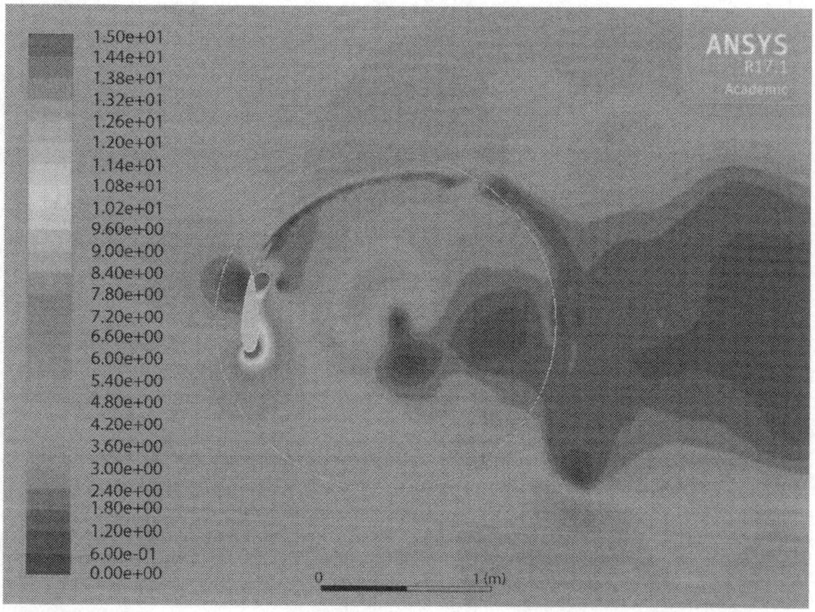

Fig. Ap1.24 Typical velocity field at the moment of jet outflow, m/s.

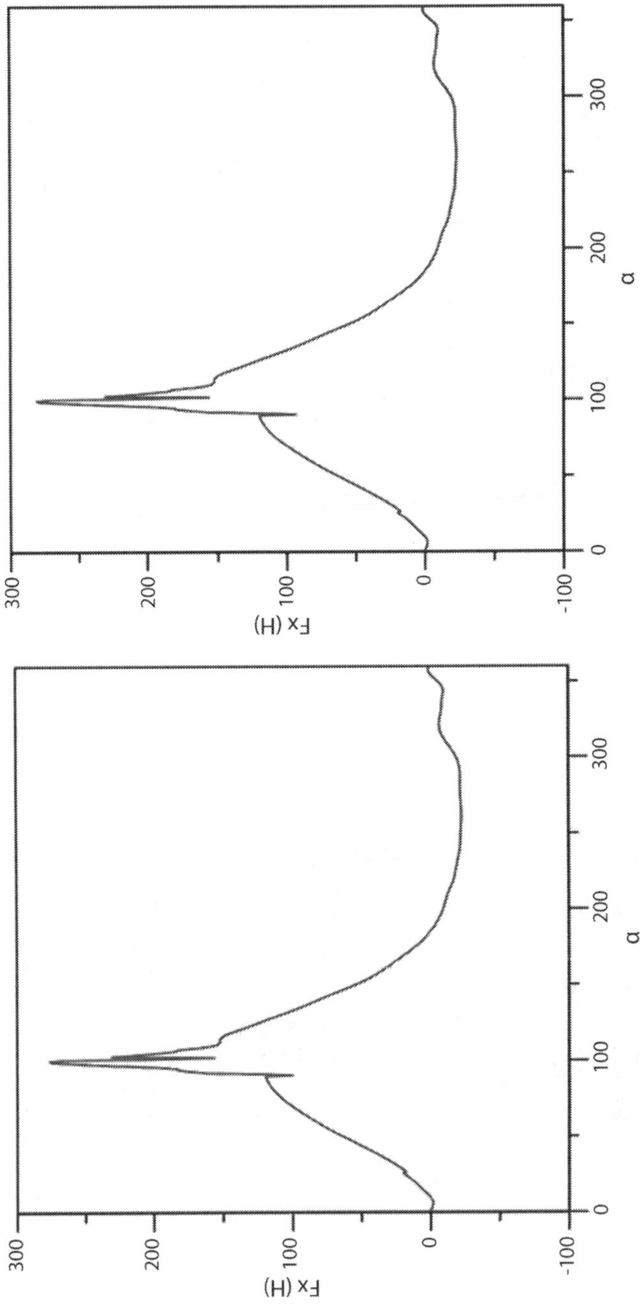

Fig. Ap1.25 Horizontal force (only aerodynamic component - on the left, with the addition of a component due to jet thrust - on the right).

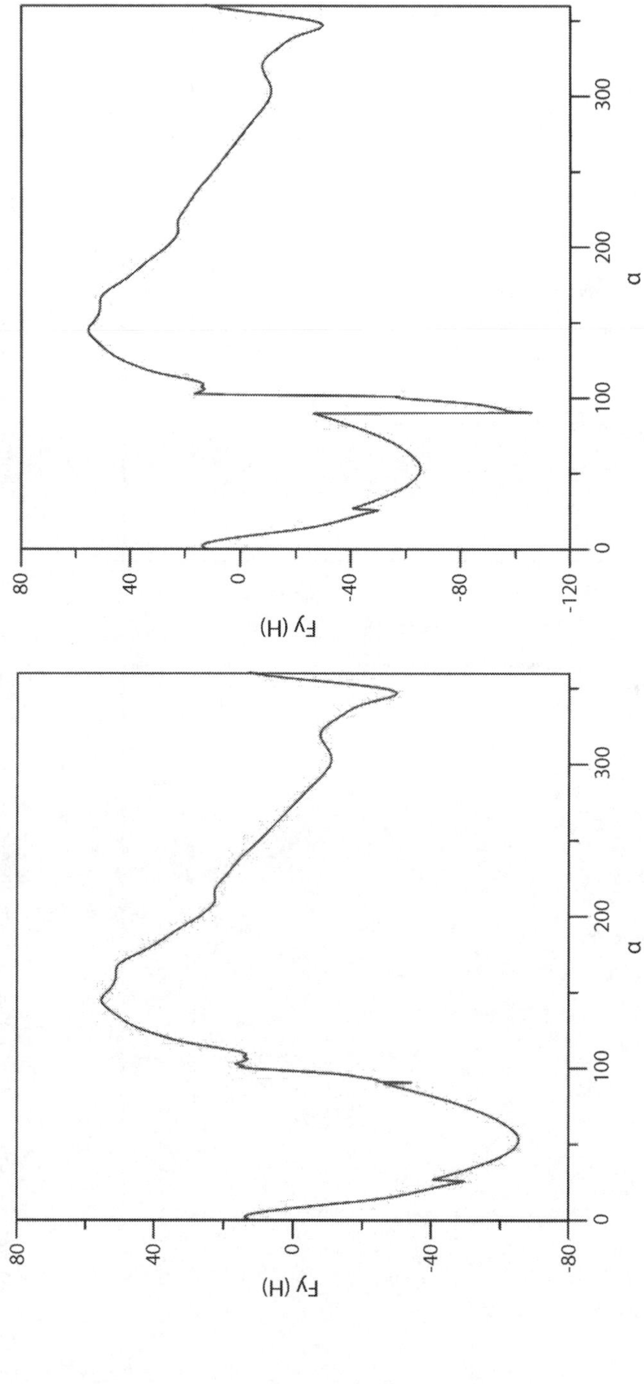

Fig. Ap1.26 Vertical force (only aerodynamic component - left, with the addition of a component due to jet thrust - right).

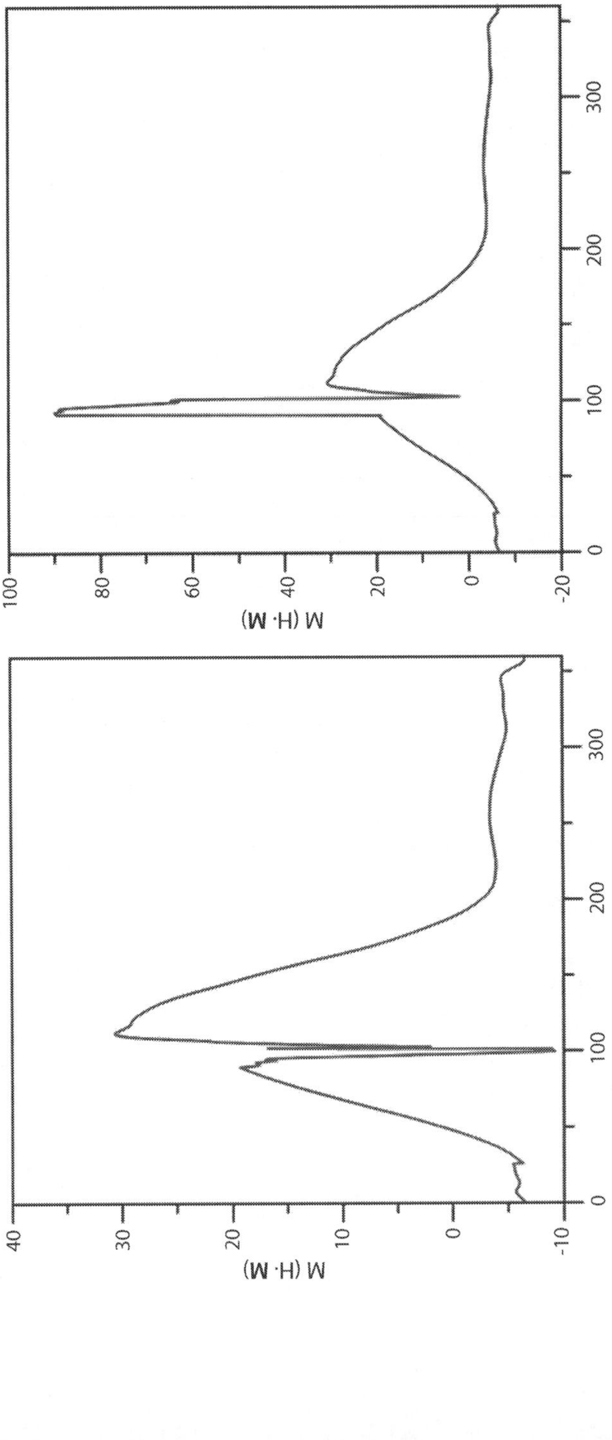

Fig. Ap1.27 Moment (only aerodynamic component) - on the left, with the addition of a component due to jet thrust - on the right.

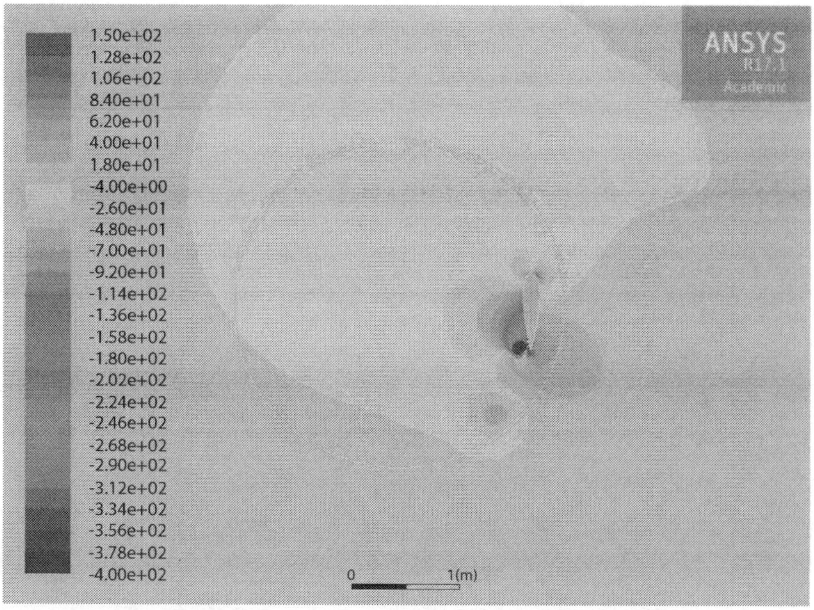

Fig. Ap1.28 Typical pressure field at the moment of jet outflow, Pa.

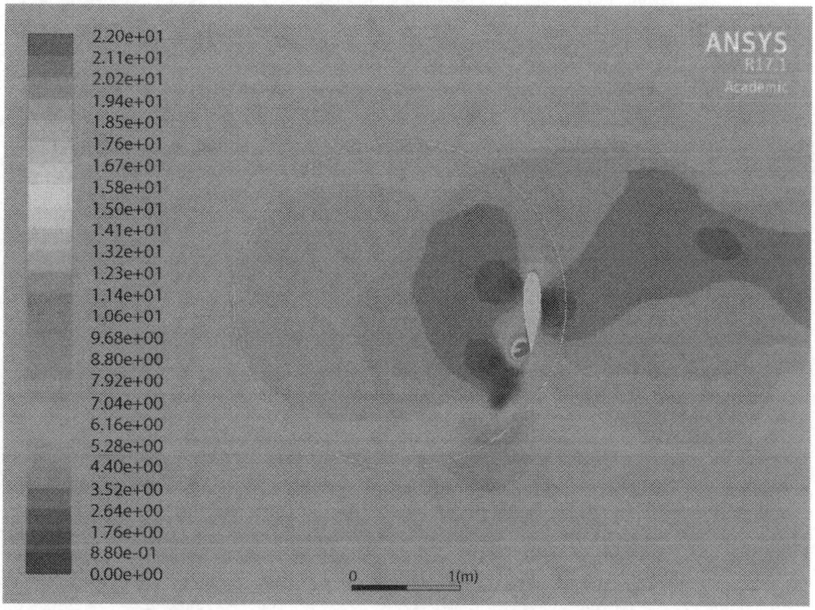

Fig. Ap1.29 Typical velocity field at the moment of jet outflow, m/s.

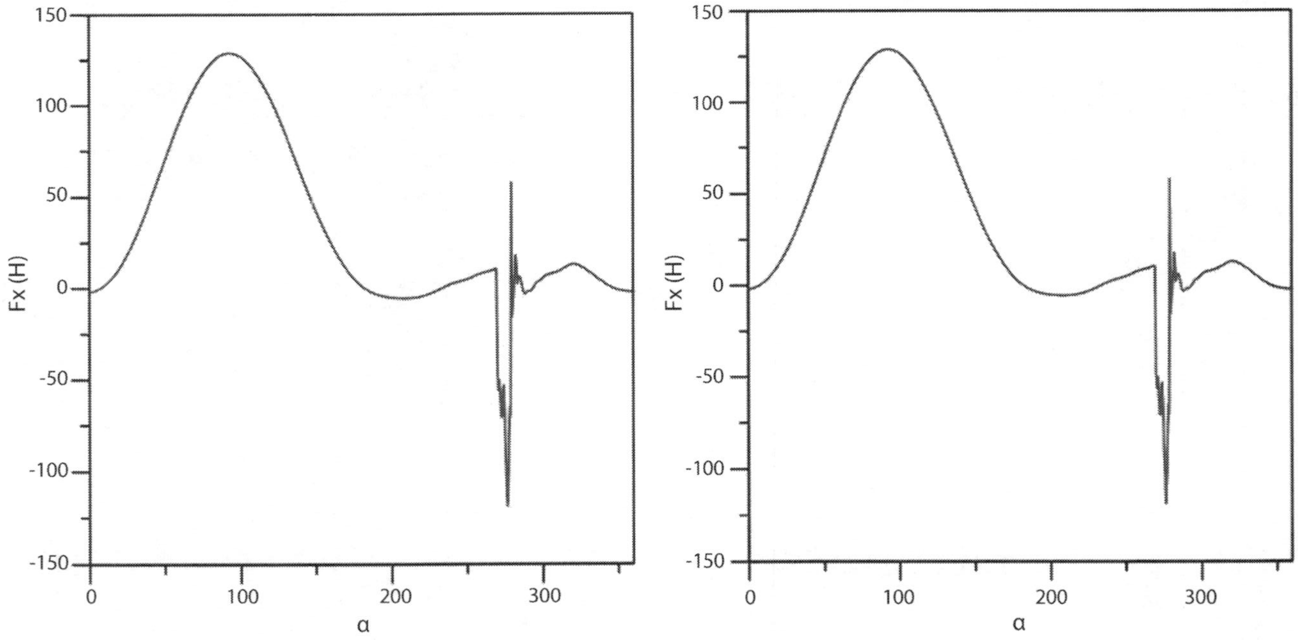

Fig. Ap1.30 Horizontal force (aerodynamic component only) - on the left, with the addition of a component due to jet thrust - on the right.

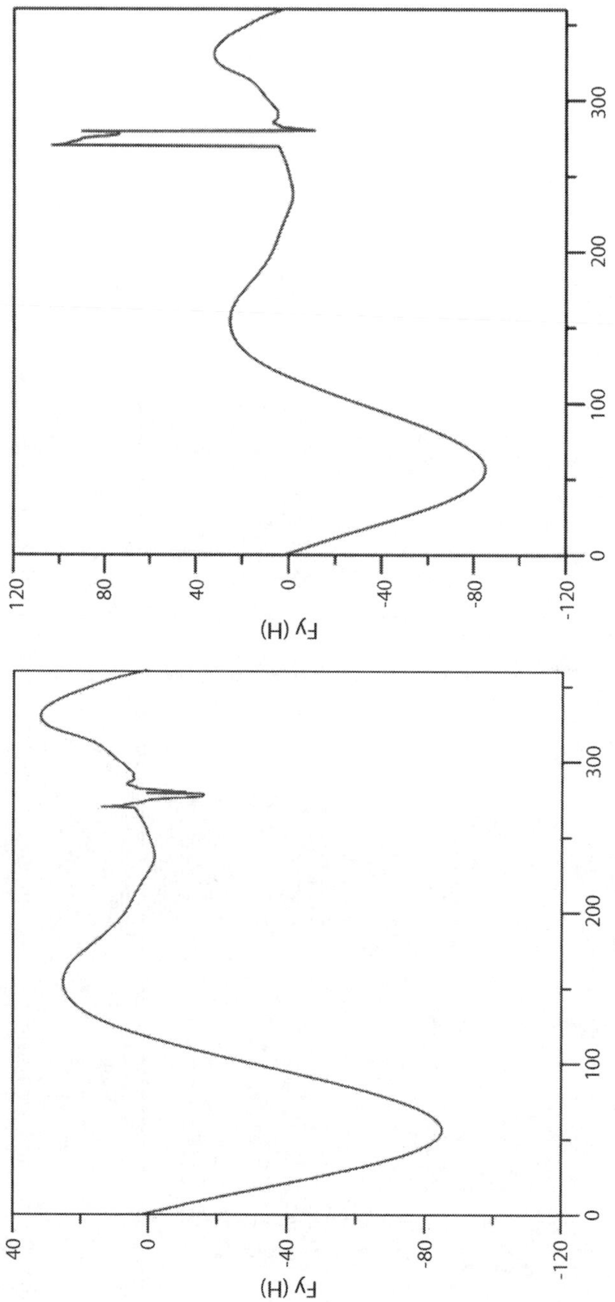

Fig. Ap1.31 Vertical force (aerodynamic component only) - on the left, with the addition of a component due to jet thrust - on the right.

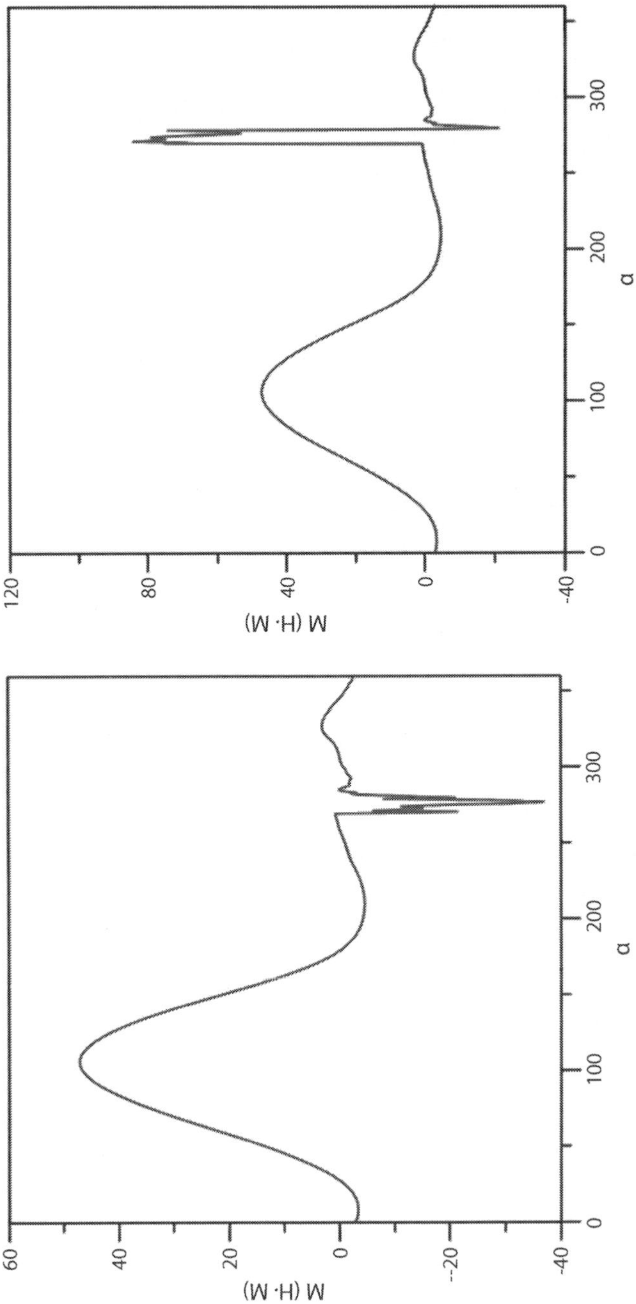

Fig. Ap1.32 Moment (aerodynamic component only) - on the left, with the addition of a component due to jet thrust - on the right.

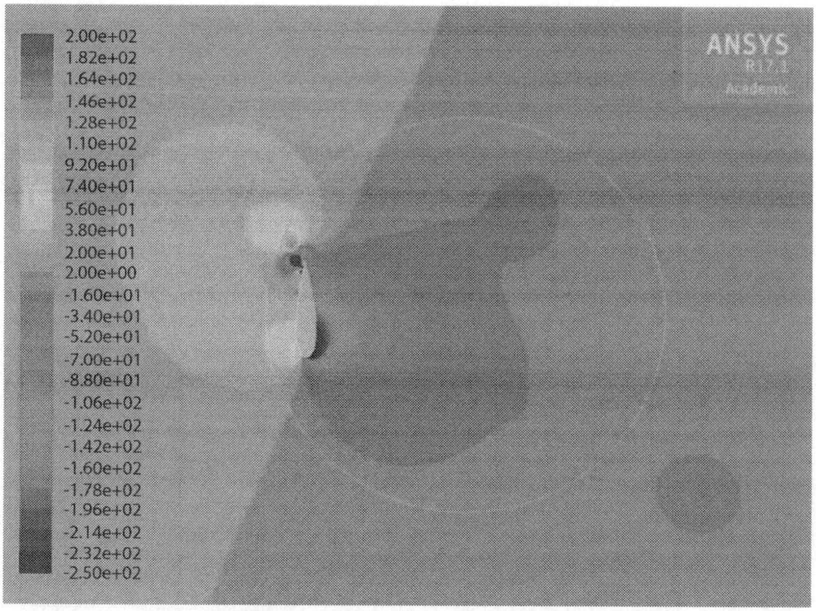

Fig. Ap1.33 Typical pressure field at the moment of jet outflow, Pa.

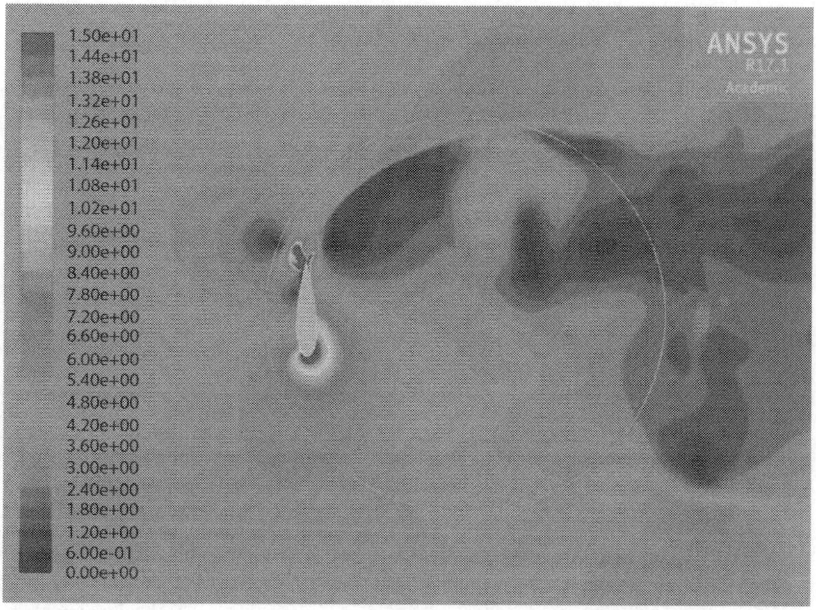

Fig. Ap1.34 Typical velocity field at the moment of jet outflow, m/s.

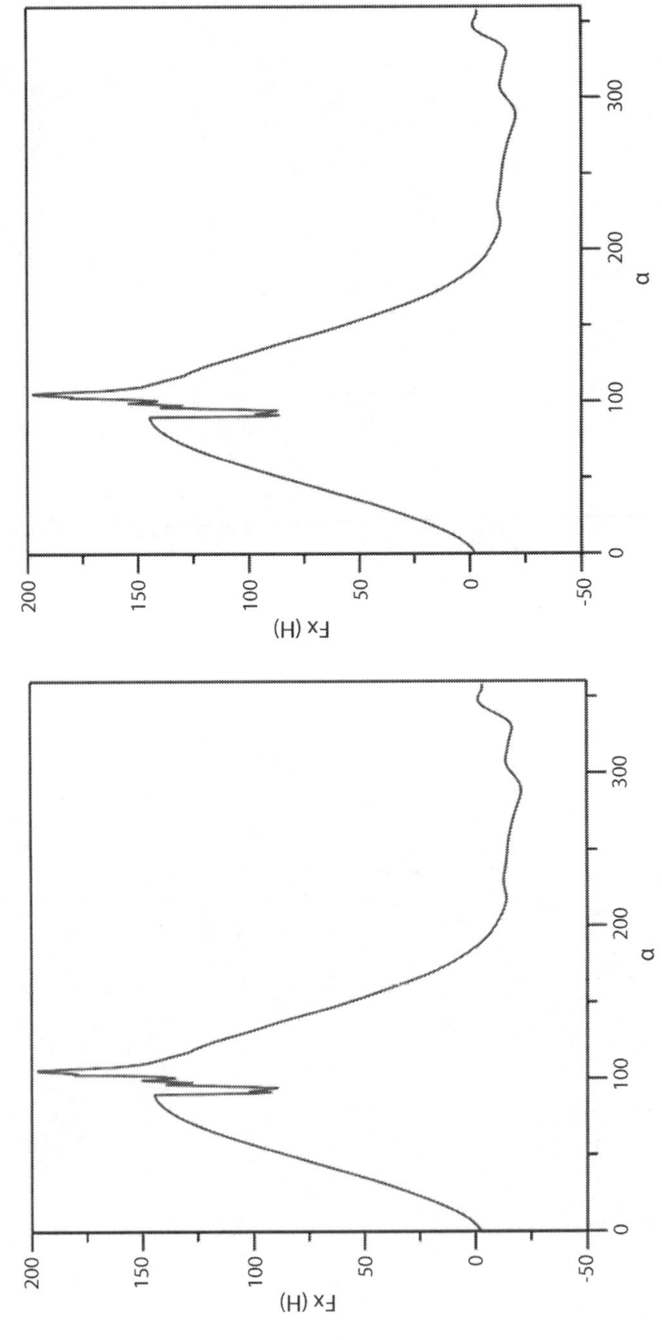

Fig.Ap1.35 Horizontal force (aerodynamic component only) - on the left, with the addition of a component due to jet thrust - on the right.

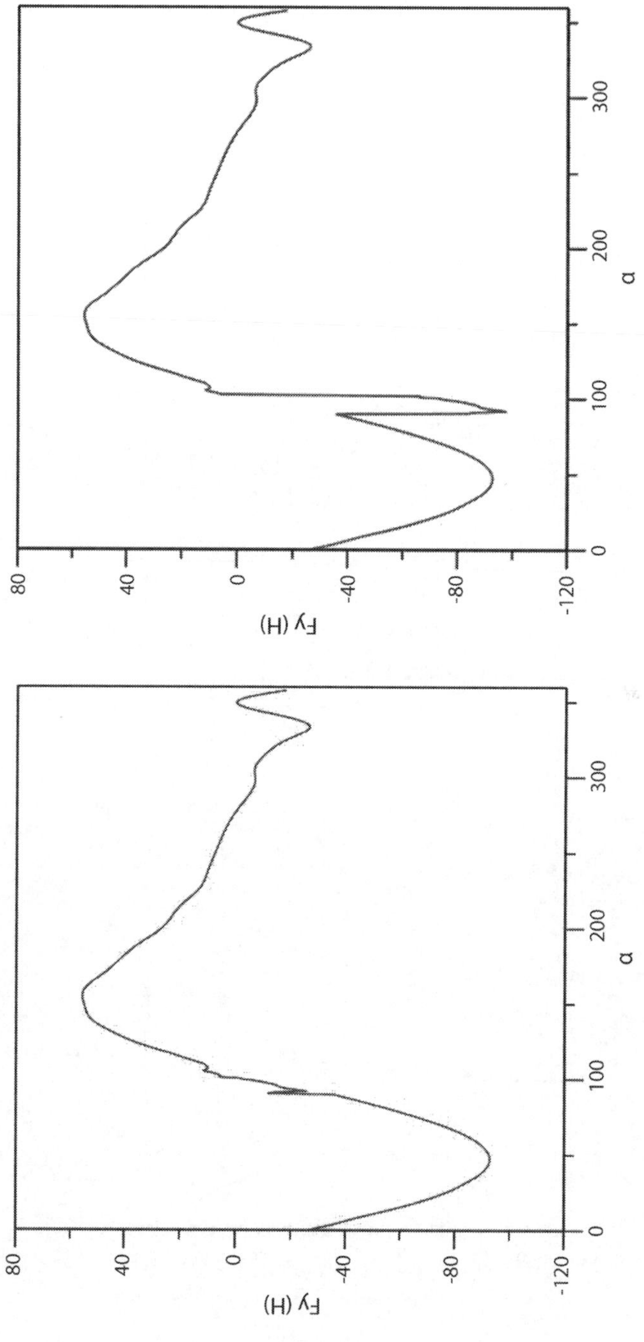

Fig. Ap1.36 Vertical force (aerodynamic component only) - on the left, with the addition of a component due to jet thrust - on the right.

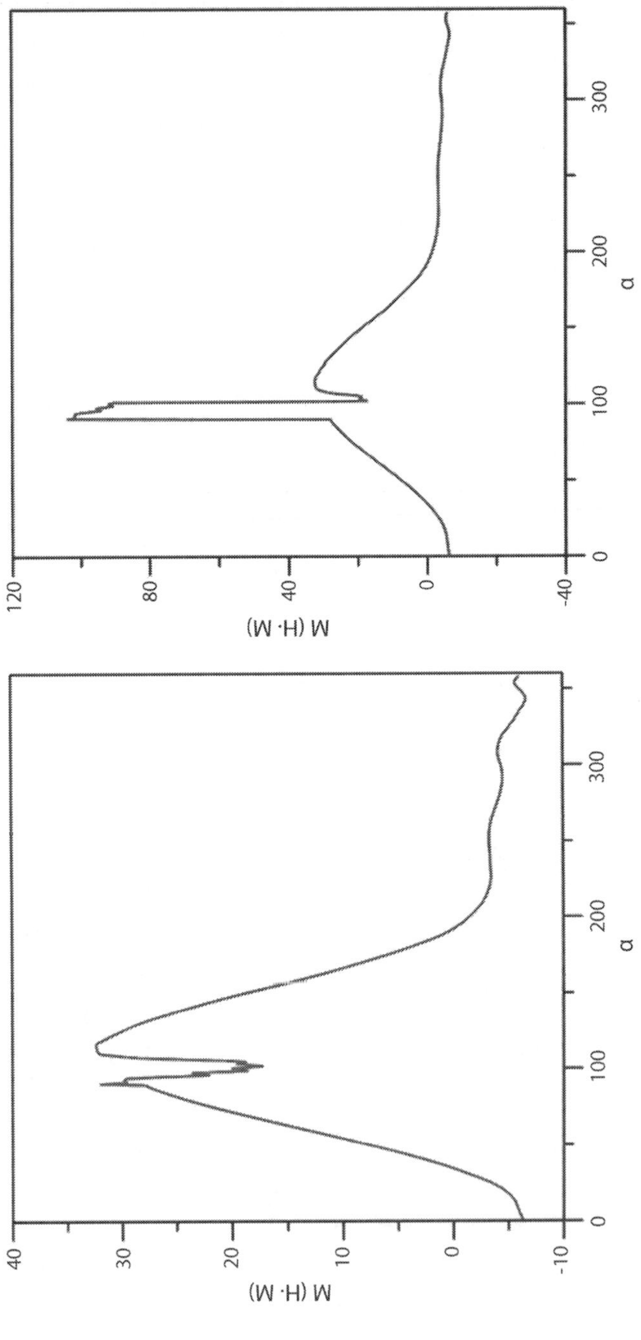

Fig. Ap1.37 Moment (aerodynamic component only) - on the left, with the addition of a component due to jet thrust - on the right.

References

1. Spalart, P. R. and Allmaras, S. R., 1992, "A One-Equation Turbulence Model for Aerodynamic Flows" AIAA Paper92-0439.
2. Patankar S.V. - Numerical methods for solving problems of heat transfer and fluid dynamics. Moscow, Energoatomizdat, 1984.152 p.
3. C Xisto, J Leger, J Páscoa, L Gagnon, P Masarati, D Angeli, A Dumas. Parametric Analysis of a Large-scale Cycloidal Rotor in Hovering Conditions. *Journal of Aerospace Engineering*, 30(1), 2017.
4. Rezaeiha, A., Kalkman, I., & Blocken, B. J. E. Effect of pitch angle on power performance and aerodynamics of a vertical axis wind turbine. *Applied Energy*, 2017, 197, 132-150.

Addition

Flow parameters:

oncoming air velocity - 3.4 m/s

blade angle - -14.5 °

kinematic viscosity 1.46E-5 m/s2

wing chord length 0.6 m (as in the project data)

nozzle diameter 0.005 m (as in the project data, in the article about 2 times less)

Cs = 0.082 (defined as in the publication)

jet speed 14.0 m/s

Because of the chord length the calculations were 3 times greater than in the publication, to preserve the Re number, the velocities of the incident flow and jets were reduced by 3 times.

The results of calculating the flow around the blade are presented below.

According to the calculation results for the variant with jet injection, the drag force decreased by 10%, and the lifting force increased 2.6 times.

RESULTS

Without jet injection

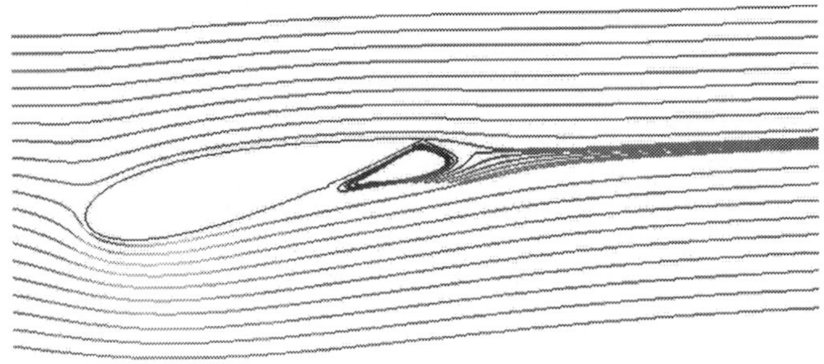

Streamlines

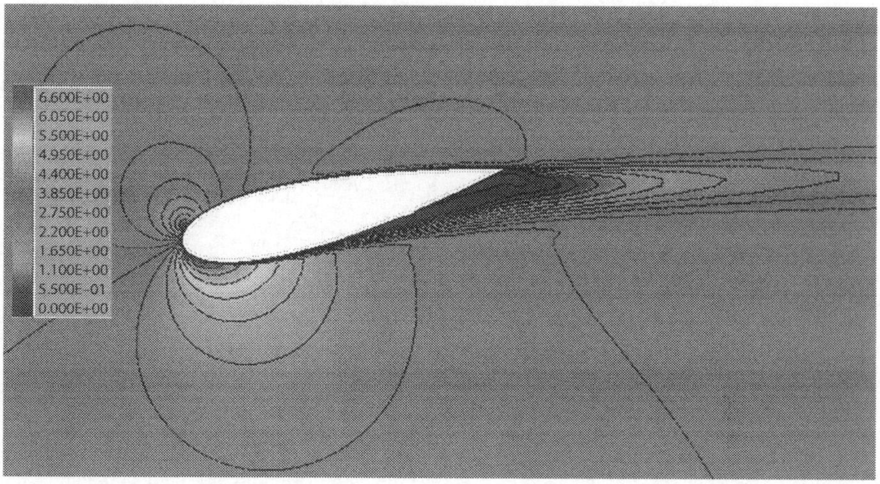

Velocity magnitude (m / s)

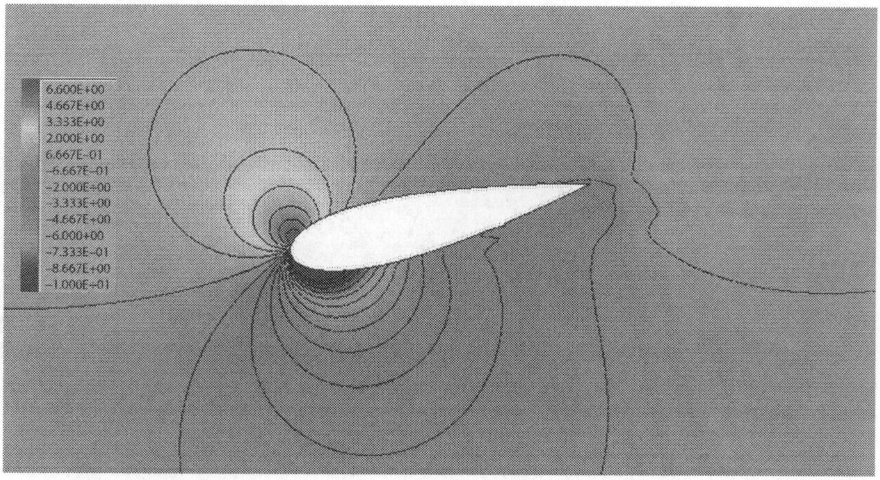

Pressure field (Pa)

With jet injection. Jet speed 14.0 m/s

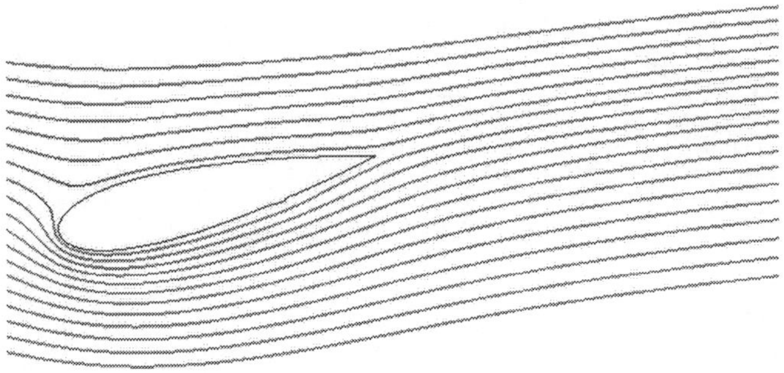

Streamlines

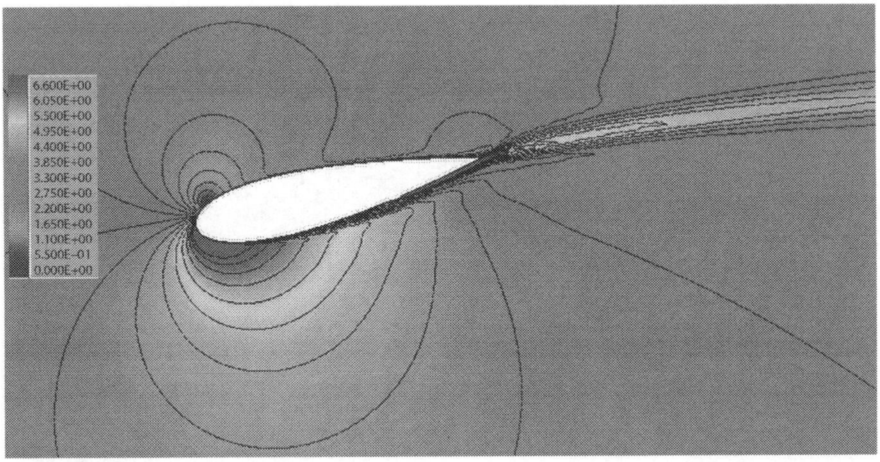

Velocity magnitude (m/s)

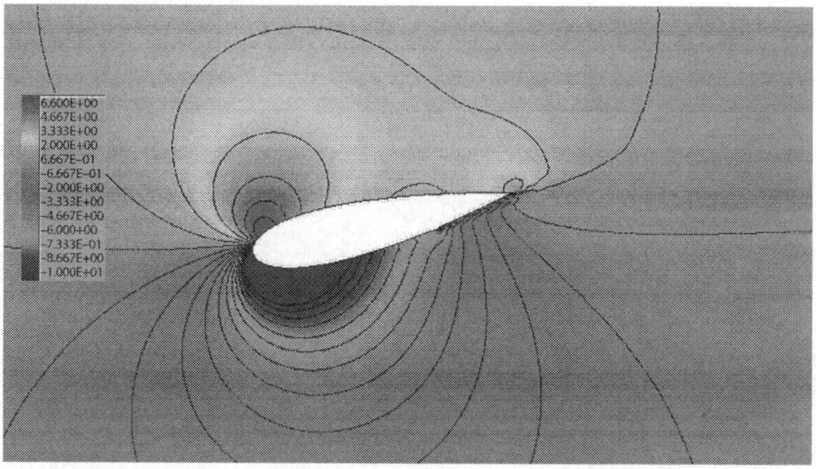

Pressure field (Pa)

Index

Also of Interest

Check out these other titles from Scrivener Publishing

From the Same Author

Hydraulic Modeling, by Victor M. Lythaker and Alexander M. Proudovsky, ISBN 9781118946190. Combining mathematical and physical modeling, the authors of this groundbreaking new volume explore the theories and applications of hydraulic modeling, an important field of engineering that affects many industries, including energy, the process industries, manufacturing, and environmental science. *NOW AVAILABLE!*

Seismic Loads, by Victor Lyatkher, ISBN 9781118946244. Combining mathematical and physical modeling, the author of this groundbreaking new volume explores the theories and applications of seismic loads and how to mitigate the risks of seismic activity in buildings and other structures. *NOW AVAILABLE!*

Wind Power, by Victor Lyatkher, ISBN 9781118720929. An up-to-date and thorough treatment of the technologies, practical applications, and future of wind power, covering the pros and cons and technical intricacies of various types of wind turbines and wind power prediction. *NOW AVAILABLE!*

Tidal Power: Harnessing Energy From Water Currents, by Victor Lyatkher, ISBN 978111720912. Offers a unique and highly technical approach to tidal power and how it can be harnessed efficiently and cost-effectively, with less impact on the environment than traditional power plants. *NOW AVAILABLE!*

Other Related Titles from Scrivener Publishing

Encyclopedia of Renewable Energy, by James G. Speight, ISBN 9781119363675. Written by a highly respected engineer and prolific author in the energy sector, this is the single most comprehensive, thorough, and up to date reference work on renewable energy. *DUE OUT IN EARLY 2022.*

Green Energy: Solar Energy, Photovoltaics, and Smart Cities, edited by Suman Lata Tripathi and Sanjeevikumar Padmanaban, ISBN 9781119760764. Covering the concepts and fundamentals of green energy, this volume, written and edited by a global team of experts, also goes into the practical applications that can be utilized across multiple industries, for both the engineer and the student. *NOW AVAILABLE!*

Microgrid Technologies, edited by C. Sharmeela, P. Sivaraman, P. Sanjeevikumar, and Jens Bo Holm-Nielsen, ISBN 9781119710790. Covering the concepts and fundamentals of microgrid technologies, this volume, written and edited by a global team of experts, also goes into the practical applications that can be utilized across multiple industries, for both the engineer and the student. *NOW AVAILABLE!*

Energy Storage, edited by Umakanta Sahoo, ISBN 9781119555513. Written and edited by a team of well-known and respected experts in the field, this new volume on energy storage presents the state-of-the-art developments and challenges in the field of renewable energy systems for sustainability and scalability for engineers, researchers, academicians, industry professionals, consultants, and designers. *NOW AVABILABLE!*

Biofuel Cells, Edited by Inamuddin, Mohd Imran Ahamed, Rajender Boddula, and Mashallah Rezakazemi, ISBN 9781119724698. This book covers the most recent developments and offers a detailed overview of fundamentals, principles, mechanisms, properties, optimizing parameters, analytical characterization tools, various types of biofuel cells, edited by one of the most well-respected and prolific engineers in the world and his team. *COMING IN SUMMER 2021*

Biodiesel Technology and Applications, Edited by Inamuddin, Mohd Imran Ahamed, Rajender Boddula, and Mashallah Rezakazemi, ISBN 9781119724643. This outstanding new volume provides a comprehensive overview on biodiesel technologies, covering a broad range of topics and practical applications, edited by one of the most well-respected and prolific engineers in the world and his team. *COMING IN SUMMER 2021*

Energy Storage 2nd Edition, by Ralph Zito and Haleh Ardibili, ISBN 9781119083597. A revision of the groundbreaking study of methods for storing energy on a massive scale to be used in wind, solar, and other renewable energy systems. *NOW AVAILABLE!*

Hybrid Renewable Energy Systems, edited by Umakanta Sahoo, ISBN 9781119555575. Edited and written by some of the world's top experts in renewable energy, this is the most comprehensive and in-depth volume on hybrid renewable energy systems available, a must-have for any engineer, scientist, or student. *NOW AVAILABLE!*

Progress in Solar Energy Technology and Applications, edited by Umakanta Sahoo, ISBN 9781119555605. This first volume in the new groundbreaking series, Advances in Renewable Energy, covers the latest concepts, trends, techniques, processes, and materials in solar energy, focusing on the state-of-the-art for the field and written by a group of world-renowned experts. *NOW AVAILABLE!*

A Polygeneration Process Concept for Hybrid Solar and Biomass Power Plants: Simulation, Modeling, and Optimization, by Umakanta Sahoo, ISBN 9781119536093. This is the most comprehensive and in-depth study of the theory and practical applications of a new and groundbreaking method for the energy industry to "go green" with renewable and alternative energy sources. *NOW AVAILABLE!*

Nuclear Power: Policies, Practices, and the Future, by Darryl Siemer, ISBN 9781119657781. Written from an engineer's perspective, this is a treatise on the state of nuclear power today, its benefits, and its future, focusing on both policy and technological issues. *NOW AVAILABLE!*

Zero-Waste Engineering 2nd Edition: A New Era of Sustainable Technology Development, by M. M. Kahn and M. R. Islam, ISBN 9781119184898. This book outlines how to develop zero-waste engineering following natural pathways that are truly sustainable using methods that have been developed for sustainability, such as solar air conditioning, natural desalination, green building, chemical-free biofuel, fuel cells, scientifically renewable energy, and new mathematical and economic models. *NOW AVAILABLE!*

Sustainable Energy Pricing, by Gary Zatzman, ISBN 9780470901632. In this controversial new volume, the author explores a new science of energy pricing and how it can be done in a way that is sustainable for the world's economy and environment. *NOW AVAILABLE!*

Sustainable Resource Development, by Gary Zatzman, ISBN 9781118290392. Taking a new, fresh look at how the energy industry and we, as a planet, are developing our energy resources, this book looks at what is right and wrong about energy resource development. This book aids engineers and scientists in achieving a true sustainability in this field, both from an economic and environmental perspective. *NOW AVAILABLE!*

The *Greening of Petroleum Operations*, by M. R. Islam *et al.*, ISBN 9780470625903. The state of the art in petroleum operations, from a "green" perspective. *NOW AVAILABLE!*

Emergency Response Management for Offshore Oil Spills, by Nicholas P. Cheremisinoff, PhD, and Anton Davletshin, ISBN 9780470927120. The first book to examine the Deepwater Horizon disaster and offer processes for safety and environmental protection. *NOW AVAILABLE!*

Biogas Production, Edited by Ackmez Mudhoo, ISBN 9781118062852. This volume covers the most cutting-edge pretreatment processes being used and studied today for the production of biogas during anaerobic digestion processes using different feedstocks, in the most efficient and economical methods possible. *NOW AVAILABLE!*

Bioremediation and Sustainability: Research and Applications, Edited by Romeela Mohee and Ackmez Mudhoo, ISBN 9781118062845. Bioremediation and Sustainability is an up-to-date and comprehensive treatment of research and applications for some of the most important low-cost, "green," emerging technologies in chemical and environmental engineering. *NOW AVAILABLE!*

Green Chemistry and Environmental Remediation, Edited by Rashmi Sanghi and Vandana Singh, ISBN 9780470943083. Presents high quality research papers as well as in depth review articles on the new emerging green face of multidimensional environmental chemistry. *NOW AVAILABLE!*

Bioremediation of Petroleum and Petroleum Products, by James Speight and Karuna Arjoon, ISBN 9780470938492. With petroleum-related spills, explosions, and health issues in the headlines almost every day, the issue of remediation of petroleum and petroleum products is taking on increasing importance, for the survival of our environment, our planet, and our future. This book is the first of its kind to explore this difficult issue from an engineering and scientific point of view and offer solutions and reasonable courses of action. *NOW AVAILABLE!*

Made in the USA
Las Vegas, NV
28 October 2025

676c5797-3be4-4dc9-ab64-2c254a7a491fR01